Fiber Network Service Survivability

For a complete listing of the Artech House Telecommunciations Library,
turn to the back of this book

Fiber Network Service Survivability

Tsong-Ho Wu

Artech House
Boston • London

Library of Congress Cataloging-in-Publication Data

Wu, Tsong-Hu.
    Fiber network service survivability / Tsong-Hu Wu.
    p.   cm.
    Includes bibliographical references and index.
    ISBN 0-89006-469-5
    1. Computer networks. 2. Fiber optics. 3. Optical
communications. I. Title.
TK5105.5.W8 1992   91-48071
621.382'75—dc20   CIP

British Library Cataloguing in Publication Data

Wu, Tsong-Ho
    Fiber Network Service Survivability
    I. Title
    004.6
    ISBN 0-89006-469-5

4ESS is a trademark of AT&T
5ESS is a registered trademark of AT&T
TIRKS is a registered trademark of Bellcore
SPARCstation is a trademark of SPARC International, Inc.

International Standard Book Number: 0-89006-469-5
Library of Congress Catalog Card Number: 91-48071

10 9 8 7 6 5 4 3 2

To
my wife, Shu-Jen, our parents,
and my children, Arthur and Mae

Contents

Preface

The increased deployment of optical fiber transmission systems with very large cross-connections supported on a few strands of fiber and the trend toward more sophisticated software-controlled networks have increased concern about the survivability of high-capacity fiber communications networks for any single point of failure. This concern arises because service disruption causes both tangible and intangible loss for users as well as for service providers.

Such losses can be avoided by using survivable network planning, which depends on architectures, technologies, and designs. Traditionally, these areas have been represented by three different disciplines. However, integrating these disciplines can help produce a cost-effective and efficient survivable network. This book focuses on network architectures and designs that use synchronous optical network (SONET) technology because SONET is a standard optical transmission technology that has been widely accepted and implemented in the telecommunications industry. In fact, network survivability is a key application of SONET technology.

The original motivation for writing this book was simple: Several Bellcore client company (BCC) network planners believed that a collection (or one-stop "shop") of Bellcore and other industrial resource works on fiber network survivability would help them plan cost-effective, survivable fiber networks. (I understood and sympathized with their concerns because I had a similar problem before I joined Bellcore in 1986.) Subsequently, this book is intended to serve as a reference tool for network planners, network architects, network engineers, and network scientists. It is not intended to be a textbook because several subjects discussed in this book are relatively new and some are still in the development and standardization processes. This book may also help some people in academia understand what SONET is and how it applies to network survivability, since SONET technology was initiated and primarily developed by the telecommunications industry.

The content of this book is primarily taken from my own research and that of my Bellcore colleagues to ensure subject accuracy. This book does not reflect any policy and position of Bellcore or the BCCs. All materials presented in this book are primarily taken from public literature and have not been reviewed by the BCCs. All ideas expressed in this book are strictly those of the author, who assumes full responsibility for the views expressed in this book.

Tsong-Ho Wu
Red Bank, New Jersey

Acknowledgments

To avoid conflicts with my company projects, I primarily wrote this book on weeknights and weekends. During this period, it became a daily routine for my son, Arthur, and daughter, Mae, to inquire about the status and progress of the book before they went to bed. It is fair to say that this book was a "family project." Without my wife's full support and understanding, and my children's expectations, it would have been virtually impossible to finish this book.

In addition to thanking my family, I would like to thank my Bellcore management and colleagues for their support of my writing efforts. Stu D. Personick supported the initial book proposal, which was later approved by the Bellcore Publications Committee. The Committee's approval allowed me to use company equipment to complete this book. I would also like to thank Joe E. Berthold and Richard H. Cardwell for their understanding and support during the preparation of this book and for their efforts in reviewing the drafts of the manuscript. (Actually, they may be the only two people who read the entire manuscript.) Additionally, I would like to thank the following Bellcore colleagues for their reviews and comments on the manuscript: Dev P. Batra, Peter J. Castaldo, Yau-Chau Ching, Richard E. Clapp, Bob Doverspike, Sarry F. Habiby, Gary A. Hayward, Vladimir Kaminsky, David J. Kolar, Dennis Kong, Joseph Lau, Richard C. Lau, Kevin Lu, Jonathan Morgan, Joseph Sosnosky, Yukun Tsai, Stu Wagner, Ondria J. Wasem, Winston I. Way, and Susumu Yoneda.

I would like to express my deep gratitude to Ruth A. Santulli for her efforts in editing the book manuscript. She made it more readable and ensured that it conformed to both Bellcore's and the publisher's quality requirements. In addition, I would like to thank Carol L. Adams, who helped prepare some illustrations for the book, and my wife, Shu-Jen, who also prepared some of the illustrations and designed the book cover. Finally, I would like to thank Mark Walsh for offering me the opportunity to publish this book.

CHAPTER 1

Introduction

1.1 OVERVIEW

Telecommunications is a technology that has significantly impacted the development of civilization. The use and integration of computers with telecommunications have created a so-called "information age" in which telecommunications has become a vital part of business's day-to-day operations. Service disruption is no longer being tolerated by industries due to the increased necessity of communications with bankers, purchasing managers, stock brokers, retailers, and so forth. At the same time, the consequences of service disruption are becoming more severe for a single point of failure, partly due to the high volume of traffic being carried by fewer fiber systems that are aggregated to a central point.

Service disruption causes both tangible and intangible losses for users as well as for service providers. To ensure service continuity, service providers have increased their efforts to alleviate such disruption. A challenging task centered on these efforts is how to ensure service continuity at affordable costs. This issue forms the core focus of this book: to offer insight and information that will help service providers and customers plan and design affordable solutions for service continuity.

The goal of this book is to provide network planners, designers, and researchers with a reference tool to understand the architectures, technologies, and design methods involved in planning and designing survivable fiber networks to support telecommunications service continuity.

This book does not discuss network restoration techniques and designs for all types of network applications. Instead, it focuses on survivability for fiber facility networks. It discusses the design principles, generic survivable network architectures, and technologies that make implementing these architectures practical and economical. The principles of architectural planning and network design are applicable to *intra-Local Access and Transport Area* (intraLATA) fiber networks, as well as to interLATA fiber networks, although design methods and case study results described in this book are primarily for intraLATA fiber networks. Due to increased acceptance and deployment of *Synchronous Optical Network* (SONET) equipment, the discussions here focus more on SONET-based survivable fiber network architectures and designs than on current non-SONET technology.

1.2 ADVANTAGES OF USING FIBER-OPTIC NETWORKS

The selection of transmission technology is determined by capacity, economics, reliability, and growth potential. The growing interest in using fiber for telecommunications networks is due to its superior attributes over existing copper and radio transmission systems for intraLATA networking applications. These attributes include higher capacity, higher reliability, longer repeater spacing, greater security, smaller size and less weight, unlimited growth potential, and lower system costs. These attributes, which are described below, have made optical fiber systems the transmission medium of choice for telecommunications networks.

- *Higher Capacity:* Today's fiber can carry gigabit-per-second data, and fiber that can carry data at 10 times that speed has been tested in research laboratories. Fiber's high capacity offers more flexible use of bandwidth for existing and new services. This attribute creates a wealth of opportunities in the field of telecommunications to transport voice, high-speed data, video, *High Definition Television* (HDTV), and so forth. It further allows new services requiring large amounts of bandwidth to be provided in a cost-effective manner.
- *Higher Reliability:* Fiber's high reliability is achieved with the following system performance factors:

  — Commercially available fiber systems have *Bit Error Ratios* (BERs) less than 1×10^{-11}.

  — The fiber-optic transmission medium is free of electromagnetic interference and is unaffected by rain, temperature, and humidity.

In addition to system reliability, fiber cable is normally buried three to four feet underground, providing substantial link protection from disasters. For example, fiber has survived hurricanes such as Hugo, which occurred in the eastern U.S. in October 1989. Whereas Hugo may have been the single largest natural disaster in U.S. history, it did not completely disrupt telephone service because fiber cables were buried under the ground and survived substantial flooding [1].

- *Longer Repeater Spacing:* Commercially available fiber-optic systems can transmit data for about 30 miles without regenerating signals. In comparison, today's copper transmission systems require a regenerator about every mile. Repeater spacing for fiber-optic systems is expected to extend to 50 miles in the very near future.
- *Greater Security:* Fiber-optic transmission systems neither emit nor induce any external energy. This feature makes fiber systems virtually untappable, provided the power level is carefully monitored, because signal loss can be detected almost

immediately. It also makes these fiber systems particularly desirable for military and banking communications applications.

- *Small Size and Less Weight:* One mile of point-to-point copper cable weighs approximately half a ton; replacing the equivalent section with fiber would decrease that weight by at least 75 percent. Because fiber-optic cable is so thin and lightweight, it can nearly always be installed in existing conduits. This attribute is particularly attractive for communications systems on naval ships where the weight is almost all above the waterline but must be compensated by weight below the waterline [2].

- *Unlimited Growth Potential:* Fiber strands have virtually unlimited growth potential because capacity is not constrained by the fiber, but rather by the optoelectronics equipment that terminates the fiber.

- *Lower System Cost:* The economic appeal of fiber systems over copper systems is primarily due to high volumes of data that can be carried by fewer fibers, longer repeater spacing, and lower maintenance costs. In fact, the driving forces behind initial fiber deployment are not new applications, but cost considerations in replacing old technologies such as T1-carrier. Future fiber deployment is likely to be used for new applications such as *Broadband Integrated Services Digital Network* (B-ISDN) and fiber-to-the-home.

1.3 PRESENT INTRALATA FIBER NETWORK ARCHITECTURES

Traditionally, the intraLATA telephone network was built primarily using copper cable. Economic facility networks are therefore designed to minimize the total circuit mileage (since circuit mileage is proportional to circuit costs). This occurs because copper cable facilities have limited bandwidth (i.e., loss and crosstalk increase as frequency increases). Copper cable has a short repeater spacing (generally about one mile) and is easily affected by electronic interference. The cost associated with line regenerators and housings is an additional factor that affects the number of circuits transmitted on a facility path and the path mileage. These design parameters make the mesh topology an efficient telephone network topology, as depicted in Figure 1.1(a).

The mesh topology is essentially a set of point-to-point links between office buildings. Network survivability is an inherent part of the mesh topology because there are usually at least two paths between two end offices. Thus, a network that uses a mesh topology can survive a single failure. In communications, network survivability is defined as the capability of a communication network to resist any interruption or disturbance of service, particularly by warfare, fire, earthquake, harmful radiation, or other physical or natural catastrophes rather than by electromagnetic interference or crosstalk [3].

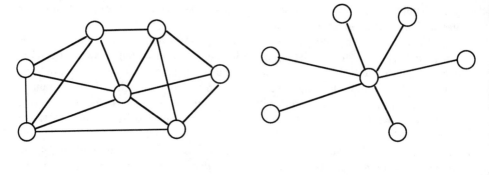

(a) Mesh Topology   (b) Star Topology

Figure 1.1.   Mesh and star topologies.

In the world of fiber technology, bandwidth may not be a constraint. Transmission capacities greater than 1.7 Gbps exist today, with large increases projected for the near future. The available bandwidth on an existing fiber link can be doubled by modifying the electronics at each end of the fiber link. Compared to smaller-capacity systems (e.g., copper systems), fiber-optic systems require more expensive terminating equipment. Equipment costs are also much higher than the fiber cost for intraLATA fiber transport networks [4]. Thus, the amount of equipment required for fiber transport systems must be reduced if they are to compete economically with networks that use conventional technologies.

An architecture that uses facility hubbing can best utilize the economical factor of high-capacity fiber systems and reduce the amount of equipment needed for signal transport. As depicted in Figure 1.1(b), facility hubbing is implemented on a star topology for fiber networks because the cost of fiber systems is relatively insensitive to distance, whereas the capacity of fiber transmission systems is very large compared to office-to-office circuit requirements. As a result, a reasonable network architecture and routing strategy is to send all the demand[1] from each office to a central point, or *hub*. Thus, demand is aggregated into the largest possible bundle to take advantage of fiber technology's economies of scale. Figure 1.2 depicts the concept of facility hubbing for

---

1. Throughout this book, we will use the term "demand" rather than "traffic" because traffic is usually represented by *erlangs*, whereas the signal unit considered in this book is *circuit* or above.

4

interoffice networks. Each *Central Office* (CO) is connected to a hub via a fiber-optic system. At the hub, a *Digital Cross-connect System* (DCS) partitions incoming traffic by destination and routes channels, over fiber, to the appropriate end office. Aggregating interoffice traffic onto a single fiber pair greatly reduces the costs for small COs that require links to other offices. Illinois Bell installed the first fiber-optic hub in the U.S. in Chicago [5].

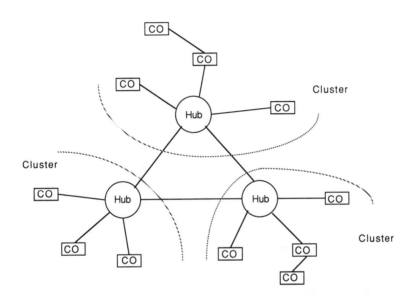

Figure 1.2.   Facility hubbing for interoffice networks.

An analysis of several alternative architectures for intraLATA interoffice networks reported in Reference [6] has suggested that (1) fiber facility hubbing is economically competitive with copper's point-to-point structure under a growth scenario, and (2) deployment of fiber only as a replacement technology for copper networks carries a significant cost penalty. In addition to capital cost advantages, hub offices create network flexibility points. These locations become the home of future equipment designed to mechanize demand switching and network reconfiguration. A hubbed network with a relatively small number of interoffice fiber links creates a simpler network than was previously possible. It also provides remotely controlled network capabilities that allow more efficient use of fiber, ease of administration and maintenance, and quick response to increased customer demand.

1.4 THE IMPORTANCE OF CONSIDERING SURVIVABILITY

The increasing deployment of interoffice optical fiber transmission systems with large cross-sections supported on a few strands of fiber and the trend toward a fiber-hubbed network architecture have increased concern about the survivability of fiber communications networks. As fewer routes carry more traffic centralized to a few points (i.e., fiber hubs), the probability increases for more users to suffer serious service disruption due to a single point of failure. For example, a fiber cable cut in the AT&T network, which occurred at Newark in January 1991, interrupted 60 percent of voice and data coming in and going out of New York City, including three major commercial airports, for about 10 hours. Three major CO fires have also been reported: two in New York City in 1975 and 1987, and a third in Hinsdale, Illinois in 1988. The building fire in Illinois Bell's Hinsdale CO, a major fiber hub, affected voice and data communications for more than one-half million resident and business customers in Hinsdale and surrounding Chicago-area communities [7]. This is because the impacted COs included not only the Hinsdale CO, but also surrounding COs. All services were completely restored four weeks after the fire [7]. In contrast, if the same fire occurred in today's copper telephone networks, which use the mesh topology, it would affect fewer users because only users associated with that CO would be affected. The CO fires that occurred in the New York telephone network in 1975 and 1987 are examples of this type of service disruption.

Service disruption causes both tangible and intangible loss for users as well as for service providers. This includes revenue loss, assets loss, restoral costs, legal costs, adverse customer relations, loss of competitive advantage, and credibility. AT&T conducted a customer survey during 1989 [8] and found that, among 1700 customers interviewed, (1) 16 percent of the companies had experienced a telecommunications failure in the last two years that impacted them financially (many reported more than one such failure), and (2) 15 percent said that, during these failures, they lost over $100,000 an hour due to service loss.

Network failures can be attributed to hardware or software problems or to natural catastrophes. Among network failures, the fiber cable cut has become a common system failure, according to a Bellcore report [9]. Most fiber cable cuts are caused by natural disasters or fiber-optic cables that are inadvertently dug up [9]. A well-known software failure occurred in January 1990 when AT&T's 4ESS switches blocked 65 million long-distance calls for about nine hours. Catastrophic failures due to fire or other natural disasters do not occur very often. However, when they do occur, the impact on user communities is significant.

1.5 SERVICE SURVIVABILITY PLANNING

Service survivability planning involves challenges, opportunities, and regulatory realities. This planning can be categorized into four phases to ensure service continuity and minimize the level of impact caused by service disruption. As shown in Figure 1.3, these phases are (1) prevention, (2) prompt detection, (3) network self-healing through robust design, and (4) manual restoration.

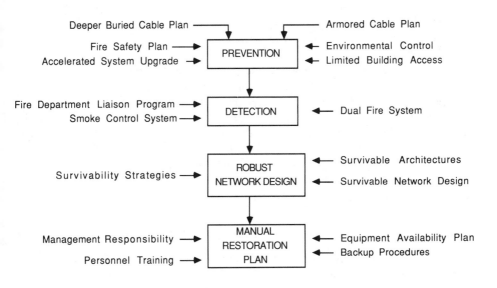

Figure 1.3. Service survivability planning.

7

The first phase focuses on preventing network failures. In this phase, efforts are placed on minimizing problems created by people and the environment (e.g., fire). Suggested guidelines for this phase include plans for limited building access, environmental control, fire safety, accelerated equipment upgrading, armored cable, and more deeply buried cable.

The second phase focuses on quick detection of network component failures. Efforts in this phase include dual fire alarm systems, fire department liaison programs, and smoke control systems.

The third phase focuses on the network self-healing capability during network component failures. In the self-healing concept, network protection is built into the network from the beginning, rather than added later. This phase emphasizes building a survivable network that can provide a self-healing capability whenever failures occur and includes survivability strategies, survivable network architectures, and survivable network design. The self-healing network design phase can represent a large investment for service providers. A big challenge in this phase is how to build a cost-effective, survivable network to provide user service assurance. This phase is the main topic of this book.

The last phase focuses on planning and practicing restoration in case the network cannot fix the problems itself. This phase emphasizes the efficient utilization of available work forces, facilities, and equipment. Thus, it requires several plans: the equipment availability plan, the backup procedures, the personnel training plan, and a plan for assigning management responsibility. Skilled personnel and a complete restoration plan can minimize damages when the network does not function properly.

In summary, engineering design, emergency power, network diversity, disaster preparedness, and skilled, dedicated people can keep the network functioning well under catastrophic and other types of failure conditions.

1.6 SELF-HEALING NETWORK DESIGN CONCEPTS

In the customer service area, serviceability (and hence customer satisfaction) needs to be designed into the network, not added later. To achieve high customer satisfaction, service assurance, which begins by incorporating sufficient diversity and capacity, should be built into fiber networks when they are originally designed. This subsequently mitigates the impact of failures and lowers the costs incurred to implement survivability. This approach provides a platform to achieve unparalleled levels of customer satisfaction with overall lower cost to both customers and service providers.

To implement a cost-effective, survivable network for service assurance, we must first understand the users' interests and requirements in service assurance. A survey [8] involving 350 large business customers showed that the top three service assurance concepts are priority restoral (63 percent), dual access (55 percent), and diverse routing (55 percent). These three requirements form a core of restoration techniques in both interoffice and loop fiber networks.

Restoration techniques are designed to make active use of available capacity and diversity after an attempt to automatically restore or maintain service fails. These techniques fall into two categories, traffic and facility restoration, as depicted in Figure 1.4. Traffic restoration is applied to switched networks, whereas facility restoration is applied to facility transport networks. A *switched network* is composed of circuit switches that switch calls carried in busy circuits. A *facility transport network* is composed of multiplexers and cross-connect systems, and transports a bundle of circuits as a whole. Note that it may not be necessary for a switched network to have a facility transport network as its subnetwork, as shown in Figure 1.4(a).

1.6.1 Traffic Restoration

Traffic restoration involves routing individual calls around a failure. A circuit switch, such as AT&T's 5ESS switch, performs traffic restoration by routing calls around failed circuits. Other techniques that can perform traffic restoration include *Dynamic Non-Hierarchical Routing* (DNHR) [10] and state-dependent routing [11], which not only reroutes traffic from failed points, but also efficiently utilizes network bandwidth.

1.6.2 Facility Restoration

Facility restoration involves rerouting transmission bandwidth in large units around a failure. It is not generally service-specific unless the units of bandwidth have been preassigned for that service. Facility restoration requires fewer operations than rerouting each call individually; thus, it has the potential to restore more services in a shorter time than traffic restoration.

For current, high-capacity asynchronous fiber facility networks, an efficient and commonly used transport signal unit is *Digital Signal level 3* (DS3), which carries 45 Mbps of data, rather than *Digital Signal level 0* (DS0), which carries a voice call of 64 kbps. Thus, facility restoration is more appropriate than traffic restoration for fiber facility transport systems. Because the subject of this book is network survivability for high-speed fiber transport networks, we will discuss only technologies, architectures, and designs that are associated with facility restoration. Strategies and some traffic restoration techniques for switched networks, *Common Channel Signaling* (CCS) networks, and other types of networks can be found in [12-14]. For convenience, restoration techniques are referred to as *facility restoration techniques* throughout this book.

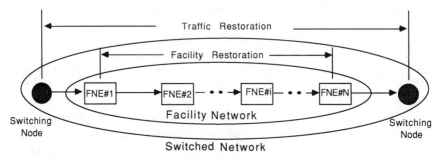

FNE: facility network element (TM, ADM, and DCS)

(a) A Layered Concept for Network Flow

Network	Equipment	Transport or Switching Unit	Dedicated Facility Restoration	Dynamic Facility Restoration
Facility Transport Network	TM, ADM, DCS	Demand (DS3 and above)	Restore demands via dedicated protection facilities	Restore demands via spare capacity within working systems
Switched Network	Circuit switch	Call (or busy circuit, 64 Kbps)	Traffic Restoration	
			Restored affected, busy end-to-end circuits (i.e., calls)	

TM: terminal multiplexer; ADM: add-drop multiplexer; DCS: digital cross-connect system

(b) A Layered Concept for Network Restoration Methods

Figure 1.4. Traffic restoration and facility restoration.

Given the service survivability requirements, facility restoration techniques can be divided into two categories: dedicated facility restoration and dynamic facility restoration. There are tradeoffs between the flexibility (and, thus, system complexity) and the additional capacity required for each category of techniques. Generally, the more sophisticated techniques require less capacity but slow the restoration procedure. The simpler techniques preassign capacity for restoration; more sophisticated techniques construct the restoration path in real time from individual restoration links.

1.6.2.1 Dedicated Facility Restoration

Restoration techniques that use dedicated facilities for protection include *Automatic Protection Switching* (APS), dual homing, and *Self-Healing Rings* (SHRs). Network design using dedicated facility assignment for restoration is sometimes referred to as *survivable physical layer network design* (see Chapter 7). The APS and dual-homing techniques are classified as conventional restoration techniques because they can be implemented using today's technology.

Conventional Restoration Techniques. Two techniques that are used today to protect facilities and office buildings are *1-for-N APS Diverse Protection* (1:N/DP) and dual homing. The 1:N/DP technique provides network protection against fiber cable cuts and multiplexing equipment failures, and dual homing provides protection against major hub failures. Chapter 3 discusses APS restoration and dual-homing techniques in more detail.

- 1:N APS Diverse Protection

  The APS approach is commonly used to facilitate maintenance and protect working services, and has the advantage of being totally automatic. The 1:N *diverse* protection structure is an alternative to the commonly used 1:N protection strategy, where N working fiber systems share one common protection fiber system. The only difference between these structures is the location of the fiber protection system; the 1:N protection structure places the protection fiber in the same route as that of working systems, and the 1:N diverse protection structure places the protection fiber in a diverse route. In a 1:N system, a cable cut may cut the protection fiber as well as the working fibers. If a fiber cable cut occurs and a 1:N diverse protection scheme is used, part of the service is lost because only one of the N working systems can be restored. Figure 1.5 shows the difference between these two structures. This diverse protection scheme is attractive because electronics costs dominate total costs and remain unchanged when attempting to achieve survivability. A 1:1 diverse protection arrangement, which provides 100 percent

11

survivability for fiber cable cuts, requires more facilities and equipment than the 1:N diverse protection arrangement.

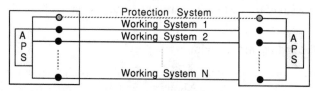

APS: automatic protection switching system

(a) 1:N APS Architecture

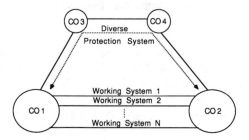

(b) 1:N APS Architecture with/without Diverse Protection

Figure 1.5.   1:N APS architecture with/without diverse protection.

• Dual Homing

In contrast to the *single-homing* approach, which aggregates demands from a CO to destination COs via an associated home hub, *dual homing* is an office backup concept that assigns two hubs to each office and requires dual access to other offices. In the dual-homing approach, demand originating from a special CO[2] is

---

2. Each CO is identified as either a special CO or non-special CO, depending upon the cost and benefit of providing survivability. Special COs, which are selected by telephone companies, are given special treatment for failure conditions.

split between two hubs: a home hub and a designated foreign hub. In the case of a home hub failure, an office that uses dual homing can still access other offices via the backup hub. Figure 1.6 shows such a dual-homing architecture. Dual homing does not automatically accomplish restoration by itself, but may be used in conjunction with dynamic restoration techniques. The dual-homing approach guarantees surviving connectivity, but it may take time to restore priority circuits via path rearrangement.

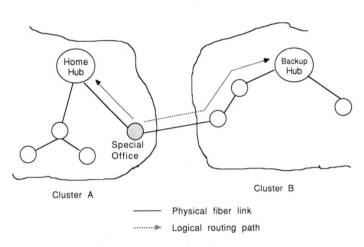

Figure 1.6.   Dual homing architecture.

Self-Healing Ring (SHR).   The SHR, like the 1:1 diverse protection structure, is totally automatic and provides 100 percent restoration capability for a single fiber cable cut and equipment failure. It can also provide some survivability for hub DCS failures and major hub failures (e.g., flooding or fires). As technology advances and competition drives the prices for higher-rate systems toward those of lower-rate systems, SHRs may become even less costly to deploy than low-cost 1:N protection systems. Chapter 4 discusses SHR architectures in more detail.

1.6.2.2 Dynamic Facility Restoration

Dynamic facility restoration techniques use DCSs to reroute demands around a failure point. These techniques are also referred to as *DCS restoration techniques*. Network design using DCS restoration techniques is sometimes referred to as *survivable logical layer network design* (see Chapter 7). DCS restoration does not require separate protection facilities dedicated to working systems for restoration. Instead, it uses spare capacities within working systems to restore affected demands. Figure 1.7 shows an example of DCS restoration. In Figure 1.7, demand between locations A and B is normally routed over a link between DCS#1 and DCS#2, but is rerouted through DCS#3 if cable cuts occur on that link. A centralized or distributed control system may optimize the use of the available spare capacities by referring to a database that contains the current status of the network (both working and spare capacities). The penalties for this flexibility are the time and complexity needed for the controller to communicate with the network DCSs, as well as maintenance of the database. Although DCS restoration does not require dedicated physical facilities for restoration, its architectural potential could be maximized by combining it with dedicated facility restoration techniques. Chapter 5 discusses DCS restoration techniques in more detail.

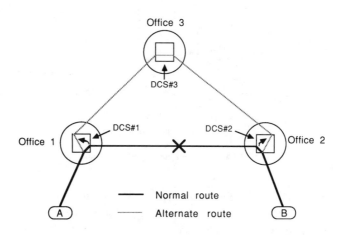

Figure 1.7. DCS restoration.

Table 1-1 compares the four restoration techniques described previously. Among these techniques, diverse protection using APS is the simplest and fastest, but may be the most costly. Dual homing, which provides protection against the major hub failure, has no demand restoration capability, but it can be used in conjunction with DCSs to restore demand. When compared to the APS approach, the SHR restoration technique can reduce costs by sharing equipment and facilities. However, it may cause an expensive system upgrade when the ring capacity is exhausted. The DCS restoration

14

technique does not require dedicated facilities for protection, but it does require a more complicated control scheme and takes longer to achieve efficient use of bandwidth and restoration after network components fail.

A cost-effective, survivable fiber network can be designed in two phases: (1) design cost-effective, survivable network architectures, and (2) optimize the network in terms of equipment and facility planning based on identified network architectures. The first phase defines survivable network architectures based on the foregoing restoration techniques and uses new technologies such as optical switching or SONET equipment (see Section 1.7) as tools to improve traditional survivable network architectures in a cost-effective way. The second phase uses network optimization techniques to allocate capacities for service and protection requirements on a minimum cost basis. Chapters 3, 4, and 5 will discuss the first phase design, and Chapters 6 and 7 will discuss the second phase design.

Table 1-1. Comparison of Facility Restoration Techniques

Restoration technique	Equipment used	Dedicated facility needed	System complexity protection	Restoration time	Major concern	Best network topology	Impacted area
Diverse Protection	OLTM/APS	yes	simple	50 ms	cost	point-to-point/ hubbed	2 points
Dual Homing	OLTM/DCS+	yes	simple	seconds-minutes%	cost	point-to-point/ hubbed	3 points
Self-Healing Ring	ADM	yes/no*	medium	50 ms	capacity exhaust	ring	smaller network
Dynamic Facility Restoration	DCS	no	complex	seconds-hours	system complexity/ restoration time	mesh	larger network

OLTM: Optical Line Terminating Multiplexer
APS: Automatic Protection Switching
ADM: Add-Drop Multiplexer
DCS: Digital Cross-connect System
+ Dual homing, in general, has no restoration technique and can be used in conjunction with DCSs to restore demand.
% The DCS is used in the dual-homing architecture.
* Some SHR architectures may not require a dedicated physical facility for protection (see Chapter 4).

1.7 TECHNOLOGY IMPACTS ON NETWORK SURVIVABILITY

Technological advancements play a crucial role in implementing cost-effective, survivable fiber networks. Among these technologies, SONET and passive optical technology have been shown to reduce survivable fiber network costs.

SONET is a standard in North America that defines both an optical interface and rate and format specifications for broadband optical signal transmission [15]. SONET is designed to transport a wide variety of signal types with a basic signal format that itself contains fixed overhead to support various operations. Network survivability is among the first applications that SONET can provide in a cost-effective manner. SONET technology, along with high-speed (e.g., 2.4-Gbps) add-drop technology, makes self-healing rings practical and economical for use in intraLATA network applications [16]. SONET DCS reconfigurable networks have the capability to restore signals easier and faster than non-SONET DCS reconfigurable networks; this is possible because the overhead communications between DCSs can be established via the SONET *Embedded Operations Channels* (EOCs) that are part of the SONET frame. More details regarding SONET can be found in Chapter 2 and Reference [17].

As mentioned previously, passive optical technology has been shown to reduce survivable fiber network costs [18]. Passive optical components of interest include optical switches, power splitters, wavelength division multiplexers, and optical amplifiers. Optical switching and optical amplifier technologies have been used to implement a cost-effective point-to-point system with 1-for-1 diverse protection against potential fiber cable cuts and a cost-effective SONET self-healing ring architecture [18]. Power splitters have been suggested for use in a cost-effective, optical, dual-homing protection architecture that may prevent major office failures [19]. Wavelength division multiplexers and optical amplifiers have been used to implement optical add-drop multiplexers for a self-healing ring application to accommodate network growth demands in metropolitan intraLATA networks [20]. Chapters 3, 4, and 8 will discuss these cost-effective survivable network architectures that use passive, optical technology.

1.8 BOOK ORGANIZATION

Figure 1.8 depicts the overall organization of this book and the relationships among its nine chapters. The arrow in the figure indicates a relationship between two chapters. For example, understanding Chapter 6 requires understanding the background information provided in Chapters 2 and 3.

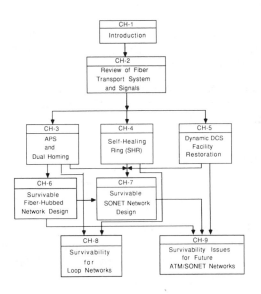

Figure 1.8. Book organization.

The remaining chapters of this book are organized as follows:

- Chapter 2 begins by reviewing architectures and functionality of fiber transport system components, and digital signal formats and hierarchy (including SONET). This chapter serves as a background review for material described in subsequent chapters.
- Chapter 3 discusses various conventional, survivable network architectures that use APS and dual-homing techniques, and shows how to reduce survivable network costs by using passive optical technology.
- Chapter 4 focuses on architectures, analysis, and planning for SONET SHRs, and also shows how to use passive optical technology to reduce ring costs.
- Chapter 5 is dedicated to architectures and analysis for dynamic facility restoration using intelligent SONET DCSs.
- Chapters 6 and 7 discuss network design methods that can be used to build cost-effective survivable networks based on survivable network architectures discussed in Chapters 3-5. (Networks considered in Chapters 3-7 are primarily for interoffice networks.)
- Chapter 8 discusses several survivable network architectures for loop networks that differ in some characteristics from interoffice networks.

17

- Finally, Chapter 9 discusses future *Asynchronous Transfer Mode* (ATM) transport network architectures and the present progress on network survivability in ATM/SONET transport networks.

Chapters 5, 7, and 9 serve as vehicles for stimulating research, development, and planning interests on SONET DCS-based self-healing networks and ATM/SONET survivable broadband switched networks.

1.9 ADDITIONAL REMARKS

Network survivability is one of the most important issues currently faced by a telecommunications industry that is eager to deploy high-capacity fiber systems. Many efforts have been explored to find affordable plans for service assurance, such as a special issue of *IEEE Communications Magazine* [21] and a special symposium of ComForum [22]. These efforts share the telecommunications industry's enthusiasm and may provide some reasonable solutions to the problems of fiber network survivability.

REFERENCES

[1] Wilson, C., and Lindstorm, A., "Survival of the Network," *Telephony*, October 23, 1989.

[2] Bailly, R., and Corporation, V., "Militarized Vehicular LANs," *FIBER OPTICS*, January/February 1988, pp. 21-26.

[3] Weik, M. H., *Communications Standard Dictionary*, Van Nostrand Reinhold Company, New York, 1983.

[4] Wu, T-H., Kolar, D. J., and Cardwell, R. H., "Survivable Network Architectures for Broadband Fiber Optic Networks: Model and Performance Comparisons," *IEEE Journal of Lightwave Technology*, Vol. 6., No. 11, November 1988, pp. 1698-1709.

[5] Timms, S., Kee, R., and Toncar, J., *Broadband Communications: the Commercial Impact*, Ovum Inc., 1987.

[6] Kerner, M., Lemberg, H. L., and Simmons, D. M., "An Analysis of Alternative Architectures for the Interoffice Network," *IEEE Journal on Selected Areas in Communications*, Vol. Sac-4, No. 9, December 1986, pp. 1404-1413.

[7] *Hinsdale Center Office Fire Final Report: Executive Summary*, Prepared by Forensic Technologies International Corporation, Annapolis, MD, March 1989.

[8] Peters, D. B., "Continuity Planning: A Study of Customer Needs," Proceeding of ComForum (Service Continuity and Disaster Preparedness), March 1990, pp. B3.1-B3.10.

[9] Anderson, B., and Yuce, H., "Optical-Fiber Reliability," *Bellcore Exchange*, September/October 1990, pp. 9-13.

[10] Ash, G. R., Cardwell, R. H., and Murray, R. P., "Design and Optimization of Networks with Dynamic Routing," *Bell System Technical Journal*, Vol. 60, No. 8, October 1981.

[11] Krishnan, K. R., and Ott, T. J., "Forward-Looking Routing: A New State-Dependent Routing Scheme," *Proceedings of the 12th International Teletraffic Congress*, Torino, Italy, June 1988.

[12] Wrobel, L. A., Jr., *Disaster Recovery Planning for Telecommunications*, Artech House, July 1990.

[13] Falconer, W. E., "Service Assurance in Modern Telecommunications Networks," *IEEE Communications Magazine*, Vol. 28, No. 6, June 1990, pp. 32-39.

[14] Cardwell, R. H., and Brush, G., "Meeting the Challenge of Assuring Dependable Telecommunications Services in the '90s," *IEEE Communications Magazine*, Vol. 28, No. 6, June 1990, pp. 40-45.

[15] Ballart, R., and Ching, Y-C., "SONET: Now It's the Standard Optical Network," *IEEE Communications Magazine*, March 1989, pp. 8-15.

[16] Wu, T-H., and Burrowes, M., "Feasibility Study of a High-Speed SONET Self-Healing Ring Architecture in Future Interoffice Fiber Networks," *IEEE Communications Magazine*, Vol. 28, No. 11, November 1990, pp. 33-42.

[17] TA-TSY-000253, *Synchronous Optical Network (SONET) Transport Systems: Common Generic Criteria*, Issue 6, Bellcore, September 1990.

[18] Wu, T-H., "Roles for Optical Components in Survivable Fiber Networks," *Digest of Optical Communications Conference (OFC'92)*, CA, February 1992, pp. ThL1.

[19] Wu, T-H., "A Novel Architecture for Optical Dual Homing Survivable Fiber Networks," *Proceedings of IEEE International Conferences on Communications (ICC)*, Atlanta, GA, April 1990, pp. 309.3.1-309.3.6.

[20] Wu, T-H., Kolar, D. J., and Cardwell, R. H., "High-Speed Self-Healing Ring Architectures for Future Interoffice Networks," *IEEE Global Communications Conference (GLOBECOM)*, November 1989, pp. 23.1.1-23.1.7.

[21] Pickholtz, R. L. (Editor), "Special Issue for Surviving Disaster," *IEEE Communications Magazine*, Vol. 28, No. 6, June 1990.

[22] ComForum on "Service Continuity and Disaster Preparedness," sponsored by National Engineering Consortium, Phoenix, AZ, March 1990.

CHAPTER 2

Fiber Transport System Components and Signals

To understand survivable fiber network architectures and their operations, it is essential to know how fiber transport systems establish an optical path. This chapter reviews fiber transport systems and signals and, thus, provides background information for readers who are unfamiliar with fiber network transport systems and the ways in which they establish optical communication channels.

This chapter first discusses an end-to-end optical path between customer premises and then addresses transport signals and component architectures associated with this end-to-end optical path.

2.1 OPTICAL PATH FOR FIBER NETWORKS

The optical path for fiber networks discussed herein is based on the fiber-hubbing network architecture, which has commonly been implemented in intraLATA fiber networks. Figure 2.1 depicts a possible network configuration for an end-to-end optical path. This configuration divides the network functions into two parts: interoffice and loop networks. The interoffice network is a set of *Central Offices* (COs) connected by high-speed fiber links. In interoffice fiber-hubbing networks, *Terminal Multiplexers* (TMs) or *Add-Drop Multiplexers* (ADMs) are placed at COs, which are connected by optic fibers, and ADMs and/or *Digital Cross-connect Systems* (DCSs) are installed at the hub, which serves as a traffic concentration point for a group of COs. Regenerators are needed if the fiber span between two offices exceeds a repeater spacing threshold. A fiber *span* is a network segment that has terminal equipment (except regenerators) at its two ends and has fibers spliced at intermediate COs if the span passes through one or more links in the network topology. The loop network is used to build interfaces between customers' premises and the interoffice transport network. Chapter 8 provides a more detailed discussion on network architectures for fiber loops.

Other network components such as wavelength division multiplexers/filters, optical switches, and optical amplifiers can also be added into the network architecture for network survivability applications. However, this chapter does not discuss these network components. Instead, later discussions about specific applications incorporate information about these components (see Chapters 3, 4, and 8).

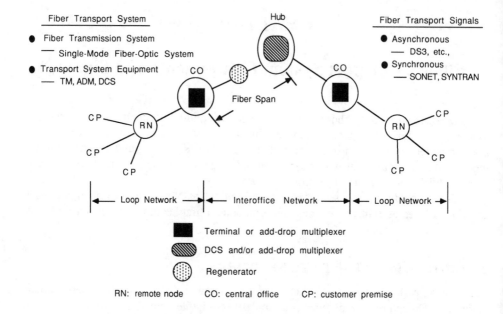

Figure 2.1. Possible configuration for an end-to-end optical path.

As shown in Figure 2.1, a fiber transport system and signals constitute an end-to-end optical communications channel. An interoffice fiber transport system includes two major elements: fiber-optic transmission systems and fiber transport equipment. Section 2.2 reviews a basic fiber-optic transmission system and associated power budget planning. Sections 2.3 and 2.4 discuss fiber transport signals including present asynchronous signal hierarchy and *Synchronous Optical Network* (SONET) hierarchy. Sections 2.5 and 2.6 discuss fiber transport equipment, which includes TMs, ADMs, and DCSs. In this book, we focus particularly on the SONET-based network components because SONET is a standard optical transmission system that has been well accepted by both service providers and equipment vendors. Because regenerators are not key components in providing network survivability, they are not discussed in this chapter. Interested readers may refer to Bellcore Technical Reference TR-NWT-000917 [1] for more details on SONET regenerators.

2.2 FIBER-OPTIC TRANSMISSION SYSTEM

2.2.1 System Components

As depicted in Figure 2.2, a fiber-optic transmission system for a fiber section includes a transmitter, a receiver, a fiber cable, splices, and connectors. A fiber span may include two or more sections if the span length exceeds a repeater spacing. The transmitter is a light source that modulates the signal and feeds it to the fiber. The fiber is a transmission medium that conveys the optical signal to the receiver. The fiber cable, which houses a number of fibers, protects the fibers from excessive strain and prevents microbending loss. The number of fibers placed in a cable depends on the application. Commercially available fiber cables for telecommunications applications may contain 12 to 144 fibers. The receiver detects and decodes the optical signal, and converts it to an electronic form, which can then be processed by equipment at the receiving end (e.g., ADM, DCS). The connector makes a temporary connection between two fiber ends, or between a fiber end and a transmitter or receiver. In contrast, splicing is a permanent junction between two fibers. Repeaters are needed when the fiber span exceeds a distance threshold that makes received signals too weak for the receiver to detect. A regenerator, which is a unidirectional device, contains a receiver and a transmitter connected in series. The receiver detects a signal from a distant transmitter, regenerates it, and produces a signal that drives the transmitter in the regenerator.

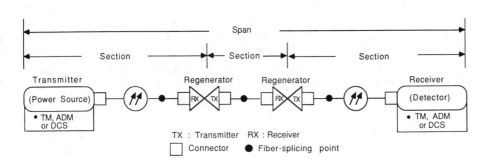

Figure 2.2. A fiber-optic system for a fiber span.

23

Optical Fibers

Two types of fiber that are used today are multi-mode and single-mode fiber. Multi-mode fiber allows multiple modes of light rays to travel through fiber, whereas single-mode fiber allows only one. Fiber's capacity to transmit an optical signal is limited by two factors: attenuation and dispersion. Attenuation, which results in power loss, occurs either when impurities in the fiber absorb light energy or when Rayleigh scattering occurs. Power loss due to attenuation is measured by comparing output power with input power in decibels (dB) as follows:

$$dB \; loss = -10 \times \log_{10} \left(\frac{Power \; out}{Power \; in} \right)$$

Attenuation depends on wavelength, and minimum attenuation occurs using wavelength windows at 1300 and 1550 nanometers (nm). Single-mode fiber, which usually operates at 1300 nm, has longer repeater spacing than multi-mode fiber; this is mainly because multi-mode fiber has serious dispersion problems. For example, the *Fiber Distributed Data Interface* (FDDI), which uses multi-mode fiber to carry signals at about 100 Mbps, may require a regenerator every 2 to 3 km, whereas single-mode fiber may carry signals at higher speeds (e.g., 600 Mbps) over a longer distance (e.g., 50 km) without using regenerators. Single-mode fiber with loss as low as 0.4 dB/km at 1300 nm is commercially available.

Interconnecting fibers using splices (permanent connections) and connectors (temporary connections) also cause attenuation. Although power losses due to splices and connectors for single-mode fiber are higher than those for multi-mode fiber, the combined loss of fiber, splices, and connectors for single-mode fiber is still considerably less than for multi-mode fiber. Splicing and connector losses for single-mode fiber at 1300 nm are typically less than 0.1 dB and 0.5 dB, respectively.

Dispersion, which results from light pulses spreading as they propagate through fiber, is a limiting factor for the data rate. Single-mode fiber, which usually operates at 1300 nm, has much lower dispersion than multi-mode fiber. The lower dispersion can be attributed to single-mode fiber having near zero chromatic dispersion at 1300 nm, where the chromatic dispersion is the sum of material dispersion and waveguide dispersion. This results in single-mode fiber being able to transmit data at a much higher rate than multi-mode fiber. The typical maximum data rate for multi-mode fiber is about 300 Mbps with repeater spacing less than 20 km. Commercially available single-mode fiber may carry data at rates up to 2.5 Gbps with repeater spacing typically beyond 35 km. Additionally, single-mode fiber that carries data at 10 times that rate has been tested in industrial research laboratories. Table 2-1 summarizes a relative comparison between single-mode and multi-mode fiber.

Table 2-1. Relative Comparison between Single-mode and Multi-mode Fiber

Fiber type	Data rate	Power loss	Repeater spacing
Multi-mode	lower (<300 Mbps)	higher	shorter
Single-mode	higher (>1.7 Gbps)	lower	longer

Single-mode fiber that operates at 1550 nm is also commercially available. In comparison to 1300-nm single-mode fiber, 1500-nm single-mode fiber has higher chromatic dispersion, but only about half the power loss.

Although multi-mode fiber cannot match the data rate of single-mode fiber, it does offer adequate capacity at lower costs for most systems[1] that are no more than a few kilometers long. For telecommunications applications in intraLATA and interLATA areas where high-speed data rates over a relatively long distance are required, single-mode fiber at 1300 nm or 1550 nm is the transmission system of choice. Thus, throughout this book, the fiber discussed is referred to as *single-mode fiber*.

Transmitters

Two types of transmitters have commonly been used in fiber-optic transmission systems: *Light Emitting Diodes* (LEDs) and semiconductor lasers. Table 2-2 shows a relative comparison between these transmitters. In general, LEDs, which are relatively inexpensive transmitters with low output power, are more appropriate for applications requiring moderate data rates (typically less than 200 Mbps) within a relatively short communication distance. In contrast, semiconductor lasers are more expensive and have higher output power. They typically transmit data up to 1 to 2 Gbps and can be used for telecommunications applications that require high-speed data rates and longer communication distances. Both LEDs and semiconductor lasers can operate at 1300-nm and 1550-nm wavelengths; however, semiconductor lasers have excellent compatibility with single-mode fiber, and LEDs usually do not.

---

1. A system includes not only fibers, but also transmitters and receivers.

Table 2-2. Relative Comparison between LEDs and Semiconductor Lasers

Transmitter type	Output power	1300-nm or 1500-nm operations	Compatibility with single-mode fiber	Repeater spacing	Maximum data rate	Unit cost
LEDs	lower	yes, but unusual	possible	shorter	< 200 Mbps	lower
Semiconductor Lasers	higher	yes	excellent	longer	> Gbps	higher

Receivers

A receiver is a device that detects incoming photonic signals and converts them to an electronic form. Signal quality is determined by two factors: signal strength and noise. Two measurements are generally used to determine the signal quality: *Signal-to-Noise Ratio* (SNR) and *Bit Error Ratio* (BER). SNR, which is the signal power divided by noise power, is normally represented in decibels (dB). The higher the SNR, the better the received signal. BER is the number of erroneous bits received, divided by the total number of bits transmitted over a predetermined detection time period. In fiber-optic systems, BER is much easier to measure than SNR. Theoretically, SNR greater than 16 dB is equivalent to BER of 10^{-9}; however, in practice, SNR greater than 20 dB may be equivalent to BER of 10^{-9}. The tolerable maximum BER depends on the service for which the system is designed. For example, the maximum tolerable BER for voice service is 10^{-3}, and this BER threshold drops to 10^{-9} for data services, such as local area network applications [2].

2.2.2 Optical Span Design (Power Budget Planning)

Fiber-optic system planners need to understand fiber span design to evaluate the impact of incorporating additional components required for new architectures (such as mechanical optical switches or *Wavelength Division Multiplexers* [WDMs]) into existing fiber-optic systems. The most important design consideration for fiber-optic systems is the power budget. By subtracting all of the system's optical losses from the power emitted by the transmitter, planners ensure that the system retains enough power to drive the receiver at the required level. The minimum, allowable input power for the receiver is called *receiver sensitivity*, and it depends on the specified BER. The difference between the transmitter's output power and the receiver sensitivity is called *system gain*. A design should also leave some extra margin above the receiver's minimum power requirement for system degradation and fluctuations, and/or for incorporating additional components into a fiber span to provide new network

26

capabilities or services. Depending on the application, the BER performance requirements, and the cost, the power margin may be 3 to 10 dB. A general equation for power budget planning is as follows:

$$(Transmitter\ Output) - (Receiver\ Sensitivity) = (\sum Losses) + Margin$$

Power losses in a fiber-optic system include light source-to-fiber coupling loss, connector loss, splice loss, fiber loss, and fiber-to-receiver coupling loss.

Repeater spacing is the maximum length that a signal can travel with a specific BER at the receiver without signal regeneration. Repeater spacing is determined primarily by transmitter power, receiver sensitivity, data rate, and system losses, where receiver sensitivity is determined by the required BER performance. Repeater spacing is calculated as follows:

$$repeater\ spacing = \frac{(allowable\ fiber\ and\ splicing\ loss)}{(fiber\ and\ splicing\ loss\ /km)}$$

$$= \frac{(transmitter\ output\ power - receiver\ sensitivity - connector\ loss - margin)}{(fiber\ and\ splicing\ loss\ /km)}$$

The following example shows how to calculate the maximum repeater spacing for a

Maximum Repeater Spacing - 155 Mbps (1300 nm)

Laser output power	0 dBm
Receiver sensitivity (for BER = 10^{-9})	-38 dBm
System gain	38 dB
Connector loss (2)	1 dB
Margin	5 dB
Allowable fiber and splicing loss	32 dB
Splicing loss (1-km section)	0.15 dB/km
Fiber loss (1300 nm)	0.5 dB/km
Fiber and splicing loss	0.65 dB/km
Maximum repeater spacing:	$\dfrac{32\ dB}{0.65\ dB/km}$ = 49 km (or 30.5 miles)

Table 2-3 shows an example of relationships between data rates and maximum repeater spacing for fiber-optic systems operating at 1300 and 1550 nm. In Table 2-3, we assume that fiber loss is 0.5 dB/km for 1300 nm and 0.3 dB/km for 1500 nm. The connector loss is 0.5 dB; the splicing loss is 0.15 dB; the power margin is 5 dB; and the

27

BER requirement is 10^{-9}. Note that the maximum repeater spacing shown in Table 2-3 represents optimistic values, which include only power loss. Other factors, such as dispersion and mode-partition effects, may become significant for high-speed systems operating above gigabit-per-second rates. However, network planners may still find this information useful because it represents an upper bound for repeater spacing.

Table 2-3. Example of Maximum Repeater Spacing vs. Data Rates for Single-mode Fiber Systems

Data rate	Receiver sensitivity $(BER=10^{-9})$	Transmitter output power	Maximum repeater Spacing	
			1300 nm	1500 nm
155 Mbps	-38 dBm	0 dBm	49 km (30.5 miles)	71 km (44.1 miles)
622 Mbps	-35 dBm	0 dBm	44 km (27.7 miles)	64 km (40.0 miles)
1.2 Gbps	-32 dBm	0 dBm	40 km (24.8 miles)	58 km (35.9 miles)
2.4 Gbps	-29 dBm	0 dBm	35 km (21.9 miles)	51 km (31.7 miles)

Note: 1 mile = 1,609.35 meters

We now consider one application using the above power budget planning guidelines. As will be discussed in Chapter 3, *Mechanical Optical Switches* (MOSs) can be used to enhance network survivability against fiber cable cuts with moderate cost increase. An architecture that uses MOSs can be implemented by upgrading existing 1:N fiber systems. Questions remain concerning the impact, in terms of power budget, of adding two MOSs to an existing fiber system. Assume that each MOS has 1 dB loss and requires two connectors with 0.5 dB each. For a 1.2-Gbps fiber-optic system operating at 1300 nm, the maximum repeater spacing is reduced from 40 km (24.8 miles) to 29.2 km (18.1 miles) to keep the received signals at the required BER level of 10^{-9}. In other words, if the considered fiber span between two COs is less than 18.1 miles, the existing 1:N system may be upgraded to the new system without adding regenerators or using transmitters with much higher output power. (Some cost penalties are associated with such upgrades.)

2.3 PRESENT DIGITAL SIGNAL HIERARCHY

Time Division Multiplexing (TDM) is a process that interleaves several low-speed digital signals to form a high-speed digital signal. The reverse process is called *time division demultiplexing*. A digital multiplexer is a device that performs not only signal *multiplexing/demultiplexing* (MUX/DMUX) functions, but also control, monitoring, and testing functions. A fundamental problem of digital multiplexing is accommodating incoming signals that are not identical in bit rate. The process of bringing input signals to a common bit rate during the multiplexing process is called

28

multiplex synchronization. Asynchronous and synchronous multiplexing essentially use different synchronization schemes.

For an asynchronous multiplexing scheme, the time slots in the high-speed signal are not preassigned to incoming low-speed signals unless these low-speed signals are only one level below the high-speed signal in the asynchronous digital hierarchy. Thus, the input signals are "invisible" within the output high-speed signal until a reverse process (i.e., demultiplexing) is performed. (An asynchronous DS3 signal is an example of an asynchronous signal.) In contrast, synchronous signals have the same average bit rate, and they can be easily justified in frequency at the point of multiplexing. *Synchronous Transmission* (SYNTRAN) DS3 is an example of a synchronous signal.

2.3.1 Asynchronous Digital Hierarchy

Today's asynchronous digital hierarchy includes five levels based on *International Telegraph and Telephone Consultative Committee* (CCITT) Recommendation G702. Table 2-4 shows the digital signal rate for each level in North America. There are five levels of digital signals used in North America: DS0 (64 kbps), DS1 (1.544 kbps), DS2 (6.312 Mbps), DS3 (44.736 Mbps), and DS4 (274.176 Mbps), where DSn denotes Digital Signal level n.

Table 2-4. Asynchronous Digital Hierarchy in North America

Digital signal level	Data rate (Mbps)	Carrier system	Mapping
DS0	0.064	-	-
DS1	1.544	T-1	24 DS0s/DS1
DS2	6.312	T-2	4 DS1s/DS2
DS3	44.736	T-3	7 DS2s/DS3
DS4	274.176	T-4	6 DS3s/DS4

A DS0 circuit (64 kbps) is used to carry a digital voice sample. A DS1 signal, as depicted in Figure 2.3, is formed from 24 DS0s by using a multiplex device, called a *Channel-Bank* (D-Bank). In this DS1 frame structure, 193 bits are repeated every 125 µs: 24 time slots (8 bits each), which are byte-interleaved together, and one bit (F-bit) for frame indication.

A DS3 signal is formed by bit-interleaved multiplexing 28 DS1 (1.544 Mbps) channels using an M13 asynchronous multiplexer. Under today's asynchronous signal multiplexing schemes, two-step multiplexing is needed to form an asynchronous DS3 from DS1s, as depicted in Figure 2.4. DS1s are first multiplexed into DS2s and then into DS3s. During each multiplexing step, overhead and stuffing bits are added. Such

29

a layered multiplexing scheme was essentially dictated by the circuit design limitations of the 1970s and the use of point-to-point configurations where each configuration had its own internal clock. Basically, this meant that phase and frequency differences existed between the various multiplexed signals.

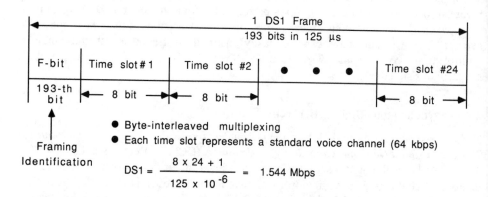

Figure 2.3. DS1 frame structure.

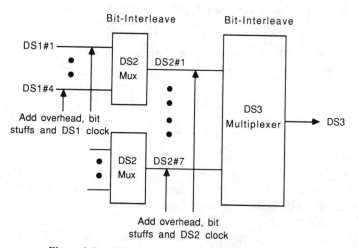

Figure 2.4. Forming an asynchronous DS3 from DS1s.

30

Bit-stuffing is used in asynchronous multiplexing to synchronize low-speed input signals and to bring the combined low-speed input signals up to the bit rate established by the multiplex high-speed pulse generator. In this process, "stuff bits" are added, as necessary, in available time slots to increase the low-speed signal rates. If stuffing is not required at the moment a stuff slot is to be filled, a valid information bit is inserted instead.

When a high-speed signal is demultiplexed, stuff bits must be removed, but information bits must be retained. For equipment to distinguish between a stuff bit and an information bit, each high-speed signal frame includes "stuff control bits," as depicted in Figure 2.5(a) (labeled as "C" bits). These bits appear in fixed positions inside each signal frame and indicate whether the stuff time slot (S) carries real or dummy data. If at least two out of three C bits are set to 1, the S bit carries dummy data; otherwise, the S bit carries real data. Bit-stuffing can accommodate normal (asynchronous) frequency variation of the multiplexed payloads. However, access to those payloads from the high-level multiplexed signal is difficult because the tributary signal must first be "destuffed" to separate real data from dummy data, and then the framing pattern of the payload must be identified if complete payload access is required.

For asynchronous signals, the DSn data rate is not equal to n times the DS1 data rate because bit-stuffing is used in forming a high-speed signal. Thus, locations of DS1s within a DS3 are not fixed, which makes direct access of DS1 signals from an asynchronous DS3 difficult. Another disadvantage of asynchronous multiplexing is the absence of spare capacity for *Embedded Operations Channels* (EOCs).

The deployment of the DS3 signal level, which has commonly been used as a basic transport unit in today's interoffice fiber telephone networks and private networks, is rapidly increasing. It is estimated that by the end of 1990, about one billion circuit miles will interface at DS3 or higher rates in intraLATA and interLATA telecommunications networks. For current private networks, the DS3 primarily carries voice traffic. However, this trend is expected to shift to DS3s carrying integrated voice, data, and video signals over the next few years. This anticipated growth in DS3 facility and terminal requirements has caused the industry to seek more efficient ways of handling DS3 signal multiplexing.

With the technology advancement of *Very Large Scale Integration* (VLSI) and the evolution toward a synchronous network,[2] Bellcore proposed a SYNTRAN DS3 format in 1984 [3].

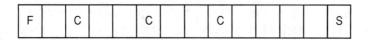

F : Framing indication bit
C : Stuff indicator bit
S : Stuff bit or real data

(a) Bit-Stuffing (Asynchronous Multiplexing)

F   : Framing indication bit
B_i : Fixed frame position for the i-th tributary signal

(b) Fixed Location Mapping (Synchronous Multiplexing)

Figure 2.5.   Payload multiplexing methods.

---

2. The synchronous network is defined as the synchronization of synchronous transmission systems with synchronous payloads to a master (network) clock that can be traced to a reference clock.

2.3.2 Synchronous Transmission (SYNTRAN)

SYNTRAN uses a restructured DS3 signal format for synchronous transmission at the DS3 level. The multiplexing method used for SYNTRAN is called *fixed location mapping,* as depicted in Figure 2.5(b). Fixed location mapping is a scheme that uses specific bit positions in a high-rate synchronous signal to carry low-rate tributary signals. For example, in Figure 2.5(b), frame position B_2 would always carry information from one specific tributary signal. This method allows easy access to the transported DS1 within a SYNTRAN DS3. However, a 125-μs slip buffer is needed for each DS1 to equalize the bit rate of incoming low-rate signals to a common rate before interleaving occurs. The slip buffer corrects possible phase-alignment problems and small frequency variation problems (by repeating or deleting a frame of information to correct frequency differences) between the high-rate signal and its tributary. Whereas this slip buffer is advantageous, it introduces undesirable delay and potential signal impairment due to slipping.

Compared with today's asynchronous DS3, SYNTRAN DS3 has the following advantages:

1. SYNTRAN provides convenient DS3 grooming[3] by directly accessing all DS0 and DS1 signals, and one-step multiplexing from DS1s, which eliminates the need for the intermediate DS2 multiplexing stage.
2. SYNTRAN frees up bandwidth used for the bit-stuffing process required in today's asynchronous DS3 signals. This additional bandwidth (64 kbps) is then used to provide EOCs, which allow the introduction of enhanced maintenance, testing, and provisioning features.

However, SYNTRAN may also introduce undesirable delay and signal impairment, which is not experienced by asynchronous signals.

2.4 SYNCHRONOUS OPTICAL NETWORK (SONET)

Increasing deployment of fiber facilities in telecommunications networks raises concerns about service efficiency on an end-to-end basis due to the lack of signal standards for optical networks. This service efficiency concern, along with the need for supporting broadband services, which require bandwidth beyond the DS3 level (e.g.,

---

3. Service (or facility) grooming is a process of consolidating and segregating a service (or facility) for special purposes.

High Definition Television [HDTV]), led to the establishment of a national standard signal format that supports present services and future broadband services. This optical signal format has been defined as SONET [4-6]. SONET is a standard in North America that defines both an optical interface, and rate and format specifications for optical signal transmission. It can support both broadband and narrowband services. Bellcore initiated SONET in February 1985 in response to MCI's proposal on interconnecting multiowner, multimanufacturer, fiber-optic transmission terminals in the *Interchange Carrier Compatibility Forum* (ICCF) in 1984 [7]. In 1988, the *American National Standards Institute* (ANSI) approved Phase I of SONET as an American standard, and CCITT approved it as an international standard. SONET Phase I specifies transmission rates, signal formats, optical interface parameters, and some payload mappings; however, it does not standardize the operations and maintenance functions that must also be exchanged between *Network Elements* (NEs). Phase II[4] of SONET, which defines the message set and protocols for using overhead channels for *Operations, Administration, Maintenance, and Provisioning* (OAM&P), is expected to be completed by U.S. standards groups (ANSI) by 1991. Phase II includes four major components: a protocol stack, a language, a message structure, and a common view of the data. References [7] and [8] provide a historical review and the standards development status for SONET.

SONET originally was envisioned as a set of standards that would encourage multivendor solutions to network engineering by allowing a "midspan meet" and positioning the network for better end-to-end service. A *midspan meet* is the ability to deploy different vendors' equipment anywhere along a fiber span and have the different pieces communicate intelligently. The transport system in today's public networks is provided by numerous point-to-point, asynchronous transmission systems using proprietary signals that do not allow the interconnection of different vendors' equipment or even of different equipment from the same vendor. Furthermore, SONET carries the SYNTRAN concept throughout the SONET synchronous hierarchy and provides single-step multiplexing for different levels of SONET signals.

Current SONET interfaces are specified for single-mode fiber using 1310-nm and 1550-nm laser sources and interconnection lengths up to 40 km. Optical interfaces for short-range connections will be specified in Phase II of the SONET standards to allow low-cost interconnections for short distances, such as intraoffice lines.

---

4. Part of Phase II has been moved to newly created Phase III.

2.4.1 SONET Benefits and Deployment

SONET provides more than increased bandwidth. Much of its value comes from being designed for synchronous transmission systems and EOCs that will reduce network operations costs and provide new services. The planning and deployment of SONET will create a transport network that is both backward-compatible with today's asynchronous networks and forward-compatible with future network architectures, such as *Broadband Integrated Services Digital Network* (B-ISDN). Table 2-5 shows a relative comparison between asynchronous systems and SONET; more detailed discussion will be provided in subsequent sections.

Table 2-5. Relative Comparison between Asynchronous Systems and SONET

Attributes	Asynchronous systems	SONET
High-Speed Interface	electrical	optical
Low-Speed Interface	electrical	optical/electrical
Maximum Data Rate	DS3 (44.7 Mbps)	OC-48 (2.488 Gbps)*
Direct Access Tributary Signals	no	yes
Payload Interleave	bit-interleave	byte-interleave
Multiplexing Method	bit-stuffing	pointer processing
Clock	local	common
Multiplexing Data Rates	DS3≠3×DS1	STS-N=N×STS-1
Midspan Meet	no	yes
Embedded Operations Channel	no+	yes
Remote Option Settings	no	yes
Applications	point-to-point	point-to-point and multipoint networks

* It can be extended to beyond 2.488 Gbps in the future.
+ Most asynchronous signals do not have EOCs; one of exceptions is the *Embedded Superframe Structure* (ECF) for DS1.

The major advantages of SONET over the asynchronous transmission systems used in many of today's public networks are as follows:

35

1. Creates a multivendor environment with a midspan meet capability; thus, it simplifies the interconnection of equipment from different vendors and allows for fast restoration in emergency situations.
2. Provides single-step multiplexing, which reduces the number of redundant terminations and interfaces, and allows greater equipment integration in terms of physical size and application functionality.
3. Provides higher network reliability because the number of network components in an end-to-end path is significantly reduced.
4. Provides a multiplexing hierarchy above DS3 that is not available today but is needed for some broadband services, such as HDTV.
5. Reduces costs for OAM&P using SONET EOCs and enhances remote provisioning and control capabilities of NEs by providing faster response time in meeting customers' needs.
6. Provides economical drop and insert by using ADMs, and eliminates the need for back-to-back terminal configurations.
7. Allows efficient and cost-effective network architectures (e.g., hubbing architectures using SONET DCSs and SHR architectures) to be implemented and operated. These architectures will position the network to provide new capabilities (e.g., survivability) and services (e.g., customer broadband network management).

Due to the above advantages, service providers have viewed SONET as their migration path to the network of the future. SONET's capabilities have also made it the first layer (the physical connection layer) in a B-ISDN network architecture hierarchy.

The benefit of operations cost reduction and network component integration may justify the slightly higher investment required to initially deploy SONET. Later deployment will be driven by new applications and new network capabilities, such as network survivability, broadband services, and customer network management. Recently, Bellcore conducted a study on the deployment forecast of SONET for intraLATA interoffice networks and loop feeders. This study was based on the information the *Bellcore Client Companies* (BCCs) provided to the *Federal Communications Commission* (FCC) regarding the number of working and equipped interoffice and loop feeder circuits from 1980 to 1989 and forecasted circuits through 1994. According to this information (estimated percentages by Bellcore varying from 15 percent in 1991 to 100 percent in 1996), SONET will be the primary fiber-optic system for new traffic growth and for replacement of non-fiber systems. Figure 2.6 shows the results of this study in terms of percentages of circuits carried on fiber-optic systems and SONET in intraLATA, interoffice networks. This study indicates that 50 percent and 99 percent of the intraLATA, interoffice circuits will be on SONET in

1997 and 2003, respectively, with a 10-year retirement planned for proprietary fiber-optic systems. Loop feeder circuits are estimated to be 50 percent and 90 percent on SONET in 2003 and 2011, respectively.

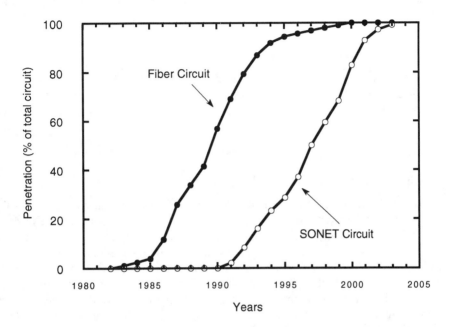

Figure 2.6.   Fiber and SONET penetration in intraLATA, interoffice networks (data courtesy of R. Clapp and R. Marzec at Bellcore).

Public network transmission products based on the SONET Phase I standard began appearing in 1989. Subsequently, field trials and deployment of SONET-compatible equipment among local telephone companies began in 1990 [9]. SONET deployment can be loosely divided into two phases. The first phase consists of initially deploying point-to-point systems, like the ones used today, in both the interoffice and feeder portions of the telephone company networks. Because of the standard interfaces, it is expected that, in the second phase of deployment, these point-to-point SONET systems will be interconnected to provide multivendor, end-to-end optical networks through high-speed SONET ADMs, and through SONET wideband and broadband DCSs. Also, more survivable network architectures, such as SONET SHRs, are expected to be built in the 1992-1993 time frame. At a much later date, SONET interfaces on digital switches, including *Asynchronous Transfer Mode* (ATM) switches, are expected to be available.

The following sections provide a brief overview of SONET that will be helpful in understanding later chapters on network restoration. Most of this information is taken from [4,6,7].

2.4.2 SONET Frame Structure and Hierarchy

Several considerations have impacted the design of the SONET frame structure and signal hierarchy. These considerations include flexibility of supporting different services, simplicity in cross-connection, benefits from synchronous networks, facility maintenance as an integral part of the network, modularity for growth, and compatibility with present networks.

2.4.2.1 STS-1 Frame Structure and Multiplexing

The basic building block (i.e., the first level) of the SONET signal hierarchy is called the *Synchronous Transport Signal-Level 1* (STS-1). The STS-1 has a bit rate of 51.84 Mbps and is divided into two portions: transport overhead and information payload, where the transport overhead is further divided into line and section overheads (see Section 2.4.3.2). The STS-1 frame consists of 90 columns and nine rows of eight-bit bytes, as depicted in Figure 2.7. The transmission order of the bytes is row by row, from left to right, with one entire frame being transmitted every 125 µs. The first three columns contain transport overhead, and the remaining 87 columns and nine rows (a total of 783 bytes) carry the STS-1 *Synchronous Payload Envelope* (SPE). The SPE contains a nine-byte path overhead that is used for end-to-end service performance monitoring (see Section 2.4.3.2). *Optical Carrier Level-1* (OC-1) is the lowest-level optical signal used at equipment and network interfaces. OC-1 is obtained from STS-1 after scrambling and *electrical-to-optical* (E/O) conversion.

38

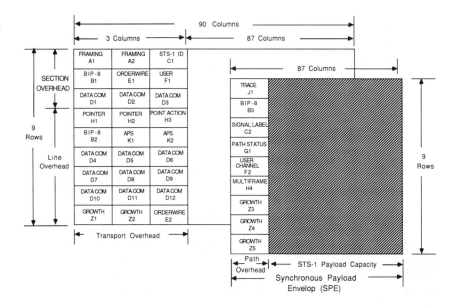

Figure 2.7. SONET STS-1 frame and overhead channels.

Higher-rate optical signals are formed by byte-interleaving an integral number of STS-1s. Figure 2.8 depicts an example of how an OC-N signal is formed. In Figure 2.8, services such as DS1 (along with path overhead) are first mapped into an SPE. Line overhead is then added to form an STS-1. A number of STS-1s are then byte-interleaved and multiplexed to form an STS-N. The frame is scrambled after section overhead (except framing and STS-N ID) is added. Framing and STS-N ID are then added into the section overhead of the scrambled STS-N, and finally, the STS-N signal is converted into optical signal OC-N.

The OC-N line rate is exactly N times that of OC-1, and all higher- rate standard interface signals are readily defined in terms of STS-1. Currently, several optical OC-N parameters are defined in SONET, as depicted in Table 2-6, where the present maximum value of N is 255 [4]. The OC-N signal may be formed from different combinations of OC-M signals ($M \leq N$). For instance, an OC-12 can be composed of 12 OC-1s, four OC-3s, or any combination of OC-1s and OC-3s that equals OC-12.

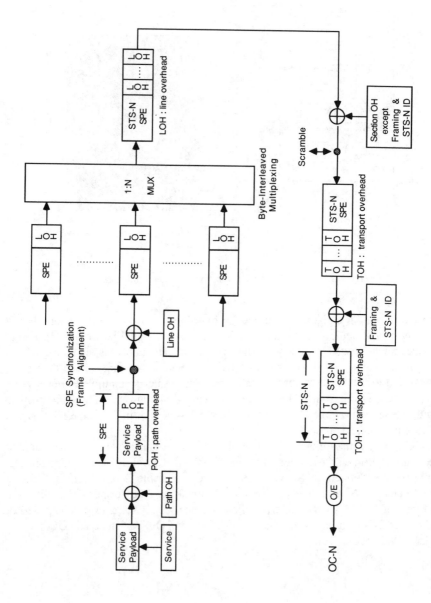

Figure 2.8. An example of forming an OC-N signal.

40

Table 2-6. Levels of SONET Signal Hierarchy

Level	Line rate (Mbps)	Digital hierarchy (ANSI)	SDH* (CCITT)
OC-1	51.84	STS-1	-
OC-3	155.52	STS-3	STM-1
OC-9	466.56	STS-9	-
OC-12	622.08	STS-12	STM-4
OC-18	933.12	STS-18	-
OC-24	1244.16	STS-24	-
OC-36	1866.24	STS-36	-
OC-48	2488.32	STS-48	STM-16

* SDH: Synchronous Digital Hierarchy
STS: Synchronous Transport Signal
STM: Synchronous Transport Module

Broadband services requiring more than one STS-1 payload capacity are transported by concatenated STS-1s. For example, HDTV signals requiring 135 Mbps can be carried by three concatenated STS-1s, denoted by STS-3c, whose transport overheads and payload envelopes are aligned. Figure 2.9 depicts an STS-3c frame overhead and information payload format. In this STS-3c, the first of three H1 and H2 bytes contains a valid SPE pointer, whereas the second and third H1 and H2 bytes contain a concatenation indicator that prevents the STS-3c signals from being demultiplexed. Concatenation specifies that these signals are considered a single unit and transports them as such through the network. Compared to an STS-3 (non-concatenated), the STS-3c can carry more information bits because only one set of path overhead (nine bytes) is required for the STS-3c. Within the STS-3 SPE, one path overhead is required for each STS-1.

2.4.2.2 Virtual Tributaries (VTs)

The STS-1 SPE can be used to carry one DS3 or a variety of sub-DS3 signals. The DS1 is a commonly used sub-DS3 signal that can be mapped into a SONET unit called *Virtual Tributary-1.5* (VT1.5). Each STS-1 SPE can carry up to 28 VT1.5s. Other types of VT signals include VT2, VT3, and VT6. Table 2-7 shows VT types, signal formats, and associated data rates.

41

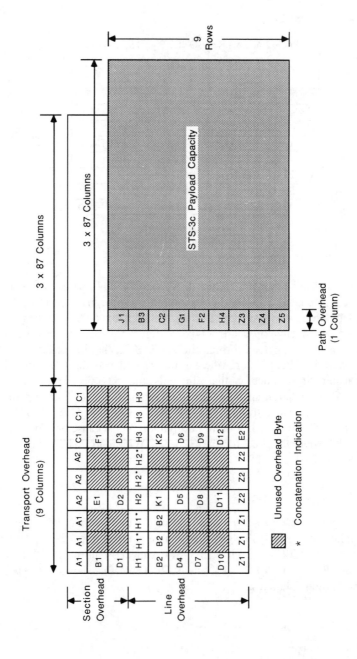

Figure 2.9. STS-3c frame structure.

42

Table 2-7. VT Types, Formats, and Data Rates

VT type	(Row, Column) within STS SPE	Data rate (Mbps)
VT1.5	(9, 3)	1.728
VT2	(9, 4)	2.304
VT3	(9, 6)	3.456
VT6	(9, 12)	6.912

Before mapping VT formats to the STS-1 SPE, one more mapping, called VT group mapping, is needed. The VT group is a fixed-size container for VT signals (approximately 6.9 Mbps). Each VT group contains up to four VT1.5s, three VT2s, two VT3s, or one VT6. When mapping services (such as DS1) into an STS-1 SPE, the service is first mapped into the appropriate VT format, and one or more VTs of the same size are then combined to form a VT group. Seven VT groups are then inserted into the STS-1 SPE. Note that the STS-1 SPE can contain VT groups of different sizes (e.g., contains four VT1.5 groups and one VT6 group), but each VT group contains VTs of the same size.

The VT structure has two possible modes of operation, floating and locked, based on the method of VT group mapping to the STS-1 SPE. Floating VTs use a flexible mapping method (e.g., the pointer method; see Section 2.4.4) to adjust locations of VT groups inside the STS-1 SPE. In contrast, locked VTs use a fixed VT group mapping method (e.g., fixed-stuffing). The floating mode minimizes delay for distributed VT switching (at least 1.5 Mbps), whereas the locked mode minimizes interface complexity in distributed DS0 (64-kbps) switching. An STS-1 payload can be structured only as all floating or all locked VTs.

2.4.2.3 SONET Payload Mapping

Based on STS and VT frame structures, the present digital hierarchy and future B-ISDN services can be readily mapped into the SONET optical hierarchy as shown in Table 2-8.

Table 2-8. Mapping between Asynchronous Digital Hierarchy and SONET Hierarchy

Service	Bit rate	SONET	VT group mapping	VT group capacity
DS0	64 kbps	-	-	-
DS1	1.544 Mbps	VT1.5	VT1.5	4 DS1s
CEPT+	2.048 Mbps	VT2	VT2	3 CEPTs
DS1C	3.152 Mbps	VT3	VT3	2 DS1Cs
DS2	6.312 Mbps	VT6	VT6	1 DS2
DS3	44.736 Mbps	STS-1	-	-
DS4NA	139.264 Mbps	STS-3	-	-

+ CEPT: Conference of European Postal and Telecommunications (administration signals)

2.4.3 Layered Overhead and Transport Functions

2.4.3.1 SONET Interface Layers

The SONET layered overhead and transport functions are divided into four layers: physical, section, line, and path. The path layer transports services between terminal multiplexing equipment at two end offices. The line layer transports STS SPE path-layer payload and overhead across the physical medium. The section layer transports STS-N frames across the physical medium, and the physical layer transports bits as optical or electrical pulses across the physical medium. No overhead is associated with the physical layer. Table 2-9 shows major functions associated with each SONET interface layer.

Figure 2.10 depicts the relationship of the layers. Each layer requires the services of all lower-level layers to perform its function. For example, suppose that two COs want to exchange DS3 services. The DS3 is first mapped into the STS-1 payload (SPE) along with the path overhead and is then delivered to the line layer. The line layer adds line overhead to the STS-1 SPE and multiplexes N STS-1s to the STS-N. The STS-N is then passed to the section layer where the section overhead is added and scrambling is performed. This scrambled STS-N is finally transmitted by the physical layer.

Table 2-9. Functions Associated with SONET Interface Layers

SONET interface layer	Payload	Layer function	Overhead layer	Overhead function
Path	service, STS-1 SPE	service mapping and transport	Path	path trace, signal label, error monitoring, user channel, path status, multiframe indication, alarm, growth
Line	STS-1	STS-1 multiplexing to STS-N	Line	pointer processing, error monitoring, automatic protection switching, data channels, growth, express orderwire
Section	STS-N	preparation of physical transport	Section	framing, channel ID, error monitoring, local orderwire, data channels, user channel
Photonic	-	optical transmission	-	pulse shape, power level, bit rate, wavelength

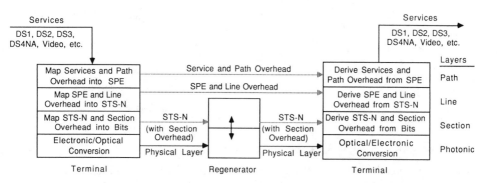

Figure 2.10.　SONET interface layers.

The layered approach to overhead allows SONET equipment to be built in a cost-effective manner according to the associated functionalities. Based on this approach, SONET-based NEs are divided into three types: *Section Terminating Equipment* (STE), *Line Terminating Equipment* (LTE), and *Path Terminating Equipment* (PTE). Figure 2.11 is a diagram that illustrates SONET STE, LTE, and PTE. The LTE, an NE that originates and/or terminates line signals, interprets and processes the transport overhead. The LTE can be any SONET equipment (such as a TM, ADM, or DCS) except regenerators. The STE, an NE that originates and/or terminates the SONET physical and section layers, interprets and processes the section overhead. The STE is either a regenerator or part of an LTE. The PTE, an NE that terminates and/or originates path overhead, can be any SONET equipment (such as TM or Wideband Digital Cross-connect System [W-DCS]) that terminates SONET STS-1 payload. (The path overhead is included in the STS-1 payload.) The VT STE can be any SONET equipment that terminates SONET VTs, such as a VT multiplexer.

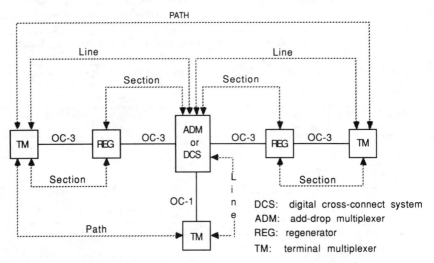

Figure 2.11.   SONET section, line, and path equipment.

2.4.3.2 Overhead Channels

SONET, which adopted a layered approach to overhead, allocates bandwidth to a layer based on the function addressed by that particular channel. SONET overhead is divided into three layers: section, line, and path layers. Figure 2.7 depicts the overhead bytes and their relative positions in the STS-1 frame structure. The line and section overheads (called *transport overhead*) are contained in the first three columns (27 bytes) of the STS-1 frame, and the path overhead is contained in the first column (nine bytes) of the SPE within the STS-1.

Section Overhead

The *section* overhead is the overhead necessary to verify reliable communication between NEs, such as terminals and regenerators. A minimum amount of overhead is

46

placed here to allow regenerators to remain cost-effective. The section overhead channels for an STS-1 include

- Two frames bytes (Bytes A1 and A2) that indicate the beginning of each STS-1 frame
- An STS-1 identification byte (Byte C1)
- An eight-bit *Bit-Interleaved Parity* (BIP-8) check for section error monitoring (Byte B1)
- An orderwire channel for craftsperson (network maintenance personnel) communications (Byte E1)
- A channel for unspecified network user (e.g., operator) applications (Byte F1)
- Three bytes (Bytes D1, D2, and D3) for a section-level 192-kbps message-based data channel that carries OAM&P information and other information between STEs.

Line Overhead

The *line* overhead was designed to verify reliable communication between more complicated NEs, such as terminals, DCSs, ADMs, and switches. The line overhead includes

- The STS-1 pointer bytes (Bytes H1, H2, and H3) (pointers will be explained in Section 2.4.4)
- An additional BIP-8 for line-error monitoring (Byte B2)
- A two-byte *Automatic Protection Switching* (APS) message channel (Bytes K1 and K2[5])
- A nine-byte, 576-kbps, message-based *Data Communications Channel* (DCC) (Bytes D4 through D12)
- Bytes reserved for future growth (Bytes Z1 and Z2)
- A line orderwire channel (Byte E2).

Note that within the section and line layers of the SONET structure, DCCs are only physical channels that can be used to transport data for any application. When DCCs are embedded in a signal format and dedicated to network operations and management, they are referred to as EOCs. More details regarding DCCs can be found in References [4,6,10].

---

5. K1 and K2 bytes are defined only for STS-1 number one of an STS-N signal. Formats for K1 and K2 will be discussed in Chapter 3.

STS Path Overhead

The STS path overhead is used to communicate operations-type functions (e.g., performance checking) from the point where a service is mapped into the STS SPE to where the point where it is delivered. The path overhead, which is part of the STS SPE, allows complete performance monitoring on an end-to-end basis. The path overhead includes

- A byte used to verify a continued connection (Byte J1)
- A path BIP-8 for end-to-end payload-error monitoring (Byte B3)
- A signal label byte to identify the type of payload being carried (Byte C2)
- A path status to carry maintenance signals (Byte G1)
- A multiframe alignment byte (Byte H4) for identifying the location of tributary signals
- Three bytes for future growth (Bytes Z3, Z4, and Z5).

2.4.4 SONET Pointers and Multiplexing Method

SONET uses a multiplexing method known as the *pointer method* to provide flexibility in (1) adjusting frequency variations for STS-1 payloads and VT signals with their corresponding SPEs, and (2) allowing for flexible frame alignment during the STS-1-to-STS-N multiplexing process. The pointer method allows the transport overhead position to be decoupled from the SPE. It also uses an STS-1 pointer, which is contained in the H1 and H2 bytes in the line overhead, to indicate the distance between the transport overhead position and the SPE. In other words, the payload pointer indicates the location of the byte where the STS-1 SPE begins. Figure 2.12 depicts an example of an STS-1 SPE that crosses two consecutive 125-μs frames. Both SPE synchronization and frame-alignment problems are handled by updating the payload pointer in every multiplexing stage.

The pointer scheme is essentially a synchronous multiplexing method with the ability to accommodate slight frequency variations (like the conventional bit-stuffing multiplexing method for asynchronous networks) and easy access to tributaries (like the fixed location mapping method used for SYNTRAN). But, unlike SYNTRAN, the SONET pointer method does not require slip buffers, which may increase network delay and potential impairment due to frame slips.

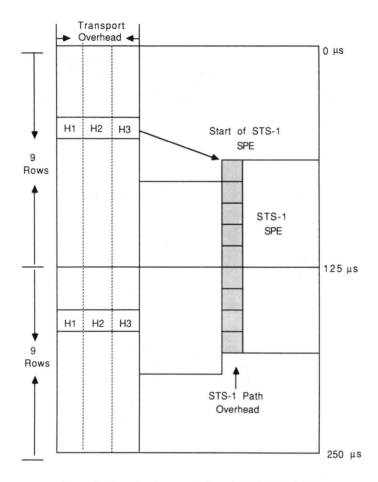

Figure 2.12. A pointer method example for STS-1 SPE.

The VT pointer, which is functionally like an STS-1 pointer, allows flexible alignment of the VT payload with respect to the STS-1 frame. The pointer method allows direct access to VT signals without demultiplexing the entire STS-1 frame because locations of VT signals can be identified using pointer values. More details on the SONET pointer method can be found in References [4,6].

49

2.4.5 SONET Protection Switching Initiation Criteria

Two SONET protection switching initiation criteria are *Signal Failure* (SF) and *Signal Degrade* (SD). An SF is a "hard" failure caused by *Loss of Signal* (LOS), *Loss of Frame* (LOF), BER threshold exceeding 10^{-3}, line *Alarm Indication Signal* (AIS), or some other protectable hard failure. An SD is a "soft" failure caused by a BER exceeding a preselected threshold that usually ranges from 10^{-5} to 10^{-9}. SF is usually given priority over SD.

The BER of an OC-N is obtained by summing the parity violations detected in each STS-1 Line BIP-8 (Byte B2) of an OC-N signal over a maximum detection time period. The maximum detection time period is a function of the BER threshold. Table 2-10 shows a relationship between BER and the number of BIP-8 violations based on a maximum detection time requirement [6]. For example, if the BER threshold is 10^{-6} for SD and the number of BIP violations detected over a 10-second period for an OC-12 signal is 6300, then the system encounters an SD situation because the detected BIP violation count is greater than 6144 (512×12).

Table 2-10. Relationship between BIP-8 Violations and BER for an STS-N Signal

BER	Maximum detection time	Number of BIP violations
$\geq 10^{-3}$	10 ms	334×N
10^{-4}	100 ms	492×N
10^{-5}	1 s	510×N
10^{-6}	10 s	512×N
10^{-7}	100 s	512×N
10^{-8}	1000 s	512×N
10^{-9}	10,000 s	512×N

The following sections describe some SONET failure states and maintenance signals based on Bellcore Technical Advisory TA-TSY-000253 [6].

2.4.5.1 Failure States

Signal failure states that persist for a specified period of time must be reported to a surveillance *Operations System* (OS). A *state* is referred to as an occurrence in the network that is required to be detected. These failure states include LOS, LOF, and *Loss of Pointer* (LOP).

Loss of Signal (LOS)

The SONET NE enters the LOS state when it detects the incoming STS-N signal within 100 μs of the onset of an all 0s pattern lasting 10 μs or longer. The SONET NE exits the LOS state if two consecutive valid frame-alignment patterns were detected and if no LOS was detected during the intervening time (one frame).

Loss of Frame (LOF)

The SONET NE enters the LOF state when an *Out-of-Frame* (OOF) on the incoming STS-N signal persists for 3 ms. The frame pattern observed by a SONET NE is a subset of the A1 and A2 bytes contained in an STS-1 signal. An OOF condition on an STS-N (or OC-N) signal is declared when four consecutive errored framing patterns have been received.

In an OOF condition, the SONET NE may declare an in-frame condition on detecting two successive error-free framing patterns. The SONET NE exits the LOF state when the incoming STS-N signal remains in-frame for 3 ms.

Loss of Pointer (LOP)

The SONET NE enters the LOP of the STS state when an STS pointer processor cannot obtain a valid pointer using the pointer interpretation rules described in the SONET standard [4,6]. It also enters this state if a pointer processor receives consecutive pointers with the first four bits of the pointer word (composed of H1 and H2 bytes) set to "1001" but not indicating concatenation, or if a valid pointer is not found in a number of eight consecutive frames.

2.4.5.2 Maintenance Signals

Maintenance (or alarm) signals notify operational conditions of transmission facilities to upstream and downstream transmission elements. SONET maintenance signals are used to alert SONET NEs of the detection and location of a failure. These signals include AIS, *Far End Receive Failure* (FERF), and Yellow alarms.

Alarm Indication Signal (AIS)

In the digital network, AIS is used to alert downstream equipment that an upstream failure has been detected. SONET provides different AIS signals for various layers of functionality. In the following list, we only describe line AIS and path AIS, which will be used for discussions of SONET SHR architectures in Chapter 4.

- Line AIS

    An STE sends Line AIS to alert the downstream LTE that a failure has been detected and to initiate protection switching when APS is provided as a feature. The STE generates Line AIS within 125 μs on entering an LOS or LOF state on the incoming signal. An STE generates Line AIS by constructing an OC-N signal that contains valid section overhead and a scrambled all 1s pattern for the remainder of the signal. An LTE detects this Line AIS as all 1s in bits 6, 7, and 8 of the K2 byte in five consecutive frames, at which point the LTE enters a Line AIS state.

    The downstream LTE detects Line AIS removal as any pattern other than all 1s in bits 6, 7, and 8 of the K2 byte in five consecutive frames. The detection of Line AIS removal causes the LTE to exit the Line AIS state within 125 μs.

- STS Path AIS

    An LTE sends STS Path AIS to alert the downstream STS PTE that a failure has been detected upstream. The LTE generates the Path AIS within 125 μs on entering a failure (e.g., LOS) or Line AIS state. STS Path AIS is an all 1s signal in bytes H1, H2, and H3, as well as in the entire STS SPE. The STS PTE detects an STS Path AIS by observing an all 1s pattern in bytes H1 and H2 (i.e., payload pointer) during three consecutive frames, at which point the STS PTE enters an STS Path AIS state.

    At STS Path AIS deactivation, the STS pointer processor constructs a valid pointer with bits 1 through 4 set to "1001", followed by normal pointer operations. The downstream STS PTE detects STS Path AIS removal when a valid STS Pointer with the "1001" code in bits 1 through 4 is observed in three consecutive frames. The detection of STS Path AIS removal causes the STS PTE to exit the STS Path AIS state within 125 μs.

Far End Receive Failure (FERF)

Line FERF alerts the upstream LTE that LOS, LOF, or Line AIS has been detected along the downstream line. Line FERF is detected when a "110" code is presented in bits 6, 7, and 8 of the K2 byte in five consecutive frames.

Yellow Alarm

STS Path Yellow alerts the upstream STS PTE that a downstream failure indication has been declared along the STS path. STS Path Yellow alarm is detected when bit 5 of the G1 byte is set to "1" in 10 consecutive frames.

Figure 2.13 depicts an example of maintenance signal operation following the detection of an LOS [10]. In Figure 2.13, STE-1 detects an LOS, generates a line AIS, and sends it to the downstream LTE-1. After LTE-1 receives a line AIS from STE-1, it initiates protection switching and tries to restore the failed link between PTE-1 and LTE-1. In this example, LTE-1 fails to complete protection switching. Thus, LTE-1 generates an STS path AIS, sends a line FERF to PTE-1, and reports the protection switching failure and the AIS received to the OS. After receiving an STS path AIS from LTE-1, PTE-2 converts that STS path AIS to a DS3 AIS for termination reporting and generates an STS Yellow signal to PTE-1.

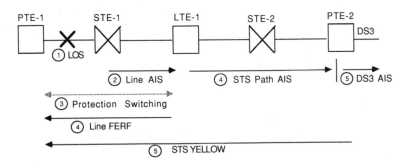

① STS-1 detects LOS caused by laser failure.

② STE-1 generates line AIS to LTE-1 and reports LOS alarm to OS via section DCC.

③ LTE-1 attemps but fails protection switching.

④ LTE-1 generates STS path AIS, sends line FERF to PTE-1, and reports switch failure and AIS received to the operations system.

⑤ PTE-2 converts STS path AIS to DS3 AIS for termination reporting and generates STS YELLOW signal to PTE-1.

Figure 2.13. An example of maintenance signal operation (© 1990, IEEE).

2.5 TERMINAL AND ADD-DROP MULTIPLEXERS (TMs/ADMs)

2.5.1 Functional Architectures

In today's fiber-optic networks, there are two types of multiplexing devices: TMs and ADMs. The M13, which provides interfaces between an asynchronous DS3 and a number of DS1s, is an example of the TM. An *Optical Line Terminating Multiplexer* (OLTM), which is another example of the TM, is a device that multiplexes a variety of low-speed signals to a high-speed signal, converts the resultant electrical signal to the optical signal, and then places it onto the fiber. Depending on an equipment vendor's specifications, the TM may also provide interfaces to synchronization, testing, maintenance, and the OS. For today's asynchronous networks, the back-to-back TM configuration is used to transport signals from the incoming fiber to the outgoing fiber, and to add-drop signals at the office via a manual or an *Electronic Cross-connect System* (DSX). A DSX provides signal cross-connects, test access, facility maintenance, and facility rearrangement. The signal add-drop operation demultiplexes the entire high-speed signal frame into tributary components. The add-drop function is usually operated manually; thus, no remote, automatic provisioning feature exists in today's asynchronous networks. Figure 2.14(a) depicts an example of a DS1 add-drop configuration in asynchronous networks. This asynchronous DS1 add-drop system requires two back-to-back OLTMs for demultiplexing the high-speed optical signal to DS3s (or vice versa), a DSX-3 which manually cross-connects DS3s for DS3 add-drop or pass-through, and several M13 multiplexing devices for DS1 add-drop from DS3s (or vice versa). If some DS1s in a DS3 are required to drop, these DS1s are dropped via an M13. After dropping these DS1s, the remaining pass-through DS1s must be remultiplexed back to the DS3 before continuing on to the next location. Managing such a back-to-back TM system requires other complex systems. Adding new services can therefore be a slow and difficult process, especially when telephone companies can choose only from proprietary fiber systems offered by different vendors.

The ADM is a device that provides not only the network multiplexing function (as does the TM), but also an automated add-drop capability and remote option settings. The ADM 3/X is a synchronous ADM that provides an interface between two full-duplex SYNTRAN DS3 signals and one or more full-duplex DS1 signals. The ADM can work in either the terminal mode or the add-drop mode. In the terminal mode, the ADM, which is functionally like a TM, interfaces only one high-speed signal and is used for the point-to-point application. In the add-drop mode, the ADM, which interfaces two full-duplex high-speed signals and a variety of low-speed add-drop signals, is used for applications such as a linear network (chain), where both add-drop and pass-through capabilities are required at each intermediate node. For the ADM 3/X in the add-drop mode, data and timing from each incoming DS3 signal are passed through the ADM 3/X and transmitted by the DS3 interface at the other end if no local DS1 access is required. Each DS1 interface reads data from an incoming DS3 stream and inserts data into an outgoing DS3 stream as appropriate. Thus, the ADM 3/X in the add-drop mode performs the same function as the back-to-back M13s in the asynchronous networks, but with fewer network components [see Figure 2.14(b)]. This functional integration makes the ADM more reliable than the back-to-back OLTM for the signal add-drop operation. For remote option settings, the ADM includes a synchronization interface for backup timing, a maintenance interface that connects the ADM to the office alarm system, and an OS for maintenance, provisioning, and testing. These ADM operations interfaces provide self-monitoring, fault detection, and isolation capabilities. For SYNTRAN networks, a 64-kbps data channel embedded in each SYNTRAN DS3 is dedicated for overhead communications between the ADM 3/Xs.

SONET ADM

A SONET ADM, as depicted in Figure 2.14(c), is an synchronous ADM (just like a SYNTRAN ADM) that directly terminates high-speed fiber. Due to the synchronous nature of SONET signals, DS3s or DS1s can be added or dropped from an STS-N signal without the need for demultiplexing all of the STS-1 or VT1.5 signals. Thus, a SONET ADM eliminates the need for back-to-back OLTMs as well as M13s as required in asynchronous networks [see Figure 2.14(c)].

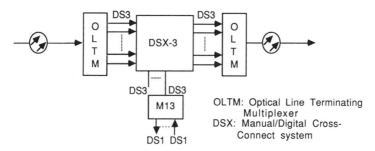

(a) DS1 Add-Drop in Asynchronous Networks

Figure 2.14. DS1 add-drop configurations in asynchronous and SONET networks.

55

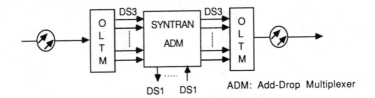

(b) DS1 Add-Drop in SYNTRAN Networks

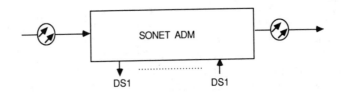

(c) DS1 Add-Drop in SONET Networks

Figure 2.14. (Continued)

Figures 2.15(a) and 2.15(b) depict generic SONET ADM functional architectures in the terminal and add-drop modes, respectively. In the terminal mode, the SONET ADM provides interfaces between a full-duplex OC-N signal and one or more full-duplex DS1, DS2, DS3, or OC-M ($M < N$) signals. DSn signal interfaces are used to support present services. In the long term, OC-M (or STS-M) interfaces may be preferable.

In the add-drop mode, the SONET ADM terminates two full-duplex OC-N signals and provides multiplex functions between the OC-N level and the DSn (n=1, 2, 3) or OC-M ($M < N$) level. The incoming (OC-N) information payloads that are not received locally are passed through the SONET ADM and transmitted by the OC-N interface on the other side. Each DSn or OC-M interface reads data from an incoming OC-N and/or inserts data into an outgoing OC-N stream as appropriate. Figure 2.15 shows a synchronization interface for a CO application with external timing and an operation interface that provides local craftsperson access, local alarm indications, and an interface to remote OSs. Detailed use of SONET overhead in the ADM can be found in Bellcore Technical Reference TR-TSY-000496 [11].

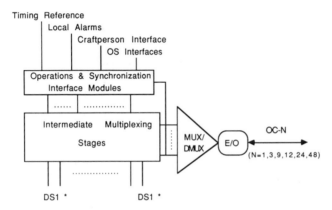

(a) ADM in the Terminal Mode

Figure 2.15.   Generic SONET add-drop multiplexer architectures.

57

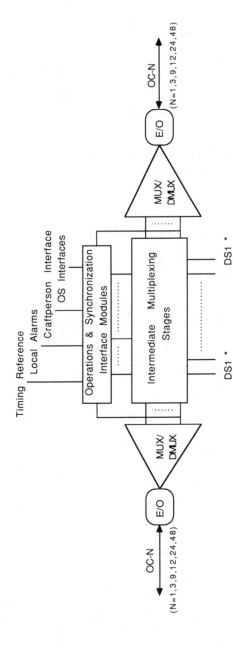

* Shown with DS1, but DS3 and OC-M (M < N) interfaces may be provided.

(b) ADM in the Add-Drop Mode

Figure 2.15. (Continued)

58

Figure 2.16 shows one possible ADM equipment architecture for interoffice applications. This two-stage ADM equipment model is proposed to provide flexibility of processing both STS-3c and STS-1 signals. In Figure 2.16,[6] an incoming optical OC-48 is demultiplexed into a maximum of 16 STS-3 (155.52-Mbps) channels. Some of the STS-3 channels may pass directly through the SONET ADM. These "through" STS-3 channels, together with other STS-3 channels added from this CO, are then multiplexed and converted to an OC-48 optical signal and sent to the next CO. If the payload of an STS-3 channel is terminated at this CO, an STS-3 MUX/DMUX is used to demultiplex the STS-3 signal to a maximum of three STS-1 (51.84-Mbps) channels and then to convert the STS-1s to DS3s (44.736 Mbps) via the STS-1 interface cards.

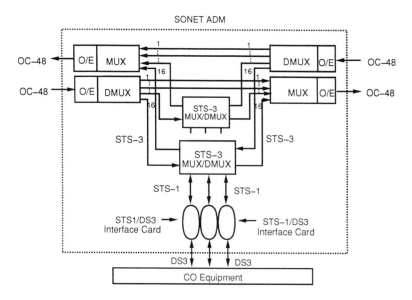

Figure 2.16.   A possible equipment model for the SONET ADM in add-drop mode.

---

6. Here we assume an optical line rate is OC-48. However, any equipment model for any OC-N with $N \geq 3$ is similar to the one described here.

Depending on the vendor's applications and product plan, SONET ADMs may or may not have the cross-connect capability (as do DCSs). The ADM without the cross-connect capability uses *Time Slot Assignment* (TSA) as its signal multiplexing scheme, whereas the ADM with the cross-connect capability uses *Time Slot Interchange* (TSI) as its signal multiplexing scheme [12,13]. Figure 2.17 depicts examples of the TSA and TSI schemes.

TSA assigns time slots to each ADM node on a dedicated basis and maps service demands (e.g., DS1s and/or DS3s) into these dedicated time slots in the high-speed, multiplexed signal. Thus, any time slot that is not assigned to this ADM node cannot be used by the local ADM node to accommodate service demands, even though this time slot is empty when passing through that ADM node. This also implies that the relative time slot positions for transit signals in the input high-speed, multiplexed data stream remain the same in the output high-speed, multiplexed data stream [see Figure 2.17(a)]. The service demand mapping in the TSA scheme can be performed in a fixed (hardwired) manner or a dynamic (programmable) manner [13]. If fixed service mapping is used, the number of time slots (e.g., VT 1.5s) assigned to the ADM node is the same as the number of equivalent capacity service demand ports (e.g., DS1 ports). For the programmable TSA scheme, the number of service demand ports can be greater than the number of equivalent capacity time slots, and the network provisioning system (e.g., Bellcore's TIRKS provisioning system) can remotely control the dynamical use of time slots.

On the other hand, TSI is a switching process that moves a time slot from one data stream to a time slot in another data stream, as depicted in Figure 2.17(b). The TSI switching matrix can be controlled by local operators or by a centralized operations system. Thus, the relative time slot positions for transit signals in the input high-speed, multiplexed data stream may be changed in the output high-speed, multiplexed data stream. Note that for ADMs with TSA, the default condition is pass-through, and a command is needed to drop a tributary signal. An ADM with TSI (or DCS) has no default configuration and needs a command to cross-connect tributary signals.

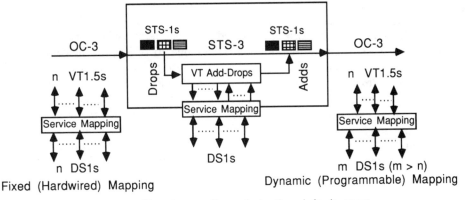

Fixed (Hardwired) Mapping   Dynamic (Programmable) Mapping

- Signal pass-through is the default case
- Economical signal add-drop

(a) ADM with Time Slot Assignment (TSA)

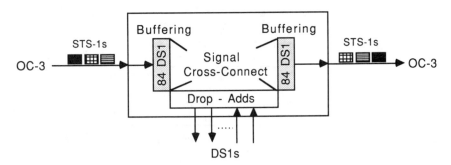

- DCS-like device
- No default for signal processing
- Flexibility within high-speed line (service and facility grooming)

(b) ADM with Time Slot Interchange (TSI)

Figure 2.17.   ADMs using TSA and TSI signal multiplexing schemes.

Table 2-11 summarizes a relative comparison between ADMs using TSA and ADMs using TSI. In an ADM, TSA does not process through traffic, whereas TSI does. TSI has flexibility with high-speed lines in terms of service and facility grooming, but experiences transmission delays due to buffering. In contrast, TSA introduces no delay and is less costly for added-dropped tributary signals than TSI; however, the flexibility is on the drop side only and is limited. For applications requiring economical signal add-drop capability, TSA is preferable for ADMs because it offers more economical signal add-drop than TSI. TSI is inherently supported by larger cross-connect systems, such as DCS 3/1, which terminates signals at the DS3 level and cross-connects signals at the DS1 level. Thus, the ADM using TSI is sometimes referred to as a "mini-DCS" because it integrates functions of signal add-drop and cross-connect, but terminates much fewer lines than a DCS. More details of the TSI scheme are discussed in Section 2.6.1.

Table 2-11. A Relative Comparison between ADMs using TSA and ADMs using TSI

Attribute	ADM (TSA)	ADM (TSI)
grooming	no	yes
add-drop cost	low	high
transmission delay	no	yes
add-drop and cross-connect integration	no	yes
flexibility	drop side only	high-speed line
SONET ring applications*	all rings except BSHR/2**	all rings

* SONET ring applications will be discussed in Chapter 4.
** BSHR/2: Bidirectional SHR using two fibers.

2.5.2 Applications

The ADM can be used in any application that requires a low-speed to high-speed signal multiplex function. One advantage of the ADM is that it requires physical access only to the low-speed signals (e.g., DS1) that need to be added or dropped from the high-speed signal (e.g., DS3). Thus, the ADM is useful in those applications requiring the adding and dropping of just a few low-speed signals. One such application is in loop networks where *Remote Terminals* (RTs) are connected using ADMs in a chain configuration. In this application, the majority of demands from each RT go to the

62

serving CO. A cluster of COs[7] with a chain configuration in interoffice, fiber-hubbed networks is another similar application. ADM chains use APS to protect the chain from fiber facility failures between nodes and do not protect the chain from node failures. Also, as will be discussed in Chapter 4, SONET ADMs can make implementing high-speed, SHR architectures possible and economical, and can provide the self-healing network capability against network component failures. Two different types of SONET SHR architectures can be implemented with ADMs using TSA or TSI.

2.6 DIGITAL CROSS-CONNECT SYSTEM (DCS)

2.6.1 Functional Architectures

A DCS is an NE that terminates digital signals and automatically cross-connects constituent (tributary) signals according to a map stored in an electronically alterable memory. The DCS, which is designed to integrate multiple functionalities such as signal add-drop, signal cross-connect, and multiplexing/demultiplexing into a single NE, can eliminate back-to-back multiplexing and reduce the need for intermediate electrical distribution frames, as well as the labor-intensive jumpers between frames. Operations cost savings are realized by DCS systems through electronically controlling cross-connections, test access and loopbacks, facility rolls, and maintenance.

Two types of DCSs are deployed in current fiber networks: *Wideband DCSs* (W-DCSs) and *Broadband DCSs* (B-DCSs). A W-DCS is a DCS that terminates DS3s and DS1s, and has the DS1 or DS0 cross-connection capability. The DCS 3/1, which only terminates DS3s and cross-connects DS1s, is one example of a W-DCS. Figure 2.18(a) depicts a functional block diagram of a DCS 3/1. The major functional components of the DCS 3/1 include the signal *Cross-Connect Network* (CCN); *Test, Monitor, and Control* (TMC) units; and MUX/DMUX units. The signal cross-connect function is performed in a non-blocking manner, where non-blocking means that a cross-connection can be made (if allowed) irrespective of other cross-connections that may already be established. The switching matrix in the CCN establishes automatic cross-connections between interface channels, DS1s on DS3s, according to a map stored in an electrically alterable memory. The alterable memory also stores all information regarding the provisioning and administration of the DCS 3/1 and terminating facilities.

---

7. In a fiber-hubbed network, several COs are grouped together to form a cluster of offices served by a central hub.

The switching function of the DCS 3/1 is performed using TSI. Figure 2.18(b) depicts a diagram of the process of TSI for a DCS 3/1. The DS1 time slots from two incoming DS3 streams are placed in buffer memory for temporary storage and frame alignment. As shown in Figure 2.18(b), the incoming DS1 time slots are stored in the buffer's locations 1 through 56 with 192 bits[8] each. Incoming DS1#7 and DS1#29 are then moved into a first outgoing DS3 (which is destined to CO-3) during the first 28 buffer reading process, based on a map controlled by the TMC unit.

The TMC units provide test access to the DS1s and to a controller that processes alarms and user commands, sets up testing, interfaces to an OS link, and programs the CCN. They also include a non-volatile, backup memory that stores sufficient data so that, during a total power failure, the DCS 3/1 can be reinitialized, and the cross-connections and status information can be reestablished. As depicted in Figure 2.18, the cross-connection for a DCS 3/1 can be between a DS1 on one DS3 and a DS1 on another (or the same) DS3, or between a DS1 on one DS3 and a DS1 connected to a drop port. The purpose of the drop port is to gain access to the individual DS1s without interfacing an entire DS3 system that must then be separately demultiplexed. DCS 3/1s have been widely deployed at hubs in today's fiber-hubbing networks to provide DS1 service and facility grooming (thus, increasing efficient use of fiber bandwidth), and to simplify operations complexity. For example, in Figure 2.18, DS1#7 (carried by the DS3 from CO-1) and DS1#29 (carried by the DS3 from CO-2) are destined to CO-3. First, DS1#7 and DS1#29 are placed on DS3s originating at CO-1 and CO-2, respectively. These DS3s terminate and are demultiplexed at the MUX/DMUX module of the DCS 3/1 at the hub. Then, DS1#7 and DS1#29 are cross-connected to two output ports that connect to a DS3 destined to CO-3.

---

8. For a DCS 1/0, each time slot uses eight bits.

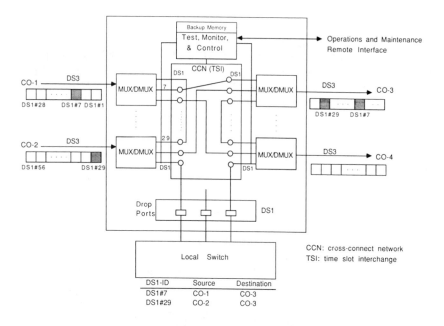

(a) DCS 3/1 Functional Diagram and Facility Grooming

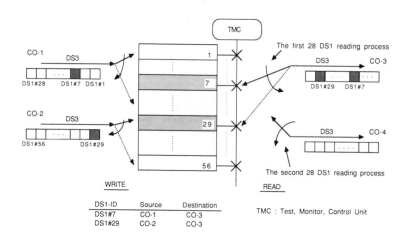

(b) TSI Process in a DCS 3/1

Figure 2.18. DCS 3/1 functional architecture and facility grooming.

65

A B-DCS, which interfaces DS3s and cross-connects DS3s, is simply a space-division switch. The DCS 3/3 is primarily deployed at the hub for facility grooming and test access. Figure 2.19 depicts an example of using a DCS 3/3 at the hub for facility grooming. For example, in Figure 2.19, DS3#7 (carried by a 1.2-Gbps fiber system between CO-1 and the hub) and DS3#28 (carried by a 565-Mbps fiber system between CO-2 and the hub) are destined to CO-3. DS3#7 and DS3#28 terminate at different input ports of the DCS 3/3 and are cross-connected to two output ports that connect to the same fiber system for CO-3. In the future, the DCSs currently being deployed in telephone companies may be expanded to capacities exceeding 2000 DS3s or 8000 DS1s [14].

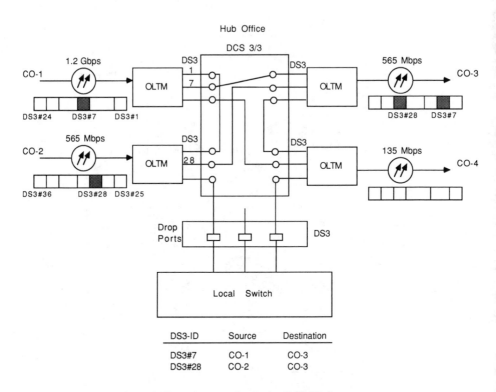

Figure 2.19. An example of using DCS 3/3 for transport.

The DCS is analogous to today's circuit switch, which is used for the telephone networks, but with some differences, as shown in Table 2-12. The major difference is

66

that the DCS switches much higher rate signals (i.e., DS1s and DS3s) than the circuit switch (i.e., DS0); however, the switching time of the DCS is much slower than that of the circuit switch. Unlike today's circuit switch, the performance requirement for higher bandwidth switching has resulted in the non-blocking switch design for the DCS. In contrast to the circuit switch, the present DCS is controlled manually. That is, data interfaces allow an operator to enter commands at a keyboard (either remotely or locally) that instruct the DCS to set up or modify its switching matrix. In effect, the operator provides the switching logic. In the near future, the DCS is expected to be controlled electrically. For the electrically controlled DCS, an external controller is needed to provide call setup functions and switching logic, which are typically part of the circuit switch. More information regarding the status of present DCSs and their applications can be found in Reference [15].

Table 2-12. Comparison between Asynchronous DCS and Circuit Switch

Attributes	DCS	Circuit switch
Switching Unit	DS1, DS3	DS0
Control	manual/automatic	automatic
Performance	Non-blocking	Acceptable blocking
Connection Duration	hours-months	minutes
Switching/ Cross-connect	not real time	real time
External Controller Needed for Switching Logic	yes	no

SONET DCS

The SONET DCS performs the same functions as the DCS 3/1 or DCS 3/3 except that it terminates and cross-connects not only present digital signals (such as DS1 and DS3), but also SONET signals (such as STS-1 and OC-N). As depicted in Figure 2.20, with the SONET interface available, optical fibers can be brought directly to the cross-connect system, thus eliminating the need for external *optical-to-electrical* (O/E) conversion. This will simplify intraoffice and interoffice transport. The SONET B-DCS terminates full-duplex OC-N signals and provides DS3 interfaces. The basic function of the B-DCS is to make two-way cross-connects at the DS3, STS-1, and concatenated STS-N (STS-Nc) levels. In an STS-Nc cross-connection, the signal is cross-connected as a single high-capacity STS-N channel, where the STS-3c cross-connection capability would be useful for broadband services such as HDTV.

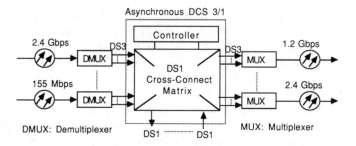

(a) A Functional Diagram for Asynchronous DCS 3/1

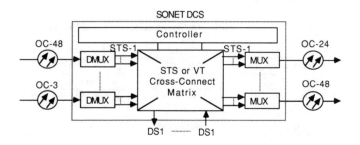

(b) A Functional Diagram for A SONET DCS

Figure 2.20.   A comparison between the asynchronous DCS and the SONET DCS.

Like SONET B-DCS, the SONET W-DCS terminates full-duplex SONET OC-N signals and provides DS3 interfaces. But, unlike the SONET B-DCS, the W-DCS cross-connects floating VT1.5 signals between OC-N terminations. It also cross-connects transparent DS1s between DS3 terminations and between DS3 and OC-N terminations [16]. The SONET W-DCS also has DS1 interface capability, which provides transparent DS1 cross-connects between DS1 interfaces, and between the DS1 interface and the DS3/OC-N termination. Table 2-13 compares SONET DCSs and present DCSs.

Table 2-13. Comparison between SONET DCSs and Asynchronous DCSs

Attribute	DCS type				
	DCS 1/0	DCS 3/1	DCS 3/3	SONET W-DCS	SONET B-DCS
Terminations	DS1	DS3	DS3	OC-N, DS3*	OC-N
Cross-connects	DS0	DS1	DS3	DS1, Floating VT1.5	DS3, STS-1, STS-Nc
Interfaces+	DS0	DS1	DS3	DS1	DS3
DS1 test-access port	-	yes	no	yes	no
Operations and Synchronization Interface	yes	yes	yes	yes	yes
Backup memory	yes	yes	yes	yes	yes

* Interface with existing digital facility. In the long run, direct interface
of all OC-Ns may be preferable.
+ Interface with DSX-1 or DSX-3

For a DS3 termination pair, a DS3 is first demultiplexed to DS1s that are cross-connected to several outgoing DS3 signals. Transparent DS1 cross-connections enable the W-DCS to handle any tributary signal at the nominal DS1 rate, like today's DCS 3/1, irrespective of the signal's nature and format. For an OC-N termination pair, an OC-N is first demultiplexed to STS-1 SPEs, and then VT1.5s are accessed and identified using VT pointers. These floating VT1.5s are then cross-connected to different outgoing STS-1 SPEs, which are finally multiplexed to form an outgoing OC-N. For an OC-N and DS3 termination pair, the cross-connect signal is at the VT1.5 as described previously for the OC-N termination pair.

Like DCS 3/1 and DCS 3/3, zero-blocking probability is required for all allowed DS1 and VT cross-connections for the SONET W-DCS, and for all allowed DS3, STS-1, and STS-Nc cross-connections for the SONET B-DCSs, irrespective of the equipment-loading and holding time of the cross-connections. EOCs for SONET provide a much more economical and flexible way to interchange overhead (required for DCS switching) between DCSs for SONET networks than today's asynchronous networks.

2.6.2 Applications and Capabilities

As DCS and SONET network deployment increases and more sophisticated operations capabilities are provided, it will be possible to offer the following new services and network capabilities: efficient network utilization, switched DS1/DS3 service, network restoration, and customer control and management. These services and capabilities are described below.

2.6.2.1 Efficient Network Utilization

DCS deployment has begun in fiber-hubbed networks and, thus, has taken advantage of the service and facility grooming and increased fiber- fill efficiency this technology offers. Network architectures that incorporate hubs can simplify the planning and provisioning of existing services and can accommodate unexpected growth more easily.

2.6.2.2 Switched DS1(VT1.5)/DS3(STS-1) Service

The electrically controlled DCSs, along with external controllers, may offer network capabilities to switch DS1/VT1.5 and DS3/STS-1. The SONET W-DCS and B-DCS can be viewed as DS1/VT1.5 and DS3/STS-1 switches, respectively, in this application, except that no inter-DCS setup function and switching logic is provided. The overhead channels in the SONET format offer a convenient method to enhance these DCSs to provide switching logic and path setup from either a central controller or a number of distributed controllers. The switched DS1/DS3 service would enable a customer to set up and tear down framed DS1/DS3 signals between customer locations in real time. Video teleconferencing is one example of a switched DS1/DS3 application.

2.6.2.3 Network Restoration

The potential of DS1/VT1.5 and DS3/STS-1 switching capabilities offers a flexible approach to using network capacity as well as enhanced survivability. The DCS provides a network restoration capability at the hub via alternate routing of DS1s and VT1.5s for W-DCS, and DS3s, STS-1s, and STS-Nc's for B-DCS. Priority restoration through DS1 and VT1.5 path rearrangement could be implemented if limited spare capacity was available in the network.

The centralized control architecture for DCS restoration has been proposed to facilitate initial implementation of the DCS restoration method. Figure 2.21 depicts an example of DCS restoration using a central controller. In Figure 2.21, the normal route between locations A and B is via a link between DCS#1 and DCS#2. Once the central controller is informed of the link failure between DCS#1 and DCS#2, it sends commands to remote DCS#1, DCS#2, and DCS#3 to change their switching matrices to reroute demand between locations A and B via a link between DCS#1 and DCS#3 and a link between DCS#3 and DCS#2. In the future, when the SONET EOC message set has been clearly defined, it may be possible to implement a distributed control architecture for fast restoration. For the distributed control architecture, one external controller is needed for each DCS. It may also be possible to integrate the function of the external controller into the DCS in the future. Details of DCS restoration methods will be discussed in Chapter 7.

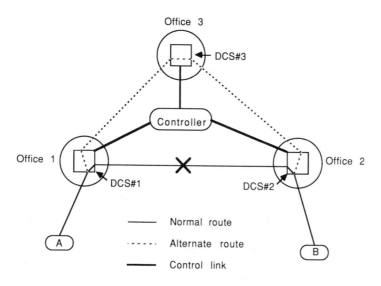

Figure 2.21. DCS network restoration using a central controller.

2.6.2.4 Customer Control and Management

Network providers can use SONET-based networks to offer wideband and broadband *Customer Control and Management* (CCM) services. The CCM capability, provided by deploying intelligent controllers in the network, enables customers to rearrange cross-connections and, thus, avoids delays currently encountered with standard private line rearrangements. CCM is being provided using the DCSs that terminate DS1s and cross-connect DS0s. The CCM capability could be extended to provide DS1

71

rearrangement capability using the W-DCS. Refer to Bellcore Technical Reference TR-TSY-000366 [17] for more details on the CCM application.

2.7 SUMMARY AND REMARKS

In this chapter, we have reviewed the major components that constitute fiber transport systems: fiber-optic transmission systems, digital signal format and hierarchy, and fiber transport equipment. Network planners need to understand performance characteristics and the power budget of fiber-optic systems to choose appropriate systems for their applications and to determine the feasibility of implementing new architectures that require adding new components on existing systems. (A book written by Hecht [18] provides an easy, but clear picture for readers who do not have an optical communications background.) The end-to-end service efficiency can be improved by introducing SONET as a standard optical signal interface. The benefit of network component integration and potential operations cost savings may drive initial SONET network deployment. Later, advanced SONET equipment, combined with standardized control and operations features, may provide opportunities for new services, such as switched DS1/DS3 services and broadband services, and a cost-effective way to enhance existing services, such as network survivability. For affordable network survivability applications, SONET equipment (such as ADMs and DCSs) that uses EOCs may offer opportunities to create new and cost-effective network architectures for restoration against network component failures. In particular, DCSs are becoming more intelligent. They can be used to restore priority demand in a cost-effective manner and to offer new opportunities for switched DS1/DS3 services, which allows business users flexibility in using these services on a per-request basis. SONET is not only serving as a transmission standard, but is also paving the way toward implementing B-ISDN. Reviews of history, development status, application potential, network transition, and technical overview for SONET can be found in References [19-22] and in papers authored by Ching [7] and Boehm [8], who both initiated the SONET concept in 1985.

REFERENCES

[1] TR-NWT-000917, *SONET Regenerator (SONET RGTR) Equipment Generic Criteria*, Bellcore, Issue 1, January 1991.

[2] Edwards, T., *Fiber-Optic Systems: Network Applications*, John Wiley & Sons Ltd., 1989.

[3] Ballart, R., "Restructured DS3 Format for Synchronous Transmission (SYNTRAN)," *Proceedings of IEEE GLOBECOM'84*, November 1984.

[4]   *American Standard for Telecommunications - Digital Hierarchy - Optical Interface Rates and Formats Specification*, T1.105/1988.

[5]   *American Standard for Telecommunications - Digital Hierarchy - Optical Interface Specifications: Single Mode*, T1.106/1988.

[6]   TA-TSY-000253, *Synchronous Optical Network (SONET) Transport Systems: Common Generic Criteria*, Bellcore, Issue 6, September 1990.

[7]   Ballart, R., and Ching, Y-C., "SONET: Now It's the Standard Optical Network," *IEEE Communications Magazine*, March 1989, pp. 8-15.

[8]   Boehm, R. J., "Progress in Standardization of SONET," *IEEE Lightwave Communication Systems*, Vol. 1, No. 2, May 1990, pp. 8-16.

[9]   Sandesara, N. B., Ritchie, G. R., and Engel-Smith, B., "Plans and Considerations for SONET Deployment," *IEEE Communications Magazine*, August 1990, pp. 26-33.

[10]   Holter, R., "SONET: A Network Management Viewpoint," *IEEE Magazine of Lightwave Communications Systems (LCS)*, November 1990, pp. 4-13.

[11]   TR-TSY-000496, *SONET Add-Drop Multiplex Equipment (SONET ADM) Generic Criteria*, Bellcore, Issue 2, September 1989.

[12]   Knapp, E. M., "SONET Technology and Applications: A User Perspective," *Proceedings of IEEE GLOBECOM'89*, Dallas, November 1989, pp. 42.1.1-42.1.7.

[13]   Bailey, D. E., "SONET Add/Drop Multiplex Equipment Administration Using the TIRKS Provisioning System," *Proceedings of IEEE GLOBECOM'89*, Dallas, November 1989, pp. 42.5.1-42.5.5.

[14]   Ince, G. C., "The Evolution of Digital Cross-Connect Systems Analysis," *Bellcore DIGEST*, July 1990, pp. 1-7.

[15]   Aveyard, R. L. (Guest Editor), "Special Issue on Digital Cross-Connect Systems," *IEEE Journal on Selected Areas in Communications*, Vol. SAC-5, No. 1, January 1987.

[16]   TR-TSY-000233, *Wideband and Broadband Digital Cross-Connect Systems Generic Requirements and Objectives*, Bellcore, Issue 2, September 1989.

[17]   TR-TSY-000366, *Customer Control and Management (CCM) Controller Requirements and Objectives*, Bellcore, Issue 2, March 1988.

[18]   Hecht, J., *Understanding Fiber Optics*, Howard W. Sams & Company, 1989.

[19]   Miki, T., and Siller, C. A., Jr. (Editors), "Global Deployment of SDH-Compliant Networks," *IEEE Communications Magazine* (Special Issue), Vol. 28, No. 8, August 1990.

[20]   Wilson, C. (Editor), "SONET Special Issue," *Supplement to Telephony*, June 1990.

[21]   Nakajima, H., "Sonet: Progress Report from Japan," *Telecommunications*, August 1990, pp. 58-61.

[22]   Warr, M., "SONET: A Status Report," *Telephony*, January 28, 1991, pp. 25-28.

CHAPTER 3

Automatic Protection Switching and Dual Homing

Automatic Protection Switching (APS) and dual homing are two restoration techniques that have been used in today's fiber networks. The survivable architectures using these two techniques are classified as *conventional* survivable architectures because they can be implemented using current technology.

APS automatically reroutes signals from a working line to a protection line during signal outage. APS has two major purposes: to provide a method to carry service for planned outages, such as new equipment installation and routine maintenance, and to provide a line restoration capability in case of unexpected outages, such as network failures. APS is the simplest and fastest facility restoration technique. The 1:N APS architecture can restore a single-fiber system[1] failure; however, it is not effective for fiber cable cuts because the protection fiber is placed in the working fiber sheath and could also be cut. With its physically diverse protection fiber, the *1:N diverse protection* (1:N/DP) architecture can partially restore services from a fiber cable cut. The 1:1/DP or 1+1/DP architecture can completely restore services from a cable cut because it offers complete protection.

Dual homing protects against major hub failures and allows partial services to remain intact if a DCS failure (at the hub) or a hub building failure occurs. Table 3.1 summarizes several conventional, survivable network architectures that defend against various types of network component failures.

---

1. A fiber system includes a fiber pair (one fiber for upstream transmission and the other for downstream transmission) and equipment that terminates this fiber pair.

Table 3-1. Network Failures vs. Survivable Architectures Using APS/Dual Homing

Survivable Architectures	Single network component failure			
	Multiplex+ equipment	Fiber system	Fiber cable	Hub building
1:N Protection	Y	Y	N	N
1:N/DP	Y	Y	Y*	N
1:1/DP	Y	Y	Y	N
Dual Homing	N	N	N	Y*

Y: Yes; N: No
* Partially restored
+ Equipment failure is not defined in the present SONET APS standard.
DP: Diverse Protection

3.1 APS ARCHITECTURES

3.1.1 1:N APS Architectures

The 1:N APS architecture, as depicted in Figure 3.1, is the simplest and most commonly used APS architecture. In the 1:N APS architecture, N working fiber systems share one protection fiber system, where working and protection fiber systems are placed in the same physical route. The 1:N APS system includes protection switching modules, a 1:N bus, and a *Protection Switching Controller* (PSC). The protection switching modules of the APS system are located within the same frame as the *Optical Line Terminating Multiplexers* (OLTMs)[2] to minimize coaxial interconnection cable congestion.

Protection switching elements are controlled externally by a PSC or internally when simple 1:1 protection is required. In practice, fiber systems with capacity above gigabit-per-second (Gbps) rates usually require 1:1 APS systems. An alternative to 1:1 APS is 1+1 APS, where the head end bridge is permanent, as depicted in Figure 3.2. The working and protection OLTMs in the 1:1 and 1+1 APS architectures may be

---

2. OLTM is one type of terminal multiplexer that terminates a high-speed optical fiber system at one end and interfaces low-speed electrical tributary signals at the other. The OLTM electronically multiplexes low-speed signals together, converts electrical signals to the optical domain, and interfaces with the optical fiber.

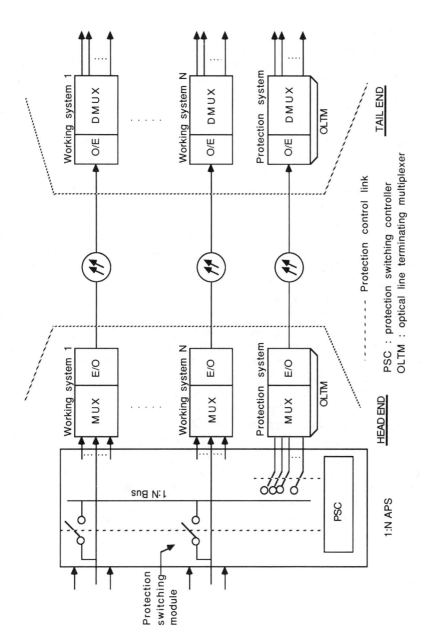

Figure 3.1. 1:N APS architecture.

77

separate or integrated as a unit with duplicated electronics sharing a common unit. These two implementations present a tradeoff between reliability and costs.

For the 1:N APS system, signals are transmitted on working lines during normal operation. The receiving terminals monitor received signals for the presence of signal, frame pattern, line code, *Bit Error Ratio* (BER), and checksum violations. If a BER or an *Out-of-Frame* (OOF) count for the working line exceeds the preset threshold, or a *Loss of Signal* (LOS) is detected on the working system, the receiver and transmitter exchange alarm control messages, thus causing the working system to switch to the protection system. After initial detection of a failed BER, OOF, or LOS, the protection switch does not activate until a predetermined time interval (usually milliseconds) expires; this ensures that the detected failure is not just a temporary interruption due to an upstream protection switch on a higher-level carrier system. Protection switching options include switching based on a specified line rate or individual *Digital Signal level 3* (DS3). For 1:N systems where two or more fiber systems have failed simultaneously, only the number of DS3s with highest priority (up to the protection line rate) can be recovered through the protection system. Each DS3 is assigned a different priority if needed. To ease system operation and administration, DS3s with the same priority are usually placed in the same fiber system.

During normal operation in the 1+1 APS system, the system transmits the signal continuously on both the working and protection systems. The receiving terminal monitors received signal quality on both working and protection systems, and the selector selects signals from the protection line if the received signals from the working line do not meet the performance requirements.

The APS system can operate in either a revertive or a non-revertive mode. In a revertive mode, services carried by the protection system are switched back to the working system once that failed working system has been repaired. In a non-revertive mode, services are not switched back to the working system even after the failed working system has been repaired.

The m:N APS architecture is an alternative to the 1:N APS architecture, where N working fiber systems share m protection fiber systems (m≤N). This architecture, which is designed to improve system availability with moderate cost increases, has rarely been implemented in present point-to-point fiber networks, although some vendors have provided 2:14 APS systems. The m:N APS architecture is also not considered part of the present *Synchronous Optical Network* (SONET) standard. Thus, we will not discuss the m:N APS architecture in this section, but rather in Section 3.6 where we discuss the nested APS architecture (in which the protection systems are shared by a chain network).

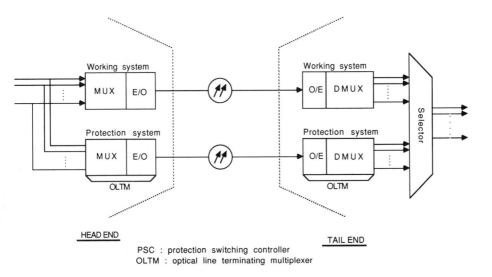

Figure 3.2.   1+1 APS architecture.

3.1.2 SONET APS Protocol

The APS switching technique is sometimes referred to as *line switching* because APS is performed on the whole optical line. The APS switching protocol uses the K1 and K2 bytes within the SONET line overhead where valid K1 and K2 bytes are only put into the first STS-1 of an STS-N signal (see Section 2.4.3.2). Figure 3.3 depicts K1 and K2 formats presently defined in the SONET standard. The K1 byte requests a channel for the switch action, and the K2 byte[3] confirms the channel that is bridged onto the protection line. Table 3.2 shows the types of requests that are specified in the K1 byte.

---

3. The K2 byte also indicates line AIS and line FERF on all lines, including the protection line. Line FERF is an indication returned to transmitting *Line Terminating Equipment* (LTE) on receipt of a Line AIS code or detection of an incoming line failure at the receiving LTE.

79

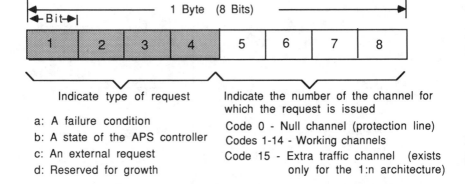

Indicate type of request

a: A failure condition
b: A state of the APS controller
c: An external request
d: Reserved for growth

Indicate the number of the channel for which the request is issued

Code 0 - Null channel (protection line)
Codes 1-14 - Working channels
Code 15 - Extra traffic channel (exists only for the 1:n architecture)

(a) SONET K1 Format

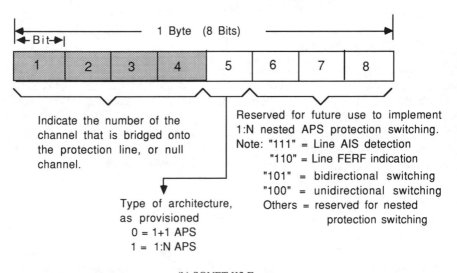

Indicate the number of the channel that is bridged onto the protection line, or null channel.

Type of architecture, as provisioned
0 = 1+1 APS
1 = 1:N APS

Reserved for future use to implement 1:N nested APS protection switching.
Note: "111" = Line AIS detection
"110" = Line FERF indication
"101" = bidirectional switching
"100" = unidirectional switching
Others = reserved for nested protection switching

(b) SONET K2 Format

Figure 3.3. SONET K1 and K2 formats.

80

Table 3-2. K1 Byte, Bits 1 through 4: Type of Request

Type of request	Bits 1-4	Action
A failure condition	1101	SF with high priority
	1100	SF with low priority
	1011	SD with high priority
	1010	SD with low priority
A state of the APS controller	0110	wait-to-restore
	0010	reverse request
	0001	do not revert
	0000	no request
An external request	1111	lockout of protection
	1110	forced switch
	1000	manual switch
	0100	exercise
Reserved for growth	1001,0111, 0101,0011	-

SF: Signal Failure
SD: Signal Degrade

Two types of APS architectures are defined in the SONET standard [1,2]. The 1:N APS architecture allows any one of the N (permissible values for N are from 1 to 14) working channels to be bridged to a single protection channel. APS protocol communication occurs via the K1 and K2 bytes that are located in the SONET Line Overhead. Referring to Figure 3.4, the protocol operation can be summarized as follows: when a failure is detected or a switch command is received at the tail end (i.e., the receiving end), the protection logic compares the priority of this new condition with the request priority of the working channel (if any) that wants to use the protection channel. The comparison includes the priority of any bridge order (i.e., of a request on the received K1 byte). If the new request is of higher priority, the K1 byte is loaded with the request and number of the channel requesting use of the protection line, and the tail end sends out the K1 byte on the protection line.

When this new K1 byte has been verified (i.e., received identically for three successive frames) and evaluated (by the priority order) at the head end (i.e, the transmitting end), the K1 byte is sent back to the tail end with a reverse request (to confirm the channel requesting use of the protection channel); a bridge is also ordered at the tail end for that channel. This action initiates a bidirectional switch. At the head end, the indicated channel is bridged to protection. When the channel is bridged, the K2 byte is set to indicate the number of the channel on protection.

At the tail end, when the channel number on the received K2 byte matches the number of the channel requesting the switch, that channel is selected for protection. This completes the switch to the protection channel for one direction. The tail end also performs the bridges, as ordered by the K1 byte and indicates the bridged channel on the K2 byte. The head end completes the bidirectional switch by selecting the channel from protection when it receives a matching K2 byte.

Note that K1 and K2 bytes always travel over the protection line in present SONET standards [2]. Protection switching, including K1/K2 operations and switch reconfiguration, must be completed within 50 ms [1].

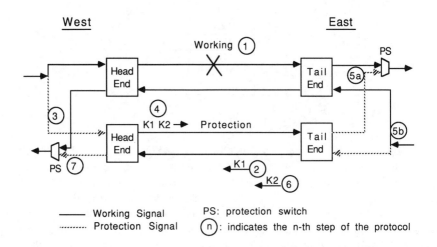

Figure 3.4.   SONET APS protocol.

Another type of APS that is also defined in the SONET standard is 1+1 protection switching (see Figure 3.2), which is a form of 1:1 APS with the head end permanently bridged. Thus, a decision to switch is made solely by the tail end. For bidirectional switching, the K1 byte is used to convey the signal condition to the other side; actual switching is decided by the tail end. A more detailed SONET APS implementation can be found in Reference [3].

Once the failure is detected, the time to restore services for bidirectional protection switching can be approximately calculated as follows:

82

$$T = (125 \times 3 + 2 \times T_{proc}) \times 3 + T_{swt} \quad (in\ \mu s)$$

where T_{proc} and T_{swt} are the K1/K2 byte processing time at each node and the switching time for the protection switching element, respectively. Assume that each node takes 125 μs of frame time to process the K1/K2 bytes; the protocol then takes 1.875 ms (i.e., T_{proc}) to process and validate the K1/K2 bytes before the protection switch performs a final bridge function at the head end. Thus, the switching time (T_{swt}) for the protection switch becomes a decisive factor in determining whether the specific SONET APS can meet the protection switching requirement of 50 ms.

3.1.3 1:N APS Diverse Protection Architecture (1:N/DP)

The 1:N APS architecture, which protects the network from a single fiber-system failure or a single multiplexing-equipment failure, may not recover from fiber cable cuts. A simple and commonly used approach is to place the protection fiber system on a physically diverse route to N working fiber systems. This approach is attractive since electronics costs, which are a dominant factor of fiber transport system costs, remain the same for 1:N APS architectures. The SONET APS protocols described in Section 3.1.2 can be applied to this diverse protection scheme, except that the K1 and K2 bytes are conveyed on the diverse protection paths.

As an alternative to the 1:N/DP architecture, full 1:1/DP, including 1:1 equipment and diverse fiber-system protection, provides 100 percent survivability for a fiber cable cut and equipment failure. The 1:1/DP structure may be preferred, particularly for fiber systems with capacities above Gbps rates, not only for promoting easy operations, but also for facilitating network transitions to ring networks where SONET high-speed SHRs may become preferable in the future.

3.2 DUAL-HOMING ARCHITECTURES

3.2.1 Dual-Homing Concept

For fiber-hubbed networks (see Figure 1.2), the failure of the hub building can isolate the area where that hub serves as a demand concentration point. A hub failure rarely occurs, but when it does, the impact on user communities is significant. The fire in Hinsdale's hub building, which occurred in 1987, is a well-known example. The 1:N/DP architecture, which protects networks from fiber cable cuts, is not effective for preventing a hub failure. To provide protection from hub failures, *dual homing* uses a hub backup concept to ensure that some services survive. In contrast to the *single-homing* approach, the dual-homing architecture designates two hubs (a home hub and a

83

foreign hub) for each special office requiring higher survivability against hub failures.

To achieve dual-homing protection, two dedicated fiber spans[4] are created from a special CO to its home hub and from that special CO to the foreign hub. Demand originating from the special CO is split between the home and foreign hubs. A dual-homing architecture, which splits demand evenly between the home and foreign hubs, is a cost-effective choice among possible demand-splitting options; it has been implemented in some telephone companies in the U.S. Under this demand arrangement, each special CO ensures that 50 percent of total demands is not affected if the home hub fails. The dual-homing architecture itself has no restoration capability; however, it can be used with DCSs in the special CO to restore more than 50 percent of demands, via the foreign hub, if the home hub fails. This scenario assumes that the special CO contains a DCS and that spare capacities exist on the fiber span connecting the special CO to the foreign hub.

3.2.2 Dual-Homing Protection for Chain Applications

Figure 3.5 depicts an *Add-Drop Multiplexer* (ADM) *fiber chain* providing connections between a group of COs and two hubs. Implementing such ADM chains between hubs is one way of providing dual homing. It appears to be advantageous in cases where COs have a significant community of interest with offices in a foreign cluster. The operation of such ADMs is described below.

Consider the ADM pair at CO-1 (Figure 3.5). One ADM transmits (adds) demand destined for Hub A and receives (drops) demand from Hub B. The other ADM works in the opposite way, i.e., it sends to Hub B and receives from Hub A. In addition to CO-hub demand, the chain provides connectivity between hubs. The principal idea behind such ADM chains is to provide connectivity from special COs to two hubs. In the event of an ADM, fiber, or hub failure, such connectivity would remain and would allow rapid restoration of priority circuits via an alternate hub. If a ring topology exists, this concept of an ADM dual-homing chain can be easily extended to a SHR architecture (see Chapter 4).

---

4. A *dedicated fiber span* has terminal equipment (e.g., OLTMs) at its two ends; it also has fibers spliced at each intermediate node if the span passes through one or more links in the network topology.

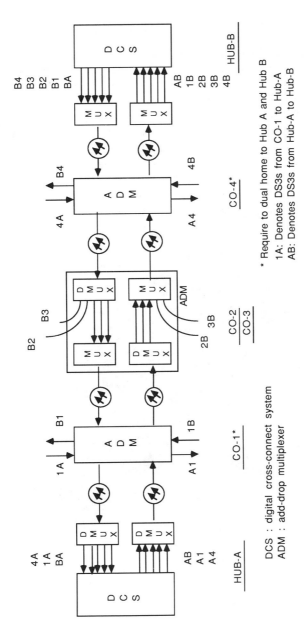

DCS : digital cross-connect system
ADM : add-drop multiplexer

* Require to dual home to Hub A and Hub B
1A: Denotes DS3s from CO-1 to Hub-A
AB: Denotes DS3s from Hub-A to Hub-B

Figure 3.5. ADM chain for dual homing.

85

3.3 COST AND SURVIVABILITY TRADEOFF ANALYSIS

The previous sections discussed several conventional, survivable architectures. A challenge that network planners face is how to plan appropriate survivable architectures to meet the network survivability requirement in a cost-effective manner. One approach is to study tradeoffs between costs and survivability among candidate survivable architectures. Reference [4] reports on a case study that shows the cost and survivability tradeoffs among several alternative architectures. The model network used in this study is a typical metropolitan *Local Access and Transport Area* (LATA) network that includes 36 nodes and 64 links (as depicted in Figure 3.6), and carries about 90 percent of the total LATA circuits. Each link is a candidate location to install the fiber system. Of these 36 nodes, four are hubs, 20 are special COs, and 12 are non-special COs. Table 3-3 summarizes this case study's results, Chapter 6 discusses the network design model used to obtain them.

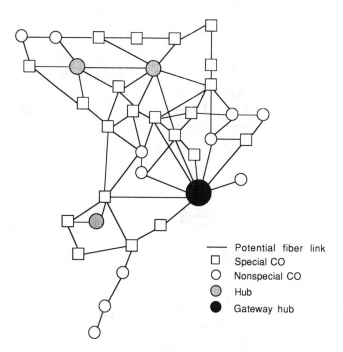

Figure 3.6. The model network.

Table 3-3. Tradeoffs between Costs and Survivability among Alternative Architectures

Network cost vs. worst case survivability				
Architecture	Cost penalty(%)	LS(%)	HS(%)	ICSR(L)
SH/1:N	0.0	50.13	30.21	-
SH/1:N/DP	6.3	72.12	30.21	11.9
SH/1:1/DP	15.2	92.57	30.21	15.0
DH/1:N	16.8	58.28	34.46	85.0
DH/1:N/DP	22.8	79.66	34.46	31.8

SH: single homing, DH: dual homing
LS = worst case survivability due to link failure (cable cut)
HS = worst case survivability due to hub failure
ICSR(L): ICSR against cable cut

The study mentioned above analyzed five alternative, conventional architectures: SH/1:N, SH/1:N/DP, SH/1:1/DP, DH/1:N, and DH/1:1/DP, where SH and DH represent single homing and dual homing, respectively. The value of N in 1:N architectures is not predetermined and is decided by a computer network design program. In this study, only special COs and hubs require diverse protection and dual-homing protection. Costs represent the fiber transport costs, which include costs for terminating equipment, fiber material and placement, APS, and regenerators; the component cost model was derived from an industrial source. Demands originating from the special CO for dual-homing architectures are split evenly between the CO's home and foreign hubs.

SH/1:N is considered a base architecture for comparison because it has been widely used for present interoffice fiber networks. *Link survivability* is defined as the percentage of total demands still intact if the fiber cable on that link is cut. *Hub survivability* uses a similar definition. Two survivability measures that are usually used are *average survivability* and *worst case survivability*. The worst case survivability is the lowest percentage of circuits surviving over all possible, singly occurring failures. The worst case survivability was used in Table 3-3 because it represents a minimum level of services that the network can support after a network component fails in the worst case. It also reflects the most sensitive (weakest) portion in terms of service protection.

As shown in Table 3-3, the SH/1:N/DP architecture, with a 6.3 percent cost increase over the base architecture, improves the worst case survivability versus link failures by 22 percent over the SH/1:N architecture. With an additional 8.9 percent cost increase over SH/1:N/DP, the SH/1:1/DP architecture further improves this worst case survivability by an additional 20 percent. The cost and survivability tradeoffs can be converted to a single measure called *Incremental Cost-to-Survivability Ratio*

(ICSR). The ICSR is the cost per additional circuit (made diverse by the architecture under consideration) compared to the base 1:N architecture. The ICSR result in the case study indicated that the diverse protection architecture is affordable in metropolitan-area LATA networks. Its incremental cost of $11.9 or $15 per circuit is insignificant compared to the total cost of a few hundred dollars for each working circuit (which includes not only the transport cost, but also multiplexing costs such as costs of D banks, M13 multiplexers, and so forth).

Dual-homing architectures improve the worst case survivability due to a hub failure by 4.2 percent, but with a significant cost penalty of 16.8 percent. SH/1:1/DP is 13 percent better for survivability due to link failures and costs 7.6 percent less than DH/1:N/DP. Dual-homing architectures are not only much more expensive than single-homing architectures, but are also more difficult to plan and administer. Thus, it is suggested that the dual-homing architectures be used only to meet special needs.

3.4 OPTICAL DIVERSE PROTECTION ARCHITECTURES

According to the cost and survivability tradeoff analysis shown in Table 3-3, the 1:N/DP architecture is attractive for metropolitan areas in terms of costs and survivability, and the 1:1/DP architecture is relatively expensive. However, in practice, fiber systems with greater than Gbps capacity usually require 1:1 protection because they carry so much service. Additionally, the 1:1/DP architecture may ease the transition to a SHR architecture, which is expected to be part of the optimum network architecture in future SONET interoffice networks. Thus, searching for an alternative 1:1/DP architecture that is affordable becomes obvious and necessary.

The primary reason that the conventional 1:1/DP architecture is relatively expensive is that it requires not only duplicate fiber facilities, but also terminating electronics equipment. The equipment costs are also a dominant factor of the total fiber transport costs [4]. This observation suggests a way to reduce survivable network costs, while retaining high survivability for cable cuts like 1:1/DP — eliminating PTE. Subsequently, this results in a cost-effective, survivable network architecture called *1:1 Optical Diverse Protection* (1:1/ODP). This architecture uses optical switches for 1:1 fiber cable protection as described in References [5,6].

3.4.1 1:1 Optical Diverse Protection Architecture (1:1/ODP)

Figure 3.7 depicts an example of the 1:1/ODP architecture. This architecture maintains 1:N electronic protection using the 1:N APS structure and 1:1 diverse fiber protection using optical switches as fiber facility protection switches. In Figure 3.7, the 1:3 APS system protects three working terminals, and three diverse protection fiber pairs protect three working fiber pairs; both pair types are connected at *the Optical Protection*

88

Modules (OPMs). Each OPM includes a 50/50 *Power Splitter* (PS) on the transmit side and a 1×2 optical switch on the receive side. The PS splits an optical signal into both working and protection systems. The 1×2 optical switch, acting as a selector, switches demand from failed fibers to corresponding diverse protection fibers when a cable containing working fiber pairs is cut. This architecture essentially uses the 1+1 APS concept.

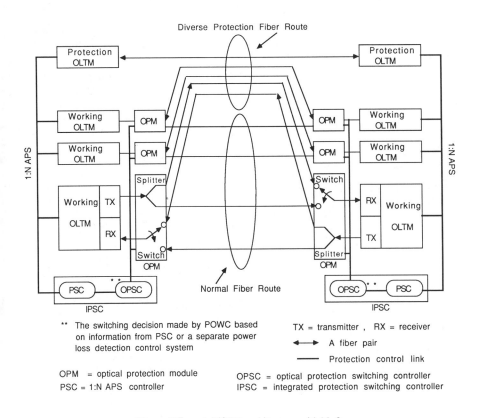

Figure 3.7. 1:1/ODP architecture with N=3.

In addition to the PSC, a scheme that controls optical protection switching is needed in the 1:N APS system. For immediate deployment of this architecture, 1×2 electrically controlled mechanical optical switches, which are commercially available, can be manually operated without changing APS design. However, this may not be acceptable for services requiring automatic (fast) reconfigurations. To overcome this reconfiguration speed problem, the electrically controlled 1×2 optical switches can be controlled remotely, rather than manually, through simple programming of a small processor. Control interfaces for these switches are also commercially available. This architecture requires an *Optical Protection Switching Controller* (OPSC) for automatic control of optical switching when protection is needed. The OPSC may be external to the PSC of the 1:N APS system yet connected to it to obtain the required link performance status data, such as BER or LOS. It may also be connected to two power loss detectors (one for working signals and the other for protection signals), as depicted in Figure 3.8 [7,8]. In Figure 3.8, the power detector monitors five or 10 percent of the optical signal, and the OPSC initiates protection switching when the detected power drops below a preset threshold. This power loss detection system works only for hard failures, such as cable cuts, not for terminal-type failures and system degradation with high BER.

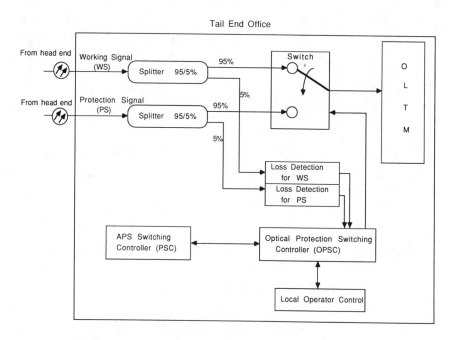

Figure 3.8. Optical power loss detection system functional diagram.

On request for switching due to detection of a failure condition, the OPSC transmits an applied voltage to transfer the incoming optical signal from the optical switch's working input port to its protection output port. Because an OPSC performs only two simple tasks (collect performance data from the PSC or a power loss detection system and send out a signal if applicable), it could be easily implemented in a simple microprocessor or integrated into the PSC features of the APS system. This 1:1/ODP architecture requires a control system only on the receive end. If one PS of 3 dB loss, four connectors of 0.5 dB loss each, and one 1×2 *Mechanical Optical Switch* (MOS) of 1 dB loss are used, the power loss for a fiber span due to optical switches and splitters is about 6 dB.

As an alternative to the 1:1/ODP architecture, a 1:1/ODP architecture having an OPM that uses two 1×2 optical switches at both ends may be used. Compared to the 1:1/ODP architecture depicted in Figure 3.7, this alternative architecture, which uses the concept of 1:1 APS protection, reduces power loss. However, it increases overall expenses because it requires control systems at both transmit and receive ends. If two 1×2 MOSs of 1 dB loss each and four connectors of 0.5 dB loss each are used, the power loss for this alternative architecture is about 4 dB.

3.4.2 1:1/ODP with 2×2 Optical Switches or WDMs

For some large areas where the protection route may require one or more repeaters, reducing the number of repeaters may be a desirable choice. The 1:1/ODP architecture depicted in Figure 3.7 can easily be modified to achieve this. As depicted in Figure 3.9, this modified architecture has the same basic structure as Figure 3.7 except that it uses 2×2 instead of 1×2 optical switches to terminate *Working Fiber-1* (WF-1) and *Diverse Protection Fiber-1* (DPF-1) systems. The use of 2×2 optical switches saves an extra fiber pair and associated repeaters (if any) placed in the physically diverse protection route. This switches WF-1 to DPF-1 during a fiber cable cut and disconnects the electrical protection node (P) from DPF-1.

As shown in Figure 3.9, the 2×2 switch is in the bar state when the WF-1 is in the normal state; it is switched to the cross state only when a fiber cable cut occurs. The controller applies a voltage to switch the 2×2 MOS. Other failures are protected based on normal APS protocols switching to the electrical protection channel when the detected LOS or BER exceeds the threshold on that working fiber system. Under this architecture, if one of the OLTMs fails, DS3s terminated at the failed OLTM are switched to the protection OLTM and sent to their destination protection OLTM through the 2×2 optical switch in the bar state. Although this alternative architecture may reduce the number of repeaters needed, it may not have 100 percent survivability for some multifailure scenarios (e.g., fiber cable cuts occur at a time when one working fiber system has been switched to a protection system due to equipment failure).

During multiple failures, this alternative architecture can possibly send control voltages only to the 1×2 switches but not to the 2×2 switches. In this case, the systems protected by the protection OLTM remain protected, but traffic from WF-1 is lost.

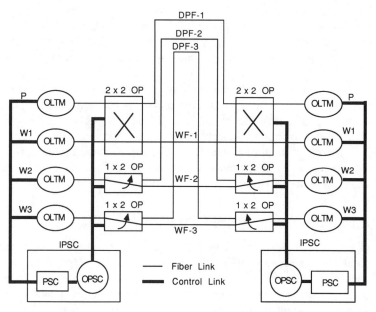

Figure 3.9. 1:1/ODP architecture with 2×2 optical switches.

The 1:1/ODP architecture can also be implemented using optical switches and *Wavelength Division Multiplexing* (WDM) devices (i.e., multiplexers and demultiplexers) as depicted in Figure 3.10 [6]. The basic operating principle for this architecture is that two input signals of two different wavelengths on separate, single-mode fiber ports of the device are multiplexed on one single-mode fiber output port. At the receiving end, the two multiplexed signals on the (one) input port are demultiplexed to two separate output fibers that can be detected separately by the optical receivers. This requires the protection system laser to be of a different wavelength than all the other working fiber systems. This is not a problem because fiber characteristics are

92

essentially equivalent, in terms of loss, for lasers operating in 1300-nm and 1550-nm ranges. Receivers sensitive to both wavelengths are available, and two wavelengths in the 1270- to 1330-nm range could also be used.

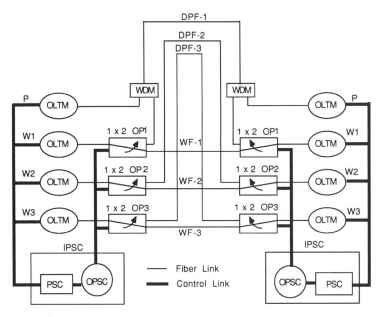

Figure 3.10. 1:1/ODP architecture with WDMs and optical switches.

The 1:1/ODP architecture using WDM allows for system survivability during multiple failures. The protected fiber system signals, and those from WF-1 that would be switched to the protection channel only during a fiber cable cut, are multiplexed by wavelength onto the same protection fiber, DPF-1. When compared to 1:1/ODP, this architecture saves the cost of an additional fiber pair and provides the same protection for all the working systems simultaneously and uniformly. The presence of the WDM devices introduces additional losses in the system, about 2.5 to 3 dB loss per span, including loss due to connectors (splices); it also requires the use of lasers at different wavelengths. These factors should be included in system performance tradeoffs when considering this architecture for a particular application.

3.4.3 Technologies for Implementation

Choosing appropriate optical switches for the 1:1/ODP architecture depends on device performance and reliability required for facility protection. Facility protection switching usually requires a long holding time with moderate reconfiguration rates, a small number of input-output ports, and a protection switching speed of less than 50 ms [1]. Table 3-4 compares optical switching technologies that can be used to implement the 1:1/ODP architecture [6]. A review of optical switching device technology can be found in Reference [9].

Table 3-4. Comparison of Optical Switching Technologies

Implementation time frames	Optical technology	Switching time	Service requirements*
immediate or near-term	mechanical optical switches	10-30 ms	low-speed protection switching
long-term	directional couplers, liquid crystal devices, acousto-optic, and photorefractive devices	ns - µs	high-speed protection switching

* All line rates (digital and analog), including rates ranging from DS0 to several Gbps (SONET STS-48, 2.49 Gbps), are acceptable.

As seen in Table 3-4, low-speed, reliable optical switching technology in the form of single-mode fiber MOSs meets most service protection switching requirements. The MOS, a unidirectional device that handles multiple-rate digital as well as analog signals, is commercially available for near-term protection of CO feeders, lower-speed telephony, and *Local Area Network/Metropolitan Area Network* (LAN/MAN) service. It has also been considered as an option to provide a redundancy capability for AT&T's SL undersea lightwave system [10].

These MOSs, which represent the first generation of optical switches, offer wavelength independence (1200 to 1600 nm), polarization independence, low insertion loss, and minimal crosstalk (60 dB or better) single-mode fiber operation while requiring low switching power. The MOS modules are independent of the optical line signal format; thus, they switch low-speed and high-speed digital, as well as analog optical signals. The MOS modules are commercially available in 1×2, 2×2, or 2×4 formats as depicted in Figure 3.11. Three port 1×2 switches and four port 2×2 switches each have two states of operation; the 2×2 switch states are referred to as the *bar* and *cross* states. Typically, the state is switched from the idle (bar) state by providing a 5 Vdc, 50 mA drive at the control terminal (allowing remote control of the switch). In one implementation, the electrically controlled movement of a prism physically switches the optical path, thus eliminating constraints for repeated fiber alignment. The prism is inserted into the beam to produce the cross state when voltage is applied at the module control terminal; this transfers the optical signal from its bar state path to the cross state path.

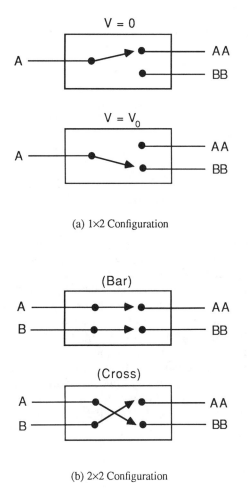

(a) 1×2 Configuration

(b) 2×2 Configuration

Figure 3.11. Mechanical optical switch configurations.

95

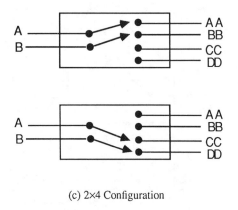

(c) 2×4 Configuration

Figure 3.11.   (Continued)

Fiber-pigtailed MOS modules have a maximum optical insertion loss of 1 dB (currently guaranteed by vendors) and have dimensions on the order of 2×2.5×1 inches. MOSs designed for operation with wavelengths between 1200 and 1600 nm have been operated at 1300 nm and 1550 nm. Their switching speed is relatively slow (10 to 30 ms is a typical range), but it is practical for protection switching applications. The switches can be obtained in both latching and non-latching modes.

High-speed or "hitless" protection switching (requiring switching speeds in the microsecond and nanosecond ranges) could be needed for some services requiring highly reliable real-time operations to minimize data loss in high-speed links. Such services include *Computer-Aided Design/Computer-Aided Manufacture* (CAD/CAM) applications, video, high-speed B-ISDN LAN/MAN data services (like *Switched Multi-megabit Data Service* [SMDS] [11]), and combat OSs in military applications. Photonic switching devices supporting high-speed protection switching, such as directional couplers, liquid crystal devices (with potential operation in the microsecond range), and photorefractive crystals [12-14] are not sufficiently mature to produce high-speed switching services, except in the laboratory [15,16]. They still exhibit tradeoffs among characteristics such as polarization and wavelength sensitivity, operating power requirements, crosstalk, and insertion loss.

3.4.4 Economical Merits and Planning Strategy

We choose to study the architecture that uses the MOS due to the MOS's availability, acceptable performance, and reliability. Comparing 1:1/ODP with the conventional 1:1/DP architecture, the cost benefit is obvious: 1×2 MOSs and PSs, plus a 1:N APS system, may cost tens of thousands of dollars, whereas two protection OLTMs at both ends may cost hundreds of thousands of dollars.

To incorporate the 1:1/ODP architecture as an option of survivable network architectures, we may need to know the conditions where this architecture is more economical than the 1:N/DP architecture. Figure 3.12 depicts an answer for this question, as reported in Reference [6], and shows a relationship between ICSR and DS3 demand requirements carried on a fiber span. In Figure 3.12, we assume the working span length is 20 miles and the ratio of protection span length to working span length is 1.5. We show two cases based on present and future fiber cost models. For the present model, when total DS3 demand is below 15, the ICSR is approximately the same for both the 1:N/DP and the 1:1/ODP architectures. However, when DS3 demand rises above 20, the ICSR of 1:1/ODP becomes better than that of 1:N/DP. The 1:1/ODP architecture will become even more attractive for these high-demand areas as fiber costs decrease. This analysis suggests that the 1:1/ODP is particularly cost-effective for high-capacity, interoffice fiber spans. The greater the number of DS3s, the more likely optical switching will be advantageous. The merits of 1:1/ODP compared to 1:N/DP are independent of span length and can be strengthened as fiber material costs decrease.

Now that we have an idea about the best locations for implementing the 1:1/ODP architecture, let us examine how much costs can be reduced in the same metropolitan area when all 1:N/DP (N>1) systems are replaced by 1:1/ODP systems. The cost of each 1×2 MOS is assumed to be $1000.

Table 3-5 shows a relative cost and survivability comparison between the 1:1/ODP architecture and several other architectures using two different costs for fiber material [5]. In Table 3-5, Study A assumes the fiber material cost to be $1700/mile/fiber pair, whereas Study B assumes $470/mile/fiber pair. Because purchasing more fibers provides diversity, two different fiber costs are used to ascertain the sensitivity of the results to fiber material. Fiber material and splicing costs are currently about $1700/mile/fiber pair, but they are expected to drop to $470 in the near future. The analysis described in Table 3-5 uses the same model network and demand data discussed in Table 3-3, but with a slightly different topology.

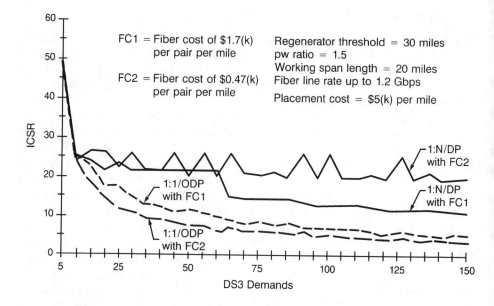

Figure 3.12. Economic comparisons between 1:N/DP and 1:1/ODP architectures.

Table 3-5. Cost and Survivability Comparison among Different Architectures

Architectures	Study A			Study B		
	CP(%)+	WLS(%)*	ICSR	CP(%)+	WLS(%)*	ICSR
1:N	0.0	55.46	--	0.0	55.46	--
1:N/DP	5.9	84.59	13.1	4.1	82.74	8.4
1:1/DP	17.7	91.05	31.9	17.8	91.05	27.8
1:1/ODP	11.2	91.05	20.1	9.4	91.05	14.7

(table header spans: *Network cost vs. worst case survivability and ICSRs*)

+ CP = cost penalty (%) for the candidate architecture when compared with the 1:N architecture

* WLS = worst case network survivability for a single fiber- link failure (e.g., cable cut)

As we see from Study A in Table 3-5, the newly proposed architecture has the same worst case survivability (for fiber cable cuts) as the 1:1/DP architecture. However, the cost penalty has dropped from 17.7 to 11.2 percent. The incremental cost for each additional survivable circuit for architecture 1:1/ODP lies between that of 1:N/DP and

1:1/DP. The ICSR for the 1:1/ODP architecture is still higher than that of the 1:N/DP. This occurs because the survivability improvement for fiber cable cuts due to 1:1/DP is not significant (with a range of 6.5 to 8.3 percent) when compared with 1:N/DP. In general, if the highest survivability is required, the 1:1/ODP architecture may be attractive to reduce costs while providing the benefits of 1:1/DP for fiber cable cuts.

There is no significant difference between the Study A and Study B results; hence, we may assume that the model is insensitive to the exact fiber cost.

3.4.5 System Development

The optical switching architecture described in Section 3.4.1 is commercially available and has been field trialed in some local telephone companies [7,8,17]. The prototypes for the 1:1/ODP architecture use PSs on the transmit end, optical switches on the receive end, and a power loss detection system as part of optical switching control (see Figures 3.7 and 3.8). These systems are classified into two categories: low-speed protection switching and high-speed protection switching. Table 3-6 summarizes their system attributes.

Table 3-6. System Attributes for 1:1/ODP Prototypes

System	Optical switch	Insertion loss	Type of switch	OS switching time	Restoration time*	Span loss/length+	Wavelength for OS
Low-Speed Switching	rotary switch	0.3 dB	mechanical	10 ms	30-40 ms	<5 dB	1300 nm/ 1550 nm
High-Speed Switching	lithium niobate crystal switch	5 dB	electro-optical	< 0.15 μs	2-4 μs	20 km**	1300 nm

OS: optical switch

* The time period that service is restored once the failure is detected.

+ Total loss/maximum repeater spacing due to additional passive components required for each fiber span

** Depends on the power budget connectors used and other factors

99

3.5 OPTICAL DUAL-HOMING ARCHITECTURES

Two approaches may be used to provide dual-homing protection: dedicated and broadcast. The dedicated approach, which was described in Section 3.2, builds dedicated fiber spans from each special CO[5] to the home and foreign hubs. When the home hub fails, each dual office associated with this affected hub restores demand via its designated foreign hub. This type of dual-homing architecture is also referred to as a *distributed* dual-homing architecture [DH(D)], since each special office is responsible for dual-homing protection.

Figure 3.13 shows an example of a DH(D) architecture with diverse routing protection, denoted by DH(D)/DP, in terms of fiber spans. This example shows two possible cases when the working span between offices (A and B) can directly span the home hub or pass through an intermediate office (i.e., Office C). As shown in Figure 3.13, implementing the DH(D)/DP architecture requires four fiber spans having four terminals at each special office: two for working and diverse protection spans to its home hub, and two for working and diverse protection spans to its foreign hub. The example of the dual-homing approach depicted in Figure 3.13 guarantees surviving connectivity against hub and DCS failures, but at a substantial cost increase compared to its single-homing counterparts (see Table 3-3 or Reference [4]). The substantial incremental cost associated with the DH(D)/DP architecture suggests that this architecture may be used only to meet special needs.

To reduce the costs associated with the DH(D)/DP architecture, a simpler, cost-effective dual-homing architecture that uses the broadcast approach has been proposed [18]. In contrast to the dedicated approach, the broadcast approach builds a fiber span from each special CO to its home hub and broadcasts signals to the foreign hub. This broadcast approach is sometimes referred to as the *centralized* dual-homing architecture, denoted by DH(C), since the dual-homing protection function is performed at a centralized location on a cluster basis. The DH(C) architecture essentially uses the SH/DP architecture, along with 1×2 PSs, to achieve the dual-homing protection if the home hub fails.

---

5. In this section, we assume each special CO requires dual-homing protection.

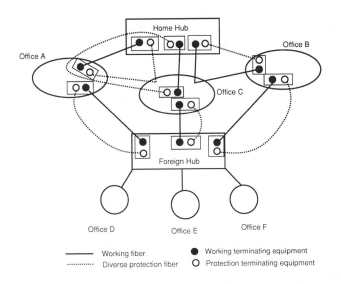

Working fiber   ● Working terminating equipment
Diverse protection fiber   ○ Protection terminating equipment

Figure 3.13.   Span layout for a DH(D) architecture.

3.5.1 Network Architecture and Control Protocol

The optical dual-homing architecture may apply to both 1:1 and 1:N protection architectures. Throughout this section, we describe the optical dual-homing architecture using only a 1:1/DP architecture. The 1:N/DP and 1:1/ODP architectures can be incorporated into this architecture in a simple manner.

The optical dual-homing architecture using a 1:1/DP structure is denoted by DH(C)/1:1/DP. Figures 3.14(a) and 3.14(b) depict an example of the DH(C)/1:1/DP architecture. Figure 3.14(a) depicts a network view for the DH(C)/1:1/DP architecture. As shown in this figure, the DH(C)/1:1/DP architecture requires a designated special CO [e.g., Office C in Figure 3.14(a)] as a *pseudo-hub* from where signals are broadcast to two hubs for dual-homing protection (for each special CO in the cluster of COs) if the home hub fails. Unlike the DH(D)/1:1/DP architecture, the DH(C)/1:1/DP architecture does not split demands between its home and foreign hubs. Instead, the DH(C)/1:1/DP architecture functions like the SH/1:1/DP architecture in normal operation and, during cable cuts, has the capability to restore all affected demands (except those that terminate at the home hub) to their destinations via the foreign hub if the home hub fails. Figure 3.14(b) shows a set of three 1×2 50/50 PS pairs at the pseudo-hub office (Office C) that are used to provide dual-homing protection if the home hub fails. For simplicity, the diversity of the protection fibers from Office C to

101

its home hub is shown in Figure 3.14(a), but not in Figure 3.14(b). Each PS shown in Figure 3.14(b) represents a PS pair because the PS is a unidirectional device.

In the pseudo-hub office, a pair of 1×2 PSs is required for each dual office in the considered cluster. To simplify our architectural description, we generalize the concept of the fiber span. We define a *Y-span* as a generalized span having three OLTMs at three ends of its Y-type structure, with a PS serving as a branch point. A *semi-span* is defined as a special span having one OLTM and one port of a PS at its two ends. Thus, one Y-span includes three semi-spans. For example, in Figure 3.14(b), a Y-span is used to connect Office A, its home hub, and its foreign hub with a PS at the pseudo-hub office (Office C). Each office except the foreign hub is equipped with an APS system that detects network faults in the architecture. Basically, the DH(C)/1:1/DP architecture can be viewed as an extension of the SH/1:1/DP architecture with the following additional requirements:

1. One fiber system constituting a fiber span (A, HH)[6] passing through the pseudo-hub office is connected through a PS at the pseudo-hub office for the DH(C)/1:1/DP architecture, rather than being spliced together as in the SH/1:1/DP architecture.

2. The DH(C)/1:1/DP architecture requires an additional semi-span connecting the pseudo-hub to its foreign hub for each dual office on a normal SH/1:1/DP network.

At the foreign hub, one protection OLTM is required for each dual office in the considered cluster. These protection OLTMs always receive signals[7] (either working signals or test/maintenance signals), but these signals are selected and processed (at the foreign hub) only when the foreign hub receives a message indicating that the home hub has failed. Note that for fiber cable cuts, each CO in the DH(C)/1:1/DP network has its own protection switching capabilities, like a normal SH/1:1/DP, to switch demands from the affected span to the diverse protection span.

To see how the DH(C)/1:1/DP architecture works for both cable cuts and hub failures, we use the example depicted in Figures 3.14(a) and 3-14(b) to describe a network protection switching control protocol (as depicted in Figure 3.15). If the current network operation mode is normal, the network operates like a *Single-Homing* (SH) network. If a cable cut occurs, each affected office initiates its own protection switching functions to switch demands from the affected working span to the diverse

---

6. Office HH is the home hub of special offices A, B, and C.
7. The protection OLTM should always receive signals so it will not activate the alarm system.

protection span; however, the protection switching function in the foreign hub is not initiated. If the home hub fails, all affected demands are rerouted to their foreign hub based on the following conditions:

1. If the working span passes through the pseudo-hub office (Office C), such as span (B, HH), demands originating from Office B are directed to the foreign hub via PS-B (i.e., no protection switching is initiated at Office B).
2. If the diverse protection span of the working span passes through the pseudo-hub office (Office C), such as span (A, HH), Office A initiates diverse protection switching to reroute its demands to the foreign hub via its diverse protection span and PS-A.
3. If the working span terminates at the pseudo-hub office, such as span (C, HH), no action is initiated.

Concurrently, the network control center passes this fault information to the foreign hub, via data channels, to start processing incoming working signals at the foreign hub.

If the pseudo-hub office (Office C) fails and the working span does not pass through the pseudo-hub office, such as span (A, HH), diverse protection switching is not initiated at Office A. If the working span passes through the pseudo-hub office, such as span (B, HH), diverse protection switching is initiated at Office B, and demands are routed to Office B's home hub via its diverse protection span.

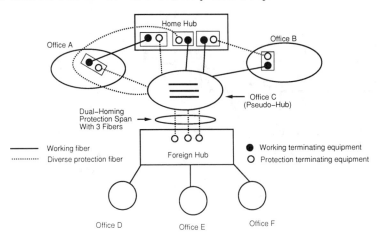

(a) Optical Dual-Homing Architecture

Figure 3.14.   An optical dual-homing architecture.

103

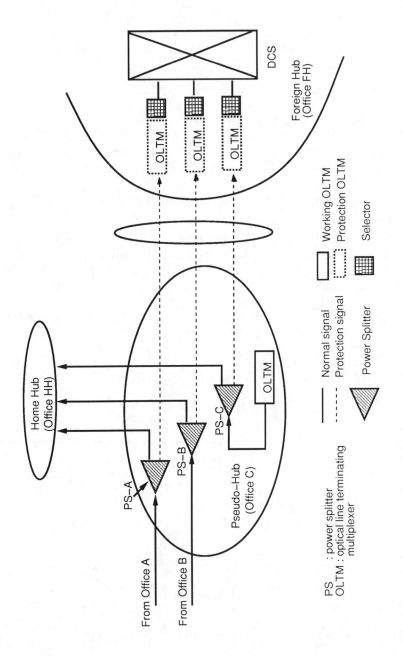

(b) Interconnection Network at the Pseudo-Hub

Figure 3.14. (Continued)

Note that if the DH(C)/1:1/ODP architecture is used (see Section 3.4.1), the *working* span between Offices A (or B) and HH should pass *through* the pseudo-hub office. Otherwise, the protection OLTM (associated with Office A) at the foreign hub will not receive continuous signals and will not remain in the working mode due to the use of 1×2 MOSs at originating COs.

The advantage of the DH(C)/1:1/DP architecture is not only that it is more cost-effective (as will be discussed next), but also that it has higher hub survivability and simpler operations than the DH(D)/1:1/DP architecture. [The DH(D) architecture is usually used in conjunction with *Digital Cross-connect System* (DCS) restoration techniques.] During normal operation or fiber cable cuts, the DH(C)/1:1/DP architecture functions like a SH/1:1/DP architecture. Should the home hub fail, the DH(C)/1:1/DP has the capability to reroute demands of special COs to their foreign hub through their normal outgoing working fiber system or protection system (depending on which span has fibers through the pseudo-hub office). A failure of the pseudo-hub office is treated as a fiber cable cut in DH(C)/1:1/DP networks.

3.5.2 Economic Merit of Optical Dual-Homing Architectures

Using the DH(C)/1:1/DP architecture can be more economical than using the DH(D)/1:1/DP architecture because savings come not only from fiber material and placement cost savings, but also from terminal (i.e., OLTMs) cost savings. For example, in Figure 3.16, using the previously proposed DH(D)/1:1/DP approach requires four fiber spans having eight OLTMs for each special office (working and diverse protection spans to the home and foreign hubs, respectively). In contrast, the DH(C)/1:1/DP architecture requires only one fiber span having two OLTMs (connecting the special CO's home hub to either the working or the protection channel) and a Y-span having three OLTMs (connecting the dual office to its home and foreign hubs) for each special CO. Thus, each office needs a total of five OLTMs and one pair of PSs for the DH(C)/1:1/DP architecture. In contrast, each needs eight OLTMs for the DH(D)/1:1/DP architecture, where each PS may cost only several hundred dollars, and each OLTM may cost about a hundred thousand dollars for gigabit-per-second systems. Because the number of fiber spans associated with the DH(C)/1:1/DP architecture is reduced, additional cost savings may result (i.e., the number of required regenerators would decrease). The regenerator requirement for the DH(C)/1:1/DP architecture can be minimized by choosing an appropriate special office as the pseudo-hub office. However, the inherent loss of the PS (and the 1×2 MOS if the 1:1/ODP architecture is used) may eventually lead to the need for more regenerators. All of these factors should be considered when planning to implement this architecture.

105

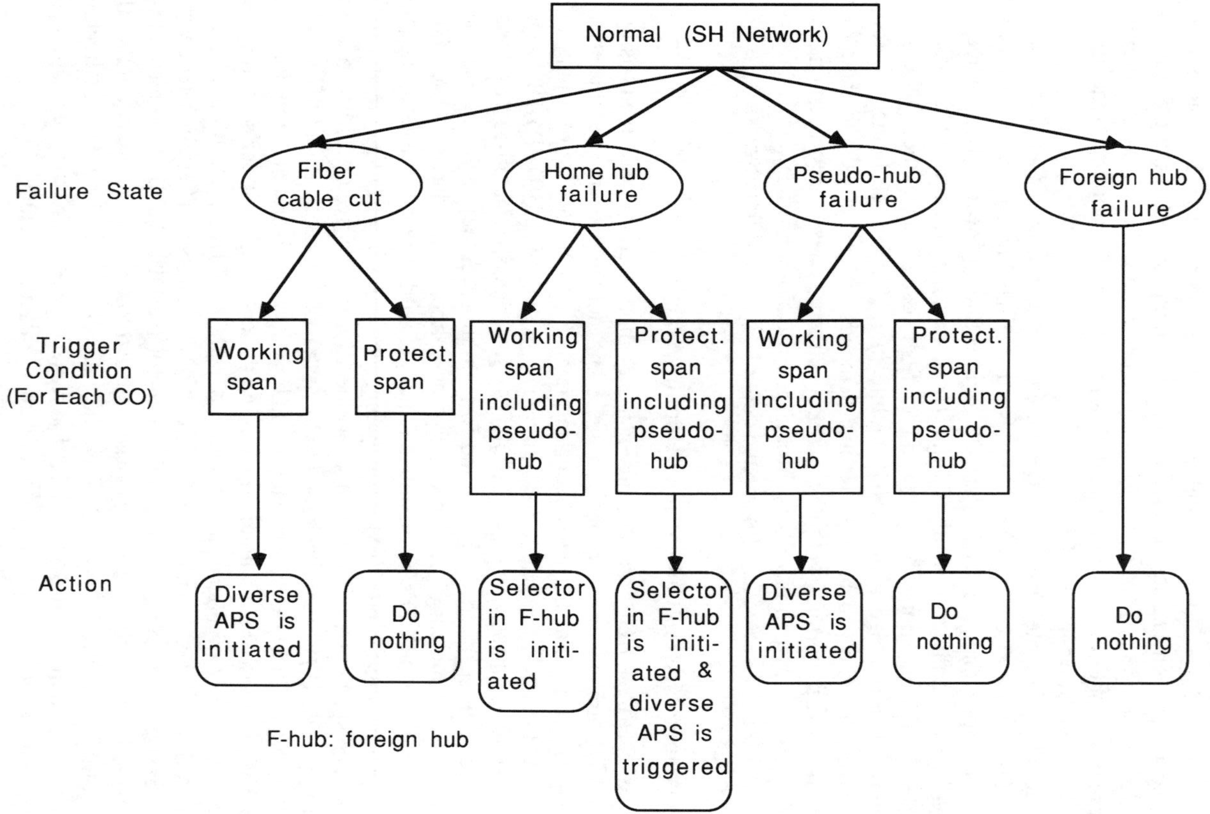

Figure 3.15. Optical dual-homing protection protocol model.

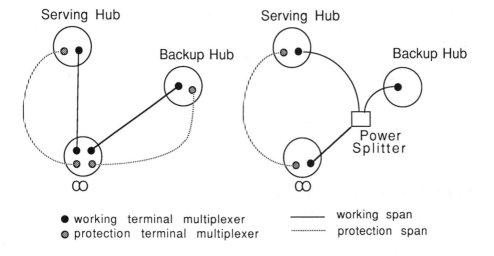

Serving Hub

Backup Hub

Serving Hub

Backup Hub

Power
Splitter

● working terminal multiplexer   ——— working span
◉ protection terminal multiplexer   ············· protection span

Figure 3.16.   Broadcast approach vs. dedicated approach.

Referring again to our example in Figure 3.14, we see that expanding a SH/1:1/DP network to form a DH(C)/1:1/DP network requires only one extra semi-span having one PS and one OLTM for each special office (from the pseudo-hub office to its foreign hub). The extra cost due to PSs is insignificant because each PS costs less than $1000. Reference [18] has reported a relative economic analysis that compares the DH(C)/1:1/ODP architecture with SH/1:1/ODP[8] and DH(D)/1:1/ODP. The purpose of this analysis was to gain some insight about a typical metropolitan area in terms of the following questions:

1. How much cost savings can be achieved using DH(C)/1:1/ODP instead of DH(D)/1:1/ODP?
2. What cost penalty is needed to expand a SH/1:1/ODP network to a DH(C)/1:1/ODP network?

---

8. The 1:1/ODP architecture, rather than 1:1/DP, is used here because it offers a promising economic benefit for protection of gigabit-per-second fiber systems against fiber cable cuts.

The answer to the second question may result in planning strategies to determine areas where the upgrade cost for an already deployed diverse protection network to a highly survivable optical dual-homing network is affordable. The case study in Reference [18] suggests that the cost savings using the broadcast approach are significant (at least 44 percent in terms of ICSR [see Section 3.3]) when compared to the dedicated approach. The extra cost to extend the 1:1/ODP network to the optical dual-homing network is affordable (about $14 per additional survivable circuit when compared to the commonly used 1:N architecture) in a light traffic environment. This extra cost may become negligible in a high-demand environment because the extra cost due to DH(C)/1:1/ODP (as compared to 1:1/ODP) comes from one extra OLTM in the foreign hub. This extra OLTM cost becomes insignificant when compared to the total transport cost in a high-demand environment.

3.5.3 Design Requirements and Limitations

When implementing the DH(C)/1:1/DP architecture in the intraLATA interoffice network, the following requirements apply:

1. All special COs under consideration, including the pseudo-hub office, should be in the same cluster (i.e., sharing the same home hub).
2. All special COs must have the same foreign hub as the pseudo-hub office.
3. Physical paths connecting the pseudo-hub to its foreign hubs must exist, and these paths should not pass through the home hub.
4. All special COs under consideration must have either working or protection spans passing through the pseudo-hub office.

Passive PSs could be realized using guided wave devices on $LiNbO_3$ [19] or using fiber couplers [20]. Power splitters with a total power-splitting loss of 4 dB, including an excess loss of 1 dB, are commercially available. Like the 1:1/ODP architecture, power loss due to splitter insertion may limit the applications that use this architecture because it makes the repeater spacing shorter. Thus, more regenerators are eventually required. However, if the cost savings of this architecture (compared to its conventional counterpart) can justify the need for more regenerators, then power loss will not be a key factor that delimits the applications for this optical dual-homing architecture. For the metropolitan-area LATA networks, where dual-homing protection and diverse protection are usually needed, the impact in PS and MOS losses could be insignificant. The typical span length in these areas is usually relatively short, and an additional 4-dB loss may be affordable.

3.6 NESTED APS SYSTEMS

3.6.1 Nested APS System Configurations

Because one protection system is dedicated to N working systems for each point-to-point span in the 1:N APS system, the APS system has traditionally been deployed in a point-to-point network and has been referred to as the dedicated APS system. A fiber *span* is defined as a segment between two terminating devices (such as OLTMs), excluding regenerators, and each fiber span may include one or more fiber systems. Two OLTMs are needed for each fiber system at both ends of each span, and fiber is spliced at intermediate COs (if any). Figure 3.17(a) depicts an example of the dedicated APS system for a point-to-point network. For example, working line W#2 includes two working point-to-point spans (one is between Nodes 1 and 2, the other is between Nodes 2 and 4), and there are two protection spans (P#2.1 and P#2.2) dedicated to these two working spans. For SONET networks, a span is a segment between two LTEs. The APS system is deployed on a span basis; therefore, the K1 and K2 bytes used for APS operations are line overhead, which is terminated and processed by SONET LTEs.

As an alternative to the dedicated APS system, the nested APS system shares protection facilities and equipment among working systems. For a nested APS system, a sector is referred to as a nested APS control area, and a segment is defined as a network portion between two adjacent nodes. The nested APS system can be applied to both point-to-point and chain networks. The chain network, which has been implemented in today's fiber networks, is a network where all nodes are connected in a linear manner and facilities are shared among nodes in the network. The ADM is used in the chain network to add and drop local demand and to pass through the transit demand from one incoming high-speed data stream to another outgoing data stream. Figures 3.17(b) and 3-17(c) depict examples of the 1:N nested APS systems for a chain network and a point-to-point network, respectively. Comparing Figure 3.17(c) with Figure 3.17(a), the protection facility (i.e., P) is not only used by two working segments in working line W#2, but also shared by working segments in other lines (e.g., W#1 and W#3). For convenience, the nested APS systems for the chain network and the point-to-point network are referred to as the C-nested and the P-nested APS systems, respectively. The SONET nested APS systems are discussed in the next section.

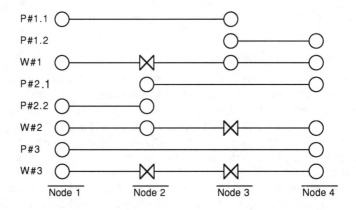

(a) Dedicated APS System for a Point-to-Point Network

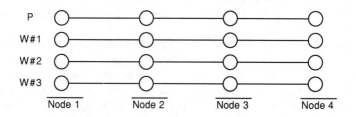

(b) 1:3 Nested APS System for a Chain Network

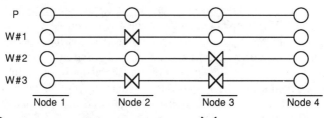

○ : ADM (Terminal or Add-Drop Mode) ⋈ : Repeater or Splicer

(c) 1:3 Nested APS System for a Point-to-Point Network

Figure 3.17.   Nested APS systems.

It is obvious that the nested APS system, when compared to the dedicated APS system, can save capital costs, but at a penalty of needing a more complicated protection control scheme. Table 3-7 shows a relative comparison between the dedicated APS system and the nested APS system. Chapter 6 will discuss network design for chain networks using the nested APS scheme.

Table 3-7. Relative Comparison between Dedicated and Nested APS Architectures

Attribute	Dedicated APS	Nested APS
Capital Cost	more	less
Protection Switching System Complexity	simple	complex
Terminating Equipment	OLTM	ADM
Number of Regenerators Needed	more	fewer
Application	point-to-point network	point-to-point or chain network

3.6.2 SONET Nested APS Proposals

3.6.2.1 1:N Nested APS

The 1:N nested APS for chain networks (i.e., the C-nested APS) is a special case of the 1:N nested APS system for point-to-point networks (i.e., the P-nested APS) in terms of the SONET K1/K2 operation. Because the protection segments of the chain network are independent of each other (i.e., the protection segment between two adjacent nodes carries demand only from working lines in the same segment during failures), the 1:N APS protocol described in References [1,2] can be applied to the 1:N C-nested APS system without any modification. The C-nested APS system is essentially the same as a set of independent dedicated 1:N APS systems. For example, the 1:3 C-nested APS system shown in Figure 3.17(b) can be viewed as three independent 1:3 dedicated APS systems. For a 1:N P-nested APS system, two protection segments may not be independent of each other. For example, in Figure 3.17(c), the protection segments between Node 1 and Node 2, and between Node 2 and Node 3, may carry demand from working lines 1, 2, and 3 during failures.

Because the 1:N P-nested APS system has non-disjoint segments, the topology information, which is needed for the P-nested APS control, is locally stored in each APS controller at each node. The ADM uses this topology information (see Table 3-8) in the protection line to determine whether to terminate/process the K1/K2 byte or pass it through to the next node. The nodal configuration mode is called an ADM mode if

111

the K1/K2 byte needs to be terminated and processed. It is called a *regenerator or splicer* (RESP) mode if the K1/K2 byte is passed through the node. For example, if working line 3 fails, the tail end (Node 4) loads a new K1 byte and sends it to Node 3 via the protection line. Because the mode of line 3 at Node 3 is the RESP mode, the protection ADM at Node 3 passes through the receiving K1 byte to Node 2. The K1 byte continues to be passed through Node 2's protection ADM and is terminated at Node 1 because the mode of line 3 at Node 1 is the ADM mode. Based on this K1/K2 pass-through/terminating algorithm, a simple extension of the SONET 1:N APS system may work well in most cases. However, current protection may be affected by another ineffective protection request with higher priority.

Table 3-8. An Example of Topology Information (ADM Mode or RESP Mode) Stored for a 1:N P-Nested APS System

Line #	Node 1	Node 2	Node 3	Node 4
P	ADM	ADM	ADM	ADM
1	ADM	RESP	ADM	ADM
2	ADM	ADM	RESP	ADM
3	ADM	RESP	RESP	ADM

RESP: Repeater or splicer (means K1/K2 pass-through)
ADM: either add-drop or terminal mode (means K1/K2 termination)

For example, in Figure 3.18, working line 3 at Node 4 issues a new request (with priority 2) to use the protection line between Node 1 and Node 4. Meanwhile, the protection segments between Node 1 and Node 2, and between Node 3 and Node 4, have been occupied by working line 2 with priority 1 and working line 1 with priority 4, respectively. Note that a lower number means higher priority (e.g., 1 has the highest priority). According to SONET 1:N APS protocol, Node 4 decides to drop the protection segment occupied by working line 1 and sends K1 through the protection line because the new request has higher priority. However, the priority in the new request is lower than the priority currently using the protection segment between Node 1 and Node 2. In this case, Node 2 does not transfer the K1 byte to Node 1, and the new request is discarded. As a result, the protection request for working line 3 is not effective; however, the protection service for working line 1 is affected.

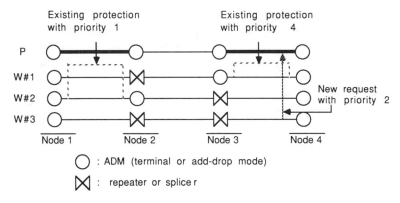

Figure 3.18. An example of ineffective protection request.

To make sure that the existing protection is not affected by such an ineffective request, the 1:N APS protocol defined in [1] needs to be modified to ensure that the existing protection cannot be dropped before the new switching request with higher priority turns out to be effective. To do that, a waiting state is introduced before the request is proved to be effective [21]. Figure 3.19 shows a state transition diagram that includes three states for each node: a protection segment idle state, a remaining state for existing protection, and a waiting state for the K2 byte. The idle state occurs when the node makes the protection line available for use. Each node executes a state machine algorithm for a protection line for each direction. Each node has the same state transition diagram despite the node mode (tail end, head end, or intermediate node). As shown in Figure 3.19, when a node in the idle state or the remaining state receives a K1 byte or a protection switching request with higher priority, it enters a state where it waits to receive the K2 byte. When a node receives a corresponding K2 byte, it drops existing protection, if any, and enters an idle state. If a node is in the waiting state (for K2), it remains in that state when the node receives the K1 byte or a switching request with higher priority. Also, any node waiting for a K2 byte returns to the remaining state when the timer expires. This rule is effective for any node; however, the actions that are taken differ according to the node mode (see Figure 3.20).

113

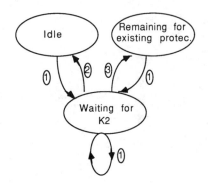

Reception of K1 or switching request with higher priority

Reception of a corresponding K2 byte

Time out

Idle state: protection line is available for use
Remaining state: remains existing protection
Waiting state: waiting for K2 byte

Figure 3.19. State transition diagram for a 1:N nested APS protocol.

According to the state transition diagram depicted in Figure 3.19, a potential security problem may occur if the head end bridges the protection line when it receives the K1 byte, as does the SONET 1:N APS protocol described in Reference [1]. The intermediate node may receive an unexpected signal from the head end via the protection line because intermediate nodes will not drop the existing protection line until they receive the K2 byte from the tail end that indicates drop of the existing protection line.

To avoid the potential security and ineffective request problems, a three-way handshake protocol was proposed [21]. In this proposal, the function of the K1 and K2 bytes is modified, but the bit sequence of these bytes remains unchanged. Moreover, the three reserved bits in the K2 byte (see Figure 3.3) are used to complete a three-way handshake protocol. The following description of a 1:N P-nested APS protocol is taken from Reference [21] with a minor modification. Figure 3.20 depicts a bidirectional operation diagram to show how this three-way handshake protocol works.

When a failure condition is detected or a switch command is received at the tail end of the system, the protection logic compares the priority of this new request for using the protection line with the priority of the line currently using the protection line. If the new request has a higher priority, the tail end loads a new K1 byte with the condition (or priority; see 'CND' in the K1 byte in Figure 3.20) and the line number ('CH #' in K1) requesting use of the protection line. At this moment, the tail end does not drop the existing protection (if any). Instead, the tail end enters the state waiting for a K2 byte from the head end.

At each intermediate node, when a K1 byte is received, the protection logic of that node compares the priority in the received K1 byte with the priority of the outgoing line using the protection line (if any). If the new request has a higher priority, the

114

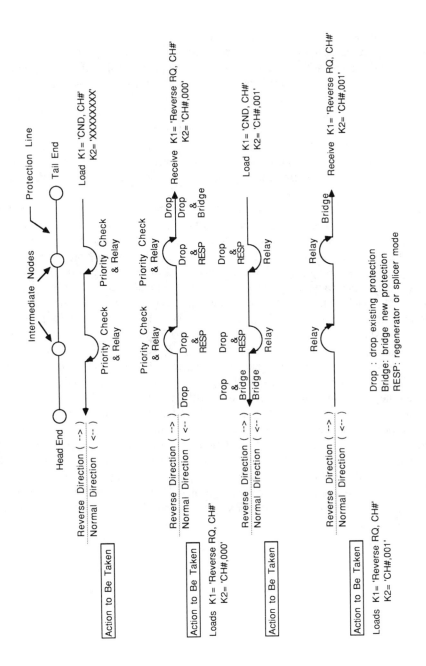

Figure 3.20. Bidirectional operation for a 1:N nested APS protocol.

115

intermediate node transfers the K1 byte to the next node. At this moment, the intermediate node does not drop the existing protection, even if the new request has a higher priority. The node then enters the state waiting for a K2 byte from the head end. If the priority of the new request is lower than the priority of the line using the protection line, the K1 byte is dropped and never transferred.

When a K1 byte from the tail end is successfully received by the head end, the head end drops the existing protection (if any) and loads a new K2 byte with the same line number as the received K1 byte and '000' in bits 6-8 of the K2 byte. The head end also loads a new K1 byte, which is used to request the reverse protection switch. Then, the head end enters the state waiting for a K2 byte with in-service bits (i.e., '001' in bits 6-8) from the tail end.

At each intermediate node, when a K2 byte from the head end is received, the node drops the existing protection and reconfigures the node mode from the ADM mode to the RESP mode, allowing subsequent K1 and K2 bytes to pass through this node. When a K2 byte reaches the tail end, the tail end drops the existing protection (if any) and bridges the line requesting use of the protection line to the protection line. The tail end also drops the existing protection for the reverse direction (if any) for the bidirectional switch. Then, the tail end loads a new K2 byte with the line number and in-service bits (i.e., '001' in bits 6-8), and sends it back to the head end. If a node is in the "waiting state" for a K2 byte and a timer expires before it receives the K2 byte, the node returns to the "remaining state," which retains the existing protection. This happens when a K1 byte is dropped and never transferred in an intermediate node because the existing protection has a higher priority than the new protection request.

When the head end receives the K2 byte with in-service bits, the line is bridged to the protection line, and the protection switch for the normal direction is completed. The operation of the reverse protection switch is similar to the one for the normal direction. The tail end completes its protection switch for the reverse direction when it receives a K1 byte with a reverse request and a K2 byte with '001' in bits 6-8. Note that if the security problem is not a concern, then a two-way handshake process is enough to avoid the ineffective request problem. More details of this three-way handshaking protocol and other proposals of 1:N nested APS protocols can be found in References [21-23].

Once the failure is detected, the time to restore services for P-segment bidirectional protection switching can be approximately calculated as follows:

$$T = (125 \times 3 + (P + 1) \times T_{proc}) \times 3 + T_{swt} \quad (in \ \mu s)$$

where T_{proc} and T_{swt} are K1/K2 byte processing time at each node and switching time for the protection switching element, respectively.

3.6.2.2 m:N Nested APS

As explained previously, the m:N APS system is rarely implemented in today's point-to-point fiber networks. However, in the nested APS system, a protection line is shared by all working lines. This may not meet system availability or survivability requirements. The number of protection lines in an m:N nested APS system depends on costs and availability/survivability requirements.

To extend the 1:N nested APS protocol to the m:N nested APS protocol, a broadcast-merger feature is added into the top of the protocol model described in Section 3.6.2.1. The basic ideas for this broadcast-merger feature are summarized as follows [21]:

1. A K1 byte is loaded and transmitted on all possible protection lines at a tail end.
2. A K1 byte is transferred to all possible protection lines at each intermediate node.
3. A head end selects one protection line on a priority basis (if more than one protection line conveys a new K1 byte), and a new K2 byte is loaded and transmitted through the selected protection line.
4. When a corresponding K2 byte is received at each intermediate node, the node transfers the K2 byte through one protection line, which is also chosen on a priority basis.

The first two items above are referred to as a *broadcast phase,* whereas the last two items are referred to as a *path-merger phase.* In a broadcast phase, a node transmits a K1 byte through all the vacant protection lines or those occupied with lower priorities. Intermediate nodes never transfer K1 bytes through the protection lines that are being used with higher priority. Hence, if at least one path is available, the K1 byte surely reaches the head end via vacant protection segments or protection segments with lower priority. In a path-merger phase, the protection line selection at each node is based on a simple rule: a vacant line is always preferred over the occupied protection line with lower priority, and a vacant line is chosen arbitrarily if more than one vacant protection line is available.

117

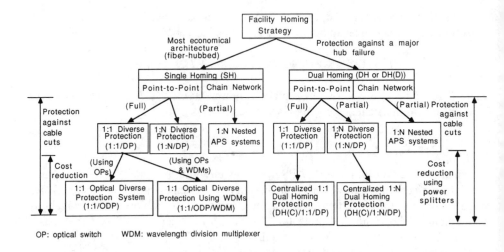

Figure 3.21. A family tree of APS and dual-homing architecture alternatives.

3.7 SUMMARY AND REMARKS

In this chapter, we have reviewed a class of survivable network architectures, as depicted in Figure 3.21, that can be implemented using today's technology. The present SONET APS standards can be applied to these survivable network architectures, except that the nested APS architecture for point-to-point networks requires a minor modification. A case study on intraLATA metropolitan areas has suggested that diverse protection offers affordable survivability for fiber cable cuts in metropolitan LATA networks. The 1:1/DP architecture is relatively expensive because it requires duplicating fiber facilities and also terminating electronics equipment, and costs of terminating equipment are a dominant factor of total fiber transport costs. In practice, the 1:1/DP architecture is preferable for fiber-system capacity above Gbps rates; it may also ease the network transition to SHR when SONET rings become economical in the future.

To reduce survivable network costs due to 1:1/DP, an alternative architecture using optical switches and power splitters was proposed. This architecture may not only reduce survivable costs for fiber cable cuts, but may also make the service restoration faster than the conventional APS system. The optical diverse protection switching architecture can be built by upgrading today's 1:N/DP networks with a minimum cost penalty, but with a significant survivability gain defending against fiber-system or fiber cable failures. In addition to cost savings, the optical protection switching system provides the following advantages:

118

1. It restores services faster than conventional APS systems.
2. It lessens the system complexity required to maintain and operate the network (because the problem is corrected at the source without a massive network reconfiguration).
3. It is independent of the transmission formats and types of equipment being protected (because the protection switching is performed at the physical layer).
4. It allows a remote demand rerouting capability from a CO.

Power splitting and insertion loss in the optical diverse protection switching system may shorten the maximum transmission distance. In the future, the loss in the optical switch can be offset and system margin can be improved using optical amplifiers, such as a semiconductor laser amplifier or an erbium-doped fiber amplifier. This optical diverse protection switching system is commercially available and has been field-trialed by some local telephone companies.

Dual-homing protection, which has been implemented in present fiber networks, is relatively expensive because it requires building separate fiber spans from the office to the home hub, and to the backup hub. The dual-homing protection scheme was not considered a survivability option until the Hinsdale fire (hub building) occurred in 1987. To obtain protection against major hub failures at a reasonable cost, an alternate optical dual-homing architecture using the signal broadcast concept was discussed. This optical dual-homing architecture uses PSs to broadcast signals to both the home and backup hubs. It also centralizes the protection function to reduce the costs due to fiber systems and terminating equipment. The optical dual-homing network can be upgraded from existing single homing with 1:1/DP networks at a moderate increase in cost, but with a full protection capability against major hub failures. The optical power loss factor of the optical dual-homing architecture may limit its application to the metropolitan LATA networks, which usually have shorter fiber spans and require dual-access protection. The power loss factor, as well as the cost factor, should be considered when planning to implement this optical dual-homing network architecture. In the near future, dual-homing protection may be provided more economically by SONET SHR technology (see Chapter 4).

REFERENCES

[1] TA-NWT-000253, *Synchronous Optical Networks (SONET) Fiber Optic Transmission Systems Requirements and Objectives*, Issue 6, September 1990.

[2] *Digital Hierarchy Optical Interface Rates and Formats Specification*, T1.105/1988.

[3] SR-NWT-001756, *Automatic Protection Switching for SONET*, Bellcore, Issue 1, October 1990.

[4]   Wu, T-H., Kolar, D. J., and Cardwell, R. H., "Survivable Network Architectures for Broadband Fiber Optic Networks: Model and Performance Comparisons," *IEEE Journal of Lightwave Technology*, Vol. 6, No. 10, November 1988, pp. 1698-1709.

[5]   Wu, T-H., Cardwell, R. H., and Woodall, W. E., "Decreasing Survivable Network Cost Using Optical Switches," *Proceedings of IEEE GLOBECOM'88*, Hollywood, FL, November 1988, Section 3.6.1-3.6.5.

[6]   Wu, T-H., and Habiby, S., "Strategies and Technologies for Planning a Cost-effective Survivable Fiber Network Architecture Using Optical Switches," *IEEE Journal of Lightwave Technology*, Vol. 8, No. 2, February 1990, pp. 152-159.

[7]   Hanson, D., Russell, D., and Roberts, H., "Optical Switching as a Facility Restoration Alternative," *Proceedings of National Fiber Optic Engineering Conferences*, April 1990, Sections 3.2.1-3.2.8.

[8]   Laos, O. C., "Fiber Optic Protection System: Architecture, Technologies, and Capabilities," *Proceedings of National Fiber Optic Engineering Conferences*, April 1990, Sections 2.3.1-2.3.12.

[9]   Marrakchi, A., Hubbard, W. M., and Habiby, S. F., "Review of Photonic Switching Device Technology," *Proceedings of the International Workshop on Digital Communications*, Italy, September 1989.

[10] Kaufman, S., Reynolds, R. L., and Loeffler, G. C., "An Optical Switch for the SL Undersea Lightwave System," *IEEE Journal on Selected Areas in Communications*, Vol. SAC-2, No. 6, November 1984, pp. 1015-1019.

[11] Hemrick, C. F., Klessig, R. W., and McRoberts, J. M., "Switched Multi-Megabit Data Service and Early Availability via MAN Technology," *IEEE Communications Magazine*, Vol. 26, No. 4, April 1988, pp. 9-14.

[12] Hinton, H.S., "Photonic Switching Using Directional Couplers," *IEEE Communications Magazine*, Vol. 25, No. 5, May 1987, pp. 16-26.

[13] Silberberg, Y., Perlmutter, P., and Baran, J.E., "Digital Optical Switch," *OFC '88 Technical Digest*, January 1988, pp. TH-A 3.

[14] Silberberg, Y., Perlmutter, P., and Baran, J.E., "Digital Optical Switch," *Applied Physics Letters*, Vol. 51, 1987, pp. 1230-1232.

[15] Special Issue on Photonic Switching, *IEEE Communications Magazine*, Vol. 25, No. 5, May 1987.

[16] Special Issue on Photonic Switching, *IEEE Journal on Selected Areas in Communications*, August 1988.

[17] Edinger, D., Duthie, P., and Prabhakara, G. R., "A New Answer to Fiber Protection," *Telephony*, April 9, 1990, pp. 53-55.

[18] Wu, T-H., "A Novel Architecture for Optical Dual Homing Survivable Fiber Networks," *Proceedings of IEEE International Conferences on Communications (ICC'90)*, Atlanta, GA, April 1990, pp. 309.3.1-309.3.6.

[19]   Findakly, T., and Chen, B. V., "Single-Mode Integrated Optical 1 X N Star Couplers," *Tech. Dig. Topical Meetings Optical Fiber Communications,* 1983, Paper ML-2.

[20]   Sheem, S. K., and Giallorenzi, T. G., "Single-Mode Fiber-Optical Power Divider: Encapsulated Etching Technique," *Optics Letter*, Vol. 4, 1979, p. 29.

[21]   Hasegawa, S., Jones, T. H., and Yang, H., "Distributed Control of Nested Automatic Protection Switch," T1X1.5/88-083 Contribution, 1988.

[22]   Ellson, J., "Nested Protection Switching," T1X1.5/90-132 Contribution, July 1990.

[23]   Easter, G., and Krause, T., "Nested Switching," T1X1.5/90-206, October 1990.

CHAPTER 4

SONET Self-Healing Rings (SHRs)

4.1 MOTIVATION FOR USING SONET SELF-HEALING RINGS

4.1.1 Benefits of Using SHRs

Present networks employing fiber-optic technology use automatic diverse protection routing and dual homing to protect networks from fiber cable cuts and major hub failures. Such networks can be evolved to SHRs if SHRs prove to be economical. A *ring network* is a collection of nodes forming a closed loop, where each node is connected via a duplex communications facility. An SHR is a ring network that provides redundant bandwidth and/or network equipment so disrupted services can be automatically restored following network failures. The multiplexing devices used in the ring[1] architectures are ADMs that add and drop local channels and pass through transit channels. Figure 4.1 depicts an example of a network evolution scenario from a diverse protection network to a ring. Figure 4.1(a) depicts a fiber-hubbed network with three COs and their serving hub. Communications between a CO and its hub are via a dedicated, point-to-point fiber span that is protected by a dedicated, physically diverse fiber span controlled by an APS system. The terminal multiplexers used in the fiber-hubbed network can be evolved to higher speed ADMs when these ADMs become available and the ring concept proves to be economical. When compared to its hubbing counterpart, the ring may need higher-speed ADMs because it shares not only fiber facilities, but also multiplexing equipment.

---

1. As used in this chapter, the term "ring" is interchangeable with the term "self-healing ring" unless otherwise specified.

123

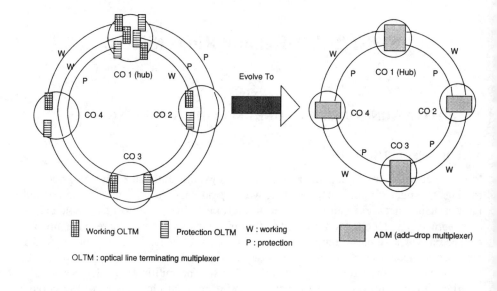

(a) Fiber-hubbed Network with Diverse Protection   (b) Fiber Ring

Figure 4.1.   Ring network architecture evolution.

As depicted in Figure 4.1[2], the ring architecture has the following advantages over the 1:1/DP hubbing architecture: (1) potential decrease in fibers, optical/electronic devices, and regenerators; and (2) complete survivability for fiber cable cuts and node failure (except the failed node). For the example shown in Figure 4.1(a), the 1:1/DP hubbing network requires 12 OLTMs, and the ring shown in Figure 4.1(b) requires only four ADMs. Depending on the line rates used, fewer ADMs for the ring may not necessarily mean lower capital costs. If we consider a case where both the ring and its hubbing counterpart use the same line rate, and the cost of one OLTM is about 80

---

2. Here, we compare the 1:1/DP hubbing network with the ring because they both have 100 percent network survivability for a single cable cut.

percent the cost of one ADM at the same line rate [1], the ring may save 58 percent of capital costs over its hubbing counterpart.

Now we consider another case where the 1:1/DP hubbing network uses the 565-Mbps system and the ring uses the 2.4-Gbps system. Assume that the relative cost of a 565-Mbps terminating equipment is 1 and the cost of 2.4-Gbps equipment is three times that cost. The ring considered here needs four 2.4-Gbps ADMs, which results in a relative cost of 12 (1×3×4=12), whereas the 1:1/DP hubbing network needs twelve 565-Mbps OLTMs, which results in a relative cost of 9.6 (1×0.8×12=9.6). Thus, in this particular example, the 1:1/DP hubbing network seems less expensive than the ring. However, regenerators may be needed for longer diverse protection spans in larger fiber-hubbed networks and may offset the advantage of using low-speed OLTMs. As the number of nodes in the ring increases to eight or more, the 1:1/DP network becomes less attractive than the ring. Note that this discussion addresses only multiplex equipment cost because it is the dominant factor of total fiber transport system costs for intraLATA networks [2].

In many cases, the ring may have economic advantages over its hubbing counterpart. However, it may be more difficult or expensive to upgrade the system when the ring capacity is exhausted. The ring may also need a more complex network control scheme than its hubbing counterpart because all nodes interact when fault conditions occur or reconfiguration is required. However, the problem of exhausted ring capacity can be reduced or solved by careful network planning (see Section 4.5) or much higher speed ADM deployment (e.g., the availability of 9.6-Gbps ADMs). Also, the control system for the ring can be simplified when SONET equipment is deployed. Table 4-1 summarizes a relative comparison between the ring architecture and the *hubbing with diverse protection network* (Hub/DP).

Table 4-1. Comparison between Diverse Protection Routing and SHR Architectures

Attributes	Hub/DP	SHR/ADM
upgradability	easy	difficult/expensive
fiber counts	more	fewer
terminal count	more	fewer
terminal speed	lower	higher
fiber cable survivability	≤100%*	100%
hub+ survivability	poor	better

Hub/DP: Hubbing with diverse protection network
* depending upon 1:1 or 1:N protection
+ assuming the entire hub building fails

125

4.1.2 Advantages of Using SONET SHRs

Two major questions arise when planning the implementation of SONET SHRs in interoffice SONET networks: (1) why use SONET SHRs? and (2) what is the appropriate SONET SHR architecture for a given application? The first question is asked because the SONET SHR is a new architecture that needs significant capital investment and new operations system development in interoffice networks. Unless the SHR architecture can show significant cost and survivability advantages over conventional counterparts, it may become only a conceptual (but not a practical) architecture. This section and Section 4.5 address the first question, and Section 4.2 discusses the second question.

In the past, the ring architecture has been restricted from interoffice applications because present metallic, low-capacity systems may make the ring uneconomical and difficult to adapt to the rapidly growing traffic in interoffice environments. Even in the present fiber world, the ring is used only in LANs, not in interoffice networks, because of its relatively low speed (on the order of hundreds of Mbps), complex control scheme (as compared to the conventional facility protection switching architectures), and complex scheme to add-drop tributary signals from a channel. Recently, standardized SONET technology and associated flexible high-speed add-drop multiplexing technology [3] have made SHR architectures practical because of SONET's simpler control scheme, ease of adding-dropping tributary signals, and high-speed add-drop multiplexing capability (e.g., OC-48 with a bit rate of 2.488 Gbps), which may meet the intraLATA interoffice demand requirement. The perceived advantages of the SONET ring in terms of cost, survivability, and simpler control have made the SHR architecture part of the target SONET network architecture.

Figure 4.2 depicts a relative comparison between the asynchronous ring and the SONET ring. Asynchronous rings may not provide a cost-effective implementation, as do SONET rings, due to the high component integration of SONET ADMs and economical signal add-drop using the SONET VT pointer processing technology. Consider a ring with DS1 (or VT1.5 for the SONET ring) requirements among ring nodes. Because the basic transport signal for fiber networks is DS3 (or STS-1), the DS1 requirement needs to be converted into a DS3 requirement. Because the ring carries demands from all ring nodes, minimizing required DS3s (or STS-1s) is desirable to minimize the ring capacity requirement. This implies that a DS3 may carry DS1s with different sources and destinations. Figure 4.2(a) depicts an asynchronous ring configuration. In this asynchronous ring configuration, two back-to-back terminal multiplexers and a wideband *Electronic Digital Cross-connect System* (EDSX 3/1)[3] or an asynchronous ADM with the TSI capability are needed to perform

126

the tributary signal add-drop and pass-through at each CO. The EDSX 3/1 is required because a local DS3 for that CO may include DS1s destined to other COs. Because DS1s within a DS3 are not "observable" from a DS3 frame, determining which DS1 should be dropped or passed through requires (1) demultiplexing an entire DS3 frame via the back-to-back terminal multiplexers, and then (2) cross-connecting DS1s to the appropriate DS3 that has the same destination via the EDSX 3/1 or asynchronous ADM with TSI. Thus, the DS1 cross-connect function is required for the asynchronous ring. To automatically restore DS1s or DS3s from a network component failure for an asynchronous ring, external controllers and an external telemetry system[4] may be needed to convey failure messages around the ring and to change the DS1 switching matrix within the EDSX 3/1 or the ADM to complete the protection switching whenever needed. The number of external controllers needed depends on whether the ring-control system is centralized or distributed. The example depicted in Figure 4.2(a) is a centralized ring-control system that uses a central controller to communicate to each node and to trigger protection switching whenever needed.

Unlike the asynchronous ring, the VT1.5 (1.5-Mbps) tributary signal add-drop function from STS-1s can be performed using VT pointer values to identify the locations of local VT signals within a STS-1 and then dropping those signals accordingly. Demultiplexing entire STS-1 frames is not necessary because VTs are "observable" within a STS-1 via their pointer values. The multiplexed signal add-drop scheme shown in Figure 4.2(b) is sometimes referred to as TSA. TSA preserves the relative STS-1 time slot positions for transit STS-1 channels in both the incoming and the outgoing STS-N signals and adds-drops tributary signals within the same time slots. The SONET ADM using TSA is much simpler and more cost-effective than the ADM using TSI in terms of tributary signal add-drop, but it may have lower network utilization. The utilization of the ring with ADMs using TSA can be improved by incorporating one or more DCSs into the ring's bandwidth management scheme at the ring interconnection point(s) (see Section 4.5.1).

---

3. The EDSX 3/3 may be needed if all signals are at the DS3 level.

4. The external telemetry system may not be needed if some DS3s are used to carry control messages.

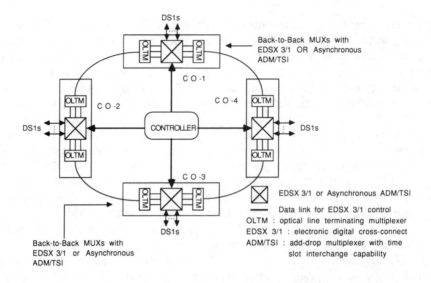

(a) Asynchronous Ring with ADMs having TSI

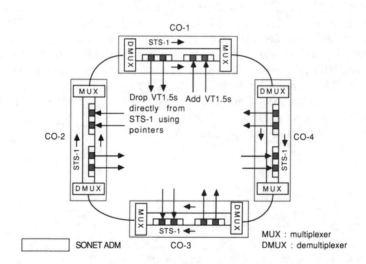

(b) SONET Ring with ADMs using TSA

Figure 4.2.   Asynchronous ring architecture and SONET ring architecture.

128

External controllers and the external telemetry system needed for asynchronous rings are not needed for the SONET ring because the failure message is conveyed and protection switching is triggered via the SONET line overhead or the path-layer signal, which is part of the SONET frame. Because communication between nodes is performed on a local basis, the distributed ring control system is preferable for SONET rings to provide a faster restoration capability and reduce vulnerability due to controller failure.

4.2 SONET SHR ARCHITECTURES

References [4,5] summarize a class of SONET SHR architectures. An SHR is a ring network that provides redundant bandwidth and/or network equipment so disrupted services can be automatically restored. A common approach to providing a self-healing capability is to provide a second communications ring parallel to the first working ring. In this case, a fault in one ring can be bypassed by transferring communications to the second ring; this is commonly referred to as a *line switch* function (or switch to protection). Moreover, if the second ring transmits in the opposite direction as the first, a break in both rings between two adjacent nodes can be remedied by the nodes on either side of the break looping back communications received on one ring onto the other ring. This is commonly referred to as a *loopback* function. Figure 4.3 depicts examples of line switch and loopback for restoration. Note that a combination of loopback and line switch functions (one on either side of the break) can be used to restore disrupted service. Therefore, an important attribute of an SHR is that, if the ring is broken at any given point, the transmission direction for signals can be reversed to prevent loss of service.

SONET SHRs can be divided into two general categories, unidirectional and bidirectional, according to the direction of traffic flow under normal conditions. SHR architectures that fall into one of these general categories may differ in the protection-control mechanism used to restore disrupted services.

Unidirectional SHR. In a *Unidirectional SHR* (USHR), working traffic is carried around the ring in one direction only (e.g., counterclockwise). Refer to Figure 4.4(a). Traffic from any Node 1 to Node 3 is routed along the working communications ring from Node 1 to Node 3 (i.e., path 1-2-3). The return traffic continues around the ring from Node 3 back to Node 1, in the same direction as from Node 1 to Node 3, using the remaining portion of the working ring (i.e., path 3-4-1). Thus, traffic arrives at Nodes 1 and 3 by different paths. Because the transmission of normal working traffic on the USHR is in only one direction, the ring capacity is determined by the sum of demands between nodes. USHRs are sometimes called "counter-rotating rings" because the second communications ring (for protection only) transmits in the direction opposite of the first (working) ring.

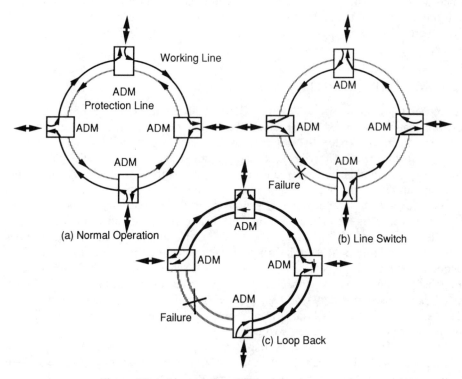

Figure 4.3. Line switch vs. loopback for ring restoration.

Because service channels are routed unidirectionally, one fiber is needed to carry them. The USHR can be implemented according to the concept of 1:1 or 1+1 protection. The 1:1 USHR uses a separate ring as the protection ring, which does not carry the service demand in the normal situation, and folds (loops) the disrupted channels onto the protection ring from the working ring when the network component fails. This 1:1/USHR is also referred to as a *unidirectional folded SHR* (USHR/L) (see Section 4.2.3). In contrast, the 1+1 USHR splits the signals onto both the working and the protection rings at the transmitting node (i.e., head end bridging), and the receiving node selects the best of two identical receiving signals based on protection switching criteria. This 1+1 USHR is sometimes referred to as a *unidirectional path SHR*

(USHR/P) (see Section 4.2.4). Note that 1:1 USHRs may be configured as 1:N USHRs, i.e., one protection communications ring is shared by N working communications rings; however, 1:N USHRs are no longer entirely self healing. Thus, this chapter does not discuss them.

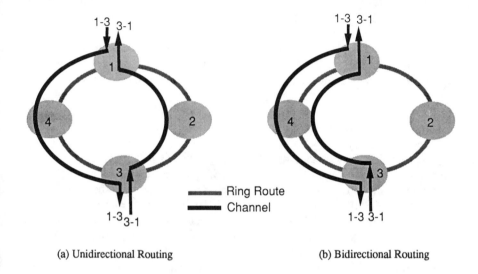

(a) Unidirectional Routing   (b) Bidirectional Routing

Figure 4.4.   Definitions of unidirectional and bidirectional SHRs.

Bidirectional SHR. In a *Bidirectional SHR* (BSHR), as shown in Figure 4.4(b), working traffic travels in both directions over a single path that uses the two parallel communications paths (operating in opposite directions) between the nodes of the ring (e.g., between Node 1 and Node 3). For example, in the normal condition, signals from Node 1 to Node 3 are routed via path 1-2-3, and the returning signals from Node 3 to Node 1 are routed via the same path (path 3-2-1). Because traffic is routed over a single path between nodes, spare capacity around the ring can be shared on a per-link basis and not dedicated to the total demand on the ring (as for a USHR). Because service channels are routed bidirectionally at two nodes, two fibers are needed to carry these service channels.

A BSHR may use two or four fibers depending on the spare capacity arrangement. In a four-fiber BSHR (or 1:1 configuration) (see Section 4.2.1), a second communications ring, separate from the first, is provided for protection, and working and protection channels use separate communications rings. The 1:1 BSHRs can also be configured as 1:N BSHRs, i.e., one protection communications ring for N working

131

communications rings. As stated above, 1:N BSHRs are not entirely self healing, and thus, this chapter does not discuss them. In a two-fiber BSHR (see Section 4.2.2), working and protection channels use the same fiber with a portion of bandwidth reserved for protection. To provide a self-healing function, half of the bandwidth is usually reserved for protection. This ring arrangement can provide line protection switching using a TSI method for merging working channels in the failed fiber with protection channels in the unaffected fiber.

USHRs and BSHRs can be further categorized into line and path protection switched SHRs, according to the SONET level used to (1) convey failure messages and (2) trigger the protection switch action that enables the ring to recover from the failure automatically.

Line Protection Switched SHR. A line protection switched SHR architecture uses the SONET line overhead (e.g., K1 and K2 bytes) to convey the failure message and to trigger the protection switching action. Switching action is performed only at the line layer for failure recovery and does not involve the path layer; it restores line demand from a failed facility. Line protection switched architectures have been defined for both USHRs and BSHRs using the principle of looping back traffic onto protection. For a break in the ring, local and remote control is necessary to effect the looping back of traffic by the nodes on either side of the break. The line protection switching scheme is a natural choice for all BSHRs because the BSHR demand routing uses the same principle as today's point-to-point systems, which use the line protection switching scheme (i.e., APS) to restore demands if a network component fails.

Path Protection Switched SHR. A path protection switched SHR architecture uses the path-layer signal (e.g., path AIS) to trigger protection switching. Unlike the line protection switching scheme, the path protection switching scheme restores an end-to-end STS or VT channel. Path switching of a specific path is independent of other paths' status. Even though a node detects a line failure, switching is done at the path layer for path switched SHRs. With path protection switching, two classes of ring levels are defined: VT and STS. A VT path protection switching SHR is defined as a ring where VT paths are switched for ring rearrangements, and an STS path protection switching SHR is defined as a ring where STS paths are switched for ring rearrangements. The STS and VT path protection switching rings are primarily used in interoffice networks and loop networks, respectively.

Figure 4.5 depicts SONET ring architecture classifications based on demand routing and the protection control mechanism.

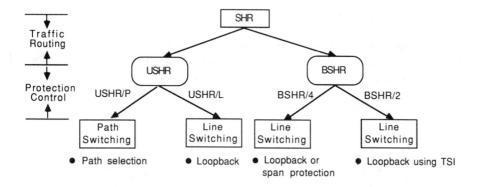

Figure 4.5. SONET SHR architectures and protection control schemes.

4.2.1 Four-Fiber BSHR with Line Protection Switching (BSHR/4)

For the four-fiber BSHR (BSHR/4), two fibers are used to carry normal services and the other two fibers are used for protection. Examples of BSHR/4 implementations have been proposed in References [6-9]. Figure 4.6 depicts an example of the BSHR/4 operation under the normal and failure scenarios. In the normal scenario, demand routing occurs as it does in today's point-to-point systems, but the service channels are looped back from working fibers to protection fibers via facility protection switches in case of network component failures. The loopback scheme is used in References [6,7] as the protection switching scheme to simplify the ring reconfiguration complexity; only two nodes adjacent to the failed network component are involved in protection switching. Another BSHR/4 implementation proposed in Reference [8] uses a combination of line loopbacks and APS span protection. The line loopback function protects against cable cuts, and the span protection protects against fiber and equipment failures. The BSHR/4 with both loopback and span protection has the highest system reliability among SONET SHR alternatives.

In the BSHR/4 architecture, the facility protection switching is triggered by detecting failure messages at the SONET line level (e.g., detection of K1 and K2 bytes). Several line overheads have been proposed to convey the failure messages for the ring application. Among them, K1 and K2 bytes used for APS in present SONET standards are a popular choice because BSHRs use the same demand-routing principle for working signals as the point-to-point systems. Therefore, it would be natural to expect that the BSHR uses a similar protection switching scheme, such as APS. However, K1 and K2 bytes defined in Reference [10] are designed for point-to-point systems and are certainly not applicable to ring applications, because an end-to-end channel may pass through one or more intermediate nodes before reaching its

133

destination. The protection ADM at any intermediate node is reconfigured as a regenerator to pass through the transit K1 or K2 byte. Also, the 1:N channel assignment and associated priority assignment are not used for the ring because the considered ring is a 1:1 SHR. Table 4-2 shows differences between point-to-point and ring applications in terms of schemes defined in K1 and K2 bytes.

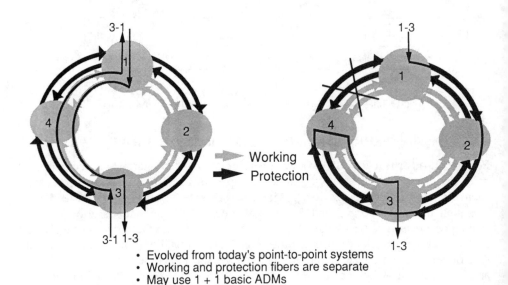

- Evolved from today's point-to-point systems
- Working and protection fibers are separate
- May use 1 + 1 basic ADMs
- Uses line protection switching

Figure 4.6. An example for BSHR/4 operation under normal and failure scenarios.

Table 4-2. Comparison between Point-to-Point and Ring Systems in Terms of K1 and K2 Bytes

Attributes	Point-to-point	Self-healing ring
1:N channel assignment	yes	no
Channel-priority scheme	yes	no
Nodal address needed	no	yes
Intermediate nodal processing	no	yes

134

Based on differences described in Table 4-2, the bits related to the 1:N channel assignment and associated priority assignment in K1 and K2 bytes for APS can be replaced by a nodal address assignment. The ring node needs this information to determine whether the incoming K1 or K2 byte is a local or transit byte in order to trigger line protection switching at the right place. Thus, the maximum number of nodes that can be supported by a SONET ring is 16 (i.e., four bits). For the ring application, the priority bit (i.e., bit number 4) of K1 will be a "span switch request" when it is set to 1 and a "loopback switch request" when it is set to 0. Similarly, the switch type bit (i.e., bit number 5) of K2 is interpreted as a "span 1:1 switch" when it is set to 1 and a "loopback switch" when it is set to 0. In addition, for four-fiber rings, bits 6-8 of K2 can be augmented to include SD and SF on the protection lines, so that each node around the ring knows the status of the protection line at any other node before performing loopback switching. Figure 4.7 shows K1 and K2 formats for point-to-point systems [10] and some possible interpretations for the ring application [8,11]. Details of SONET K1/K2 operations for the ring application can be found in References [7,8,11,12]. Note that the signaling scheme proposed in Reference [8] can work for both the BSHR/4 and the BSHR/2 architectures.

For BSHR/4, two basic ADMs are needed at each node: one for working channels and the other for protection channels. A basic ADM is an ADM that uses the hard-wired or programmable TSA (see Chapter 2, Section 2.5) as its signal multiplexing scheme. The two ADMs at each node may be implemented as a single ADM (defined by *1+1 ADM*) that terminates four fibers and shares a common unit for working signals (e.g., control software and control circuits) or as two independent ADMs connected by external protection switches. These two implementations have tradeoffs between the cost and the degree of survivability for a single ADM failure per node. The BSHR/4 using two independent ADMs is more costly but provides complete survivability for a single ADM failure per node. In fact, the BSHR/4 with two independent ADMs per node is the most survivable architecture and has the highest capacity among all SONET ring alternatives. The BSHR/4 using 1+1 ADMs costs less than the first implementation but has no protection capability for the failure of the common unit of this 1+1 ADM (e.g., software failure). In addition to the two implementations previously described, the other possible BSHR/4 implementation uses one SONET regenerator as the protection component at each node (i.e., equipped with one ADM plus one regenerator). However, an additional protection control is needed for this implementation because the SONET regenerator does not terminate the line overhead, which is a necessary part of the BSHR line protection switching protocol (i.e., to determine the failure messages, such as K1 and K2 bytes, that should be passed through to the next node or terminated and processed).

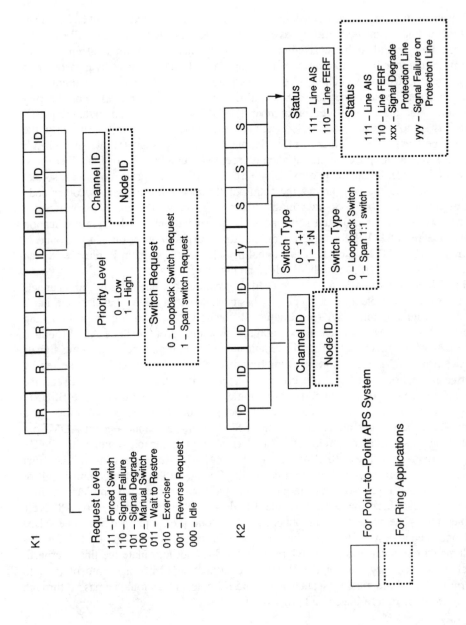

Figure 4.7. K1 and K2 formats for point-to-point systems and line protection switching rings.

136

The BSHR/4 has essentially evolved from today's point-to-point systems (or hubbed protection systems) with working and protection signals that are carried by separate fibers, as depicted in Figure 4.1. Thus, in comparison to other SONET SHR architectures, the BSHR/4 can be used with minimal changes to today's OSs.

For areas where only a limited amount of fiber is in the present fiber cables and deploying new fiber cables is expensive, the BSHR/4 may be implemented using the matured and inexpensive WDM technology to reduce the required number of fibers from four to two. Figure 4.8 depicts such a special network configuration for the BSHR/4 using two fibers with WDMs. In Figure 4.8, two wavelengths, λ_1 and λ_2, and two pairs of 1×2 WDMs are needed for each node. The protection line, which is dedicated to one working line, uses the same wavelength as that working line. The 1×2 WDM, a low-loss (about 0.7 dB) passive device that multiplexes two optical wavelengths onto a fiber, is commercially available. In this special BSHR/4 configuration, two *Electrical-to-Optical* (E/O) conversions in each ADM are driven by two different wavelengths (typically 1300 nm and 1550 nm). This differs from the conventional ADM design, which is driven by only one wavelength (typically 1300 nm).

4.2.2 Two-Fiber BSHR with Line Protection Switching (BSHR/2)

The BSHR architecture can also be implemented using two fibers [6,11]. Unlike the BSHR/4, working and protection channels for BSHR/2s are routed on the same fiber with a portion of bandwidth reserved for protection. To provide a self-healing capability (1:1 protection) and simplify ring control system complexity, half of the fiber system bandwidth is reserved for protection (e.g., for an OC-12 BSHR/2, STS-1 channels 1 to 6 may be assigned to working STS-1s, and channels 7 to 12 may be dedicated to protection).

In the normal situation, traffic is evenly split into the outer ring and the inner ring by filling the first half of the STS-1 time slots (or even and odd numbers of time slots, respectively) [see Figure 4.9(a)]. During a fiber break or equipment failure, traffic is automatically switched into corresponding vacant time slots in the opposite direction to avoid the fault [see Figure 4.9(b)]. Thus, the ADM for the BSHR/2 needs to have a TSI capability to move working channels from the failed fiber to the corresponding protection time slots in the other fiber.

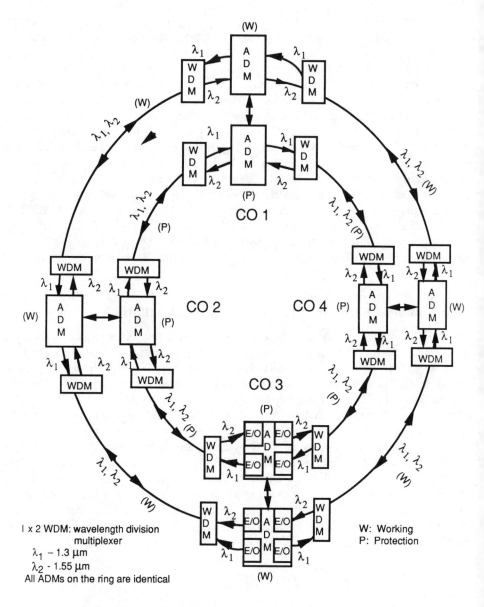

Figure 4.8. A BSHR/4 implementation using two fibers and WDMs.

138

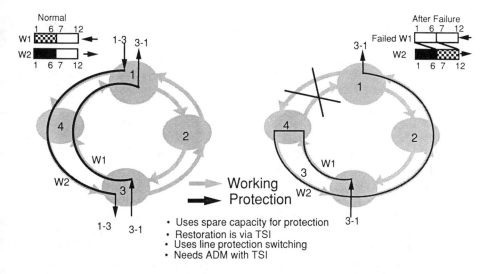

Figure 4.9. An example of BSHR/2 operation under normal and failure scenarios.

The simplest implementation of the BSHR/2 architecture can be achieved using rings with a capacity of STS-N, where N is an even number. In this case, traffic would be loaded only onto n/2 of the STS-Ns. Each ADM would then incorporate a modified loopback switch that can loop back loaded STS-1s from the failed fiber into the unused STS-1 time slots in the other fiber. As a result, all traffic that was normally on the working channels on the link between Nodes 1 and 3 [see Figure 4.9(a)] is now on the protection channels in the reverse direction around the ring [see Figure 4.9(b)]. The protection channels can then be reused by other links when the protection channels become available again. During a fiber break, local action by the ADMs adjacent to the break is sufficient to restore service.

For an STS-n, where n is odd, two possible implementations exist; they have different tradeoffs between ring protection control complexity and ring utilization. If traffic utilization is less than or equal to 50 percent, the principle just described can be applied. If traffic utilization is greater than 50 percent, one of the STS-1 time slots (e.g., the last odd-number STS-1) may be used to carry lower-priority services. For network failures, this STS-1 carrying lower-priority services is simply dropped. Of course, in this case, the considered ring is no longer completely self healing. For ring utilization greater than 50 percent, a priority scheme is needed. Thus, the priority bit in the K1 byte should be reserved, which results in a maximum number of 16 nodes that can be carried in the K1 byte.

As explained previously, the ADM used for the BSHR/2 architecture requires a TSI capability for protection. The TSI capability, typically used in large-scale DCSs, is a multiplexing scheme that adds and/or drops channels by rearranging the relative time slot positions within a frame and reconfigures the ring by changing the switching matrix if a network component fails. The BSHR/2 restoration method is similar to the DCS path rearrangement scheme except that the DCS restoration scheme restores individual STS-1s or VTs, and the BSHR/2 scheme restores all demands from the failed facility simultaneously. With TSI, the ADM's complexity can be much simpler than that of the DCS because the ring topology used in the BSHR/2 makes complete interchange freedom (required for DCS applications) unnecessary. Also, the BSHR/2 protection control system is much simpler than the DCS restoration scheme because of the ring topology involved. Under the ring topology, only two paths exist, and the node sequences in these paths are predetermined. Thus, the BSHR/2 can use the SONET line overhead (such as K1 and K2 bytes) to convey the necessary protection switching message, unlike the DCS rearrangement, which may use the SONET *Data Communications Channel* (DCC) to convey similar messages. Although the demand routing for the BSHR/2 is similar to today's point-to-point systems under normal situations, it is very different from today's operations environment under failure situations. Today's OSs always route the protection channels in separate, standby fibers.

4.2.3 Folded USHR Architecture with Line Protection Switching (USHR/L)

The folded USHR proposed in References [7,12] is a 1:1 USHR that folds (or loops) the disrupted line signal [i.e., the optical line signal or N line *Synchronous Payload Envelopes* (SPEs)] onto a separate protection fiber ring. Figure 4.10 shows an example of USHR/L operation under normal and failure scenarios. In the normal situation, incoming and returning signals are routed unidirectionally around the ring, as depicted in Figure 4.10(a). During a fiber break [e.g., a break between Nodes 1 and 4 as shown in Figure 4.10(b)], the nodes adjacent to the break (i.e., Nodes 1 and 4) perform the fold or looping function. Because of the loopback capabilities at the nodes adjacent to the break, the folded SHR remains in a ring topology under a single failure situation. The STS-1 time slot locations for working channels in the failed fiber remain unchanged when they are moved to the protection fiber during the restoration process. Thus, the ADM used for the USHR/L can be a basic ADM. The K1/K2 format and associated line protection switching scheme can be similar to those used in the BSHR/4 because they both use the same 1:1 protection concept.

140

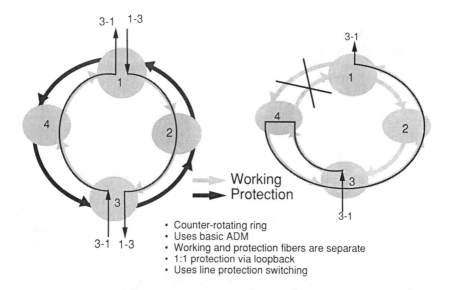

Figure 4.10.   An example of USHR/L operation under normal and failure scenarios.

4.2.4 USHR Using Path Selection (USHR/P)

A USHR/P ring architecture is defined as an architecture that operates under a ring topology in the normal situation but changes to a linear network when a network component fails. The SONET USHR/P architecture discussed in this section was proposed in References [13,14] and is based on a concept of signal dual-feed or 1+1 protection. The dual-feed concept has also been used in similar ring architectures described in References [15-17]. In this architecture, one ADM is equipped at each node with a pair of fibers with traffic going in opposite directions. Figure 4.11 shows an example of USHR/P operation under the normal and failure scenarios. In the normal state, one or more STS-M (M<N) signals or different services that can be carried inside the STS path-layer signals are transmitted onto both the clockwise and the counterclockwise directions of the ring from the transmitting node [e.g., Node 1 in Figure 4.11(a)]. These two identical signals propagate along the ring and are finally dropped at one of the offices. Thus, at the receiving node [e.g., Node 3 in Figure 4.11(a)], two identical tributary signals with different delays are observed. Suppose these two signals are designated as the primary and secondary signals. During normal operation, only the primary signal is used, although both signals are monitored for alarms and maintenance signals. If the ring is broken because of a catastrophic failure

141

(e.g., fiber cut or hardware failure), it is possible to resume the service by performing proper tributary protection switching to select the secondary signal. Thus, to perform tributary protection switching, ADMs only need to identify which of the two tributary signals is valid.

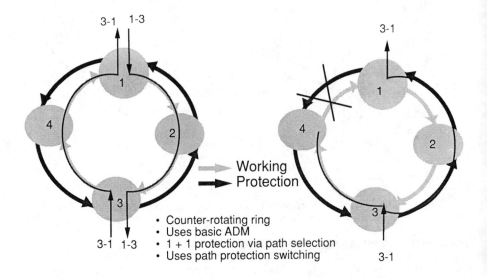

Figure 4.11. An example of USHR/P operation under normal and failure scenarios.

A simple control mechanism of the USHR/P architecture for failure recovery is given below [13,14]:

1. Detection of LOS or Line AIS triggers the insertion of Path AIS[5] (STS or VT level, depending on ADM application) onto all the downstream tributary paths (STS- or VT-level)
2. Detection of Path AIS on one of the two tributaries initiates protection switching to the other tributary

---

5. The operation of inserting path AIS is simple because it is indicated by STS pointers having "all 1s" values.

3. Detection of Path AIS on both identical tributaries signifies a multiple failure situation and triggers the generation of an AIS in the dropped signal

Note that in this control mechanism, the protection switching can be non-revertive (i.e., the system does not switch back to the original state even after the failure has been removed) with no performance penalty. After the failure is removed and the ring is brought back to the normal working condition, the identification of the primary and secondary loops is irrelevant. Each ADM low-speed interface monitors the Path AIS and chooses the valid signal to perform protection switching. Because TSI is not needed for this path protection switching, the USHR/P can use a basic ADM.

Figure 4.12 shows an example of how the SHR recovers from a fiber cut. For simplicity, we assume the ring capacity to be OC-12 with four ADMs in the ring to add-drop STS-1 signals. Furthermore, we assume that each ADM communicates with all other ADMs via a single STS-1 channel (the numbers in Figure 4.12 indicate communication between nodes, e.g., 2-1 means Node 2 to Node 1). These assumptions are given to simplify the description of the ring operation; they have no effect on the generality of the protection switching scheme.

As shown in Figure 4.12, during normal operation, STS-1 signals are inserted onto both the clockwise and the counterclockwise loops with the latter designated as the primary loop. When the fiber pair between Nodes 1 and 4 fails, the receiver at Node 1 detects the failure and declares a red alarm state. Because the control mechanism applies to both the clockwise and the counterclockwise directions, the following discussion refers to the counterclockwise loop but is applicable to both loops. After ADM #1 declares LOS or detection of Line AIS, it inserts STS Path AIS onto all tributary channels. At the STS-1 interfaces of Node 1, the status of the tributary signal is monitored. The detection of STS Path AIS in one of the two tributary STS-1 signals triggers tributary protection switching, which allows the 2:1 selector to select the valid tributary signal. Because Node 1 is adjacent to the fiber cut, all the STS-1 interfaces of Node 1 perform protection switching. The output of Node 1 on the counterclockwise loop thus has STS Path AIS on all of its STS-1 channels, except those that are added at Node 1.

At Node 2, the high-speed demultiplexer will not detect any alarm because SONET framing is never interrupted. After extracting the dropped channels, the STS-1 interfaces of Node 2 (2a, 2b) detect the STS-1 Path AIS and perform protection switching. Interface #2c corresponds to communication between Nodes 1 and 2 and is not affected by the failure; thus, no protection switching is required. For those channels that propagate through Node 2, only the 4-3 (communication from Node 4 to Node 3) channel carries an STS-1 Path AIS. Both the 1-3 and 1-4 channels originate from Node 1; thus, both are carrying valid data.

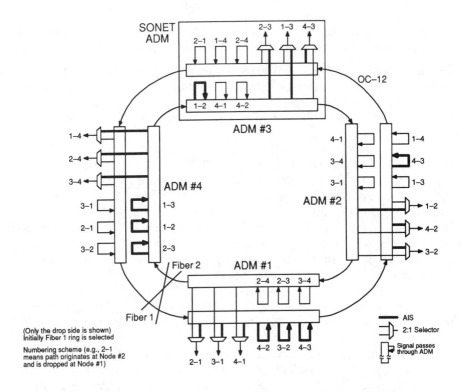

Figure 4.12. An example for the self-healing function for USHR/P.

If the same algorithm is applied to Node 3, we notice that only the 4-3 channel is required to perform protection switching. Moreover, all three pass-through channels (2-1, 1-4, 2-4) are carrying valid data that has been inserted onto the ring from an ADM that is downstream from the failure location. Using the same reasoning, none of the STS-1 interfaces in Node 4 is required to perform protection switching. The ADMs in the clockwise loop operate in exactly the same way as those in the counterclockwise loop. Because the USHR/P is a form of channel switching, and the indication of switching occurs using the path-level AIS signal, no APS protocol (i.e., K1 and K2 operations) is required. As a result, the USHR/P architecture is totally independent of the development or standardization process of the APS protocol.

4.3 SONET RING ARCHITECTURAL ANALYSIS

Determining which ring architecture is appropriate for network providers depends on several factors and their relative weight in strategic planning. These factors include

1. Applications (capital cost and capacity analysis)
2. SONET standards

    a. Multivendor compatibility
    b. Implementation time frame

3. Protection switching time requirement
4. Operations impacts.

4.3.1 Applications: Capital Cost and Capacity Analysis

Capital cost and capacity is the first and probably the most important factor in network applications. The network application is represented by the network demand requirement and is a tradeoff analysis between the network cost and the ring capacity. Two methods are usually used to evaluate merits for ring alternatives. The first method is referred to as a *consumer* approach; it ensures that the system is built in a cost-effective manner with acceptable performance. For this approach, the most inexpensive ADM is used to support the candidate ring architecture. The network costs for implementing these ring alternatives are then compared under the same demand requirement. In this approach, any applicable new technology is used to reduce the system cost, rather than to improve system performance. The second method is referred to as a *possible vendor* approach; it uses a particular type of ADM to support all ring architectures (if the equipment vendor manufactures only that particular type of ADM). The network costs for different rings are then compared under the same demand requirement. Because the primary concern is to evaluate relative merits among ring alternatives, the first method is assumed here.

For the first approach, the basic ADM is used for USHRs and the BSHR/4, and the ADM with TSI is used for the BSHR/2. Two basic ADMs (or a 1+1 basic ADM) and four fibers are needed for the BSHR/4, and one ADM and two fibers are needed for the USHRs and the BSHR/2. If any regenerators are needed between ring nodes, the number of regenerators for the BSHR/4 is twice the number for the USHR and the BSHR/2.

Two major questions in the application factor are (1) is there one ring architecture that is always better than all other ring architectures under all possible demand requirements? and (2) if no ring architecture is dominant, what is the best application for each ring architecture? Before discussing these two questions, it is necessary to discuss how to determine the ring capacity requirement and its required line rate,

145

provided the demand requirement is given. Note that the demand requirement discussed here is the STS-1 (or DS3) demand requirement because it is commonly used in fiber transport systems for interoffice networking applications.

The *ring capacity requirement* is defined to be the largest STS-1 (or DS3) cross-section in the ring. The line rate of the ring is selected based on its capacity requirement. The ring capacity requirement for BSHRs is determined by the particular ring demand assignment algorithm. This section briefly explains how this demand assignment algorithm works. Chapter 7 discusses a heuristic demand assignment algorithm for BSHRs in more detail. Among the six possible OC-N line rates defined in the present SONET standard (OC-3, OC-9, OC-12, OC-24, OC-36, and OC-48), the OC-12, OC-24, and OC-48 rates are popular for interoffice networking applications. Thus, throughout this section, we assume that only OC-12, OC-24, and OC-48 are used.

Figure 4.13 illustrates how to calculate the ring capacity requirement. In this example, we assume that demand is not split for BSHRs. In other words, all demand from one CO to another CO uses the same routing path from BSHRs. For example, in Figure 4.13(b), four STS-1s between CO-2 and CO-4 are routed through a single path: Path 2-3-4. For the considered demand requirement, the ring needs an OC-12 USHR because the ring capacity requirement for a USHR is the sum of all demands carried by the ring, that is 12 STS-1s [see Figure 4.13(a)]. To calculate the ring capacity requirement for BSHRs, we first sort the demand pair so that demand is in decreasing order. Then we distribute demand as equally as possible to minimize the number of STS-1s passing through each link, as depicted in Figure 4.13(b). The STS-1s carried on each link in this example are (2, 6, 7, 6, 3, 5), and the ring capacity requirement is the largest STS-1 cross-section, that is seven STS-1s. Thus, the BSHR/4 requires an OC-12 line rate, whereas the BSHR/2 needs an OC-24 rate because it uses only half the capacity of the ring (see Section 4.2.2).

Demand splitting is an option in BSHR demand assignment algorithms. In terms of planning and OS design, BSHRs with demand splitting are certainly more complicated than BSHRs without demand splitting. However, in some cases where the ring capacity requirement exceeds the candidate OC-N rate by only a few STS-1s (e.g., 13 STS-1s or 25 STS-1s), demand splitting may lower the capital cost significantly. Because the BSHR/2 uses only half of the available bandwidth, a load balancing (demand-splitting) traffic arrangement may be used to increase ring utilization. Figure 4.14(a) depicts an example of a BSHR/2 without demand splitting that needs an OC-24 BSHR/2 to carry demands because the ring capacity requirement is seven STS-1s. When demand is allowed to split, the ring capacity requirement is reduced from seven to four STS-1s [see Figure 4.14(b)]. Thus, only an OC-12 BSHR/2 is required for this particular demand requirement.

Demand Pair	STS-1s
(1,4)	3
(2,4)	4
(3,5)	3
(3,6)	2
Total	12 ◄ —— USHR ring capacity requirement

Assumptions:

(1) Only OC-12, OC-24 and OC-48 are considered.
(2) Demand is not split for the BSHR.

(a) USHR

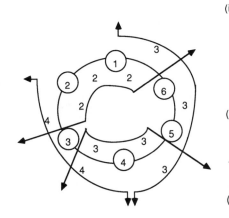

(i) Sort demand requirement

Demand Pair	STS-1s
(2,4)	4
(1,4)	3
(3,5)	3
(3,6)	2
Total	12

(ii) Assign demand in the clockwise and the counterclockwise directions as balanced as possible.

--> Link capacity requirement vector
= (2,6,7,6,3,5).

(iii) Capacity requirement for BSHR = 7 STS-1s.
For BSHR/4 --> OC-12 ring required
For BSHR/2 --> OC-24 ring required

(b) BSHR

Figure 4-13. Calculate the ring capacity requirements.

147

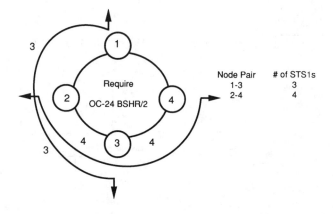

(a) BSHR/2 without Demand Splitting

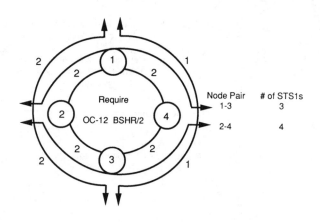

(b) BSHR/2 with Demand Splitting

Figure 4.14.   Benefits of demand splitting for BSHR/2.

4.3.1.1 Is There a Superior Ring Architecture?

As mentioned earlier, the first question that needs to be addressed is whether a ring architecture exists that is more cost-effective than all other types of ring architecture for all possible applications. This question can be easily answered by examining some counter-examples that suggest a negative solution. Before we discuss these counter-examples, we need to make the following assumptions:

1. The USHR uses a basic ADM at each node.
2. The BSHR/4 uses a 1+1 basic ADM at each node, and the BSHR/2 uses an ADM with TSI at each node.
3. A basic OC-N ADM is less expensive than an OC-N ADM with TSI.
4. An OC-12 ADM with TSI (used in the BSHR/2) is less expensive than a basic OC-48 ADM.
5. A 1+1 basic OC-N ADM is more expensive than one basic OC-N ADM but is less expensive than two basic OC-N ADMs.

The example depicted in Figure 4.13 shows that, for that particular demand pattern (all nodes' demands home to two central nodes), the USHR needs an OC-12 ring, whereas the BSHR/4 and the BSHR/2 need OC-12 and OC-24 rings, respectively. Thus, according to Assumptions 1, 2, and 3, the USHR is the least costly ring alternative, providing demand is not split for BSHRs. Even allowing demand splitting for BSHRs, the USHR is still the least costly ring alternative [see Figure 4.15(a)]. Figure 4.15(b) depicts an example where both the BSHR/2 and BSHR/4 require the OC-12 line rate and, thus, are less expensive than the USHR, which requires an OC-48 rate. Figure 4.15(a) also shows that the BSHR/4 requires six 1+1 basic OC-12 ADMs, whereas the BSHR/2 needs six OC-24 ADMs having TSI capability. Assuming that one basic OC-24 ADM costs twice as much as a basic OC-12, a 1+1 basic OC-12 ADM will be less expensive than an OC-24 ADM with TSI based on the three assumptions mentioned previously. Thus, the BSHR/4 is more economical than the BSHR/2 for this example.

On the other hand, referring to Figure 4.15(b), both the BSHR/2 and BSHR/4 require the OC-12 rate. If an OC-12 ADM with TSI is less expensive than a 1+1 basic OC-12 ADM, the BSHR/2 is more economical than the BSHR/4 for this example. Thus, the relative economics between the BSHR/2 and the BSHR/4 really depend on the relative cost between the basic ADM and the ADM with TSI, the nodal implementation for the BSHR/4 (i.e., two independent ADMs or 1+1 ADM), and the relative cost factor for a 1+1 basic OC-N ADM compared to a basic OC-N ADM. In addition to these traditional BSHR/4 implementations, the BSHR/4 can also be implemented using passive optical technology to reduce the protection cost. Section 4.4 discusses this passive protected BSHR/4 implementation.

Demand-Routing Table

Demand Pair	STS-1s	Link(1,2)	Link(2,3)	Link(3,4)	Link(4,5)	Link(5,6)	Link(6,1)
(1,4)	3	1	1	1	2	2	2
(2,4)	4	2	2	2	2	2	2
(3,5)	3	1	1	2	2	1	1
(3,6)	2	1	1	1	1	1	1
Link Cross-Section (STS-1s)		5	5	6	7	6	6

** USHR Needs OC-12; BSHR/4 Needs OC-12; BSHR/2 Needs OC-24.

(a) Example for Allowing Demand Splitting for BSHRs

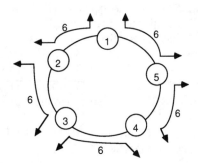

** USHR Needs OC-48; BSHR/4 Needs OC-12; BSHR/2 Needs OC-12.

(b) Best Demand Pattern for BSHRs

Figure 4.15. Counter-examples for the first main question.

4.3.1.2 Applications Areas for SHR Alternatives

Because no single ring architecture can show its dominant role in terms of applications, the next question is: what is the best application for each type of ring architecture? This question is very complex and difficult because so many factors impact the solution (e.g., the network size, the demand pattern and associated demand requirements, ADM cost factor, and so on). However, we can still examine several examples that may show promising applications for SHR alternatives. To achieve that, we assume a uniform demand requirement, i.e., all demand pairs have the same STS-1 demand requirement.

Figure 4.16(a) depicts seven possible demand patterns that can be generated from a six-node ring network. As shown in Figure 4.16(a), the demand pattern is gradually decentralized: demand pattern 1 is essentially a *centralized demand pattern* where all demands are homing to a central node; demand pattern 2 is a *dual-homing pattern* where all demands are homing to two nodes; and demand pattern 5 is a *mesh demand pattern* with 15 total demand pairs. The second portion of the demand patterns starts from the mesh demand pattern (pattern 5), gradually changes to the *braided demand pattern* (pattern 6), where each demand pair has one intermediate node between two ends of that demand pair, and finally progresses to the pure *point-to-point demand pattern* (pattern 7), where demand passes only from one ring node to the next. Such a pure point-to-point demand pattern is sometimes called a *cyclic demand pattern*. For example, the cyclic demand pattern has six demand pairs: (1-2), (2-3), (3-4), (4-5), (5-6), and (6-1). These demand patterns may represent a trend of showing economical merits for ring alternatives under different applications.

Figure 4.16(b) depicts results for the uniform demand requirement of five STS-1s for all demand pairs under different demand patterns. The relative ADM cost for the OC-48:OC-24:OC-12 line rate is 2.7:1.6:1, where the relative cost of a basic ADM(OC-12) =1.0. The OC-N ADM cost for BSHR/2 is (p×basic OC-N ADM cost), where p is the parameter used to indicate the relative cost factor of the ADM with TSI to the basic ADM (without TSI) at the same line rate. The value of p may vary from 1 to 1.5 depending on the equipment vendor's marketing plan. The OC-N per-node cost for the BSHR/4 is computed as (q×basic OC-N ADM cost), where q indicates the relative cost factor per node for the BSHR/4 to the basic ADM cost at the same rate. If two independent basic ADMs are used in each node, q=2; if the 1+1 basic ADM (two ADMs share a common unit) or one basic ADM plus a regenerator[6] are used in each

---

6. The cost of one SONET regenerator is estimated as 50 to 70 percent of the basic ADM cost at the same line rate.

node, q=1.5 to 1.7. Two cost models are used in Figure 4.16(b): the higher-cost model using p=1.5 and q=2, and the lower-cost model using p=1 and q=1.5. A *Cost Ratio* (CR) greater than 1 indicates that the USHR is more economical for that demand pattern.

As shown in Figure 4.16(b), for the higher-cost model, the USHR is always economical, except for a pure cyclic demand pattern, P7. Both of the BSHR architectures are more economical in this case. However, comparing the two BSHR architectures, the BSHR/4 is more economical than the BSHR/2 for centralized demand patterns P1 and P2; nearly equal for patterns P3 and P4; equal in costs with the USHR for a pure mesh demand pattern, P5; and more economical for pattern P6. It is less economical for a pure cyclic demand pattern, P7.

For the lower-cost model, the BSHR/4 architecture is more economical than the USHR for all demand patterns except P3. The BSHR/2 architecture is as economical as the USHR for centralized patterns P1 and P2, and equal in costs for P6. Comparing the two BSHR architectures, the BSHR/2 is more economical in four out of seven patterns; however, costs are relatively close for three out of the four patterns.

References [4,18] contain more details on the traffic analysis of different demand patterns for SHR alternatives. In these studies, two demand patterns are particularly interesting to SONET network applications. These two demand patterns are centralized and mesh demand patterns. The centralized demand pattern is a traffic pattern in which all the traffic on a ring goes to a single central site. This is typical for rings deployed in the periphery of some metropolitan networks. The mesh demand pattern is a traffic pattern where traffic is uniformly distributed between sites. Every site has approximately the same level of traffic as every other site on the ring. This is typical for rings in metropolitan and intraLATA applications.

The traffic studies in References [4,18] suggest that, in general, the USHR may be economically attractive for the centralized demand pattern if the higher-cost model is assumed. If the lower-cost model is assumed for the centralized demand pattern, the BSHR/4 becomes economical because it has twice the capacity for service on each span, and the BSHR/2 and the USHR are both economically equivalent for most cases.

For the mesh demand pattern with the higher-cost model, the BSHR/4 may be more economical than the USHR in many cases, depending on the demand requirement, particularly for the case of larger ring size. Under the same condition, the BSHR/2 may not have the economical advantage over the USHR (as does the BSHR/4) because only half of the capacity can be used in the BSHR/2. If the lower-cost model is assumed for the mesh demand pattern, the BSHR/4 and BSHR/2 architectures are economical candidates, particularly the BSHR/4, among three possible ring architectures.

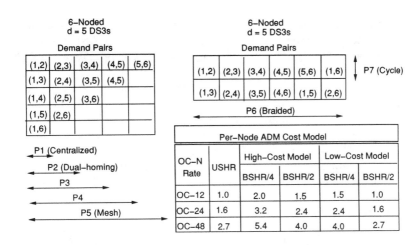

6–Noded
d = 5 DS3s

Demand Pairs

(1,2)	(2,3)	(3,4)	(4,5)	(5,6)
(1,3)	(2,4)	(3,5)	(4,5)	
(1,4)	(2,5)	(3,6)		
(1,5)	(2,6)			
(1,6)				

P1 (Centralized)
P2 (Dual–homing)
P3
P4
P5 (Mesh)

6–Noded
d = 5 DS3s

Demand Pairs

(1,2)	(2,3)	(3,4)	(4,5)	(5,6)	(1,6)
(1,3)	(2,4)	(3,5)	(4,6)	(1,5)	(2,6)

P7 (Cycle)

P6 (Braided)

Per–Node ADM Cost Model				
OC–N Rate	USHR	High–Cost Model		Low–Cost Model	
		BSHR/4	BSHR/2	BSHR/4	BSHR/2
OC–12	1.0	2.0	1.5	1.5	1.0
OC–24	1.6	3.2	2.4	2.4	1.6
OC–48	2.7	5.4	4.0	4.0	2.7

(a) Seven Possible Demand Patterns

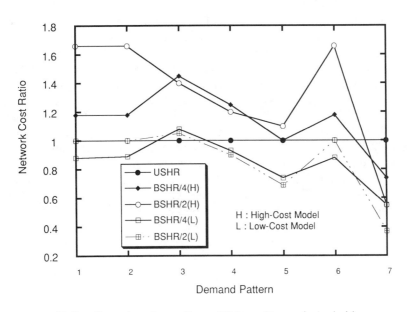

(b) Cost Comparison Among Demand Patterns (Economic Analysis)

Figure 4.16. Economic trend analysis for six-node rings.

153

The next logical question is: which demand pattern is more practical? This question is difficult to answer because it depends on many factors, such as the ring deployment strategy, the demand requirement, and the existing network architecture. The demand pattern for the loop network is an example of the centralized demand pattern. If the ring is planned in the existing hubbing network architecture, the STS-1 demand pattern for suburban areas (which usually include smaller COs) is most likely to be the centralized or near-centralized demand pattern. Because the VT1.5 (or DS1) demand requirement between two smaller COs is typically not significant enough to build a direct STS-1 (or DS3), this VT1.5 (or DS1) demand is usually aggregated to the hub DCS for grooming. Also, a concern about ring capacity exhaust may force the VT1.5 demand requirement to the centralized or near-centralized demand pattern (using hub DCS) to minimize the number of STS-1s needed to carry the VT1.5 demand requirement. Hub-to-hub subnetworks and networks with mutually heavy STS-1 demand requirements may be examples where the mesh demand pattern is used.

4.3.2 SONET Standards

The USHR using path protection switching (USHR/P, see Section 4.2.4) does not require changes in present SONET APS standards because it uses the SONET path layer to convey the failure message and to trigger protection switching. Thus, the USHR/P may be implemented as soon as its SONET ADM becomes available in the multivendor environment. On the other hand, the ring control scheme using line protection switching cannot be implemented until present SONET standards for protection switching change. This is because the present SONET APS system (see Chapter 3) is designed for point-to-point systems rather than for ring applications (see Sections 4.2.1-4.2.3). The difference between the point-to-point application and the ring application is that the line overhead associated with restoration (e.g., K1 and K2 bytes) for the ring application may pass through one or more intermediate nodes. The SONET line overhead for the ring's proposed line protection switching scheme includes the K1 and K2 bytes (used in present SONET APS), the F1 byte, or other growth bytes. Among these candidate line overhead bytes, the K1 and K2 bytes are common choices for line switching because they have been used in the present SONET line protection switching standard for point-to-point systems. If the line protection switching scheme applies to the ring network, the node needs the following additional capacities and different signal formats:

- The ring node must have intelligence to decide whether the incoming K1/K2 byte should be passed through or terminated.
- The K1 and/or K2 format used in the SONET APS must be redefined to incorporate the ring nodal address scheme.

- The ring line switching protocol must be redesigned to meet the 50-ms protection switching objective (see Section 4.3.3).

Until the universal K1/K2 formats and associated protocol for the ring application are defined in the standards group, all SHR architectures using the line protection switching scheme may not meet the requirement of intervendor compatibility. Consider a scenario where a company's SONET transition plan is first to deploy SONET point-to-point systems (ADMs in terminal mode) using multivendor equipment, and then to connect those offices to form a SONET SHR, provided connectivity exists and economic advantages are proven. In this scenario, it may not be possible to connect the offices to form an SHR using line protection switching (e.g., BSHR/2 or BSHR/4) if line protection switching for the ring application is not standardized.

4.3.3 Protection Switching Time Requirement

The restoration process in current asynchronous rings may take approximately 200 to 300 ms. The restoration time for asynchronous rings is dominated by the inter-node signaling delays. The signaling scheme used is very similar to the SONET DCC overhead channel for signaling, which is a relatively slow signaling scheme that can introduce several milliseconds per node in signaling delay. This inter-nodal signaling delay can be significantly reduced when SONET line overhead is used for signaling.

The SONET line overhead for the ring's proposed line protection switching scheme includes the K1 and K2 bytes (used in present SONET APS), the F1 byte, or other growth bytes. Among these candidate line overhead bytes, the K1 and K2 bytes are common choices for line switching because they have been used in the present SONET line protection switching standard for point-to-point systems. Based on possible line protection switching protocols, the protection switching time after a failure is detected can be approximately calculated. The loopback function is assumed here for channel restoration, i.e., only protection switches at two nodes adjacent to the failed network component are involved in protection switching.

Because the ring-signaling protocol uses K1 and K2 byte signaling, the pass-through function becomes the most critical factor in the protocol delay. In a ring, intermediate nodes (not affected by the failure) must be able to read and pass through the K1 and K2 bytes. It was assumed that the processing time and signaling delay take at least two to three 125-µs frames of delay because the K1 and K2 bytes are required to be terminated and regenerated at each intermediate node. In the line protection switching scheme, three consecutive K1/K2 bytes are required to trigger the protection switching, as does the current SONET APS protocol [10].

155

Assuming the maximum number of ADM nodes on a ring to be 16, the worst signaling delay from one end to the other end (sometimes it is referred to as a signaling pass) is approximately calculated as follows:

$$(16-2) \ nodes \times (3 \ frames) \times 125 \ \mu s/frame = 5.25 \ ms$$

If a three-pass signaling protocol like the one used in the present APS standard for point-to-point systems [10] is considered, the total protocol processing delay will be about 15.75 ms. If the protection switching for ring restoration is required to be completed within 50 ms as specified for SONET point-to-point systems, the time for each protection switch (at both ends of the failed link) to trigger and complete protection switching must be within 17.125 ms [(50-15.75)/2], where two bidirectional protection switches are involved. This may impose a stringent requirement for protection switch implementation (and thus be more expensive). Therefore, all current line switching proposals [7,8,11,12] suggest that a two-pass signaling scheme should be used to meet the 50-ms objective without imposing a stringent delay requirement for switch implementations. Details of proposed two-pass signaling schemes can be found in References [7,8,11,12].

In contrast to the line protection switching scheme, the path protection switching scheme uses the SONET path-layer signal (e.g., path AIS) to convey the failure messages and restore channels (STS-1 or VT) on an end-to-end basis. The restoration time for the path protection switching scheme can be as fast as the 1+1 point-to-point system because the path AIS carried by transit STS-1s (or VTs) does not stop at intermediate nodes of the ring.

4.3.4 Operations Analysis

The operations issues for SONET ring alternatives are very complex and include network *Operations, Administration, Maintenance, and Provisioning* (OAM&P). Reference [18] provides a preliminary view of operations impacts in different types of rings. From the previous operations discussion on USHRs and BSHRs, it is apparent that rings, in general, may indeed impact day-to-day operations.

From a provisioning standpoint, both the USHR and BSHR can be supported with today's assignment algorithms, but cannot necessarily maximize the capacity of the BSHR. Enhanced assignment algorithms may need to be developed to support BSHRs (particularly the BSHR/2 architecture) efficiently.

Both the USHR and BSHR architectures are susceptible to multiple failures. The algorithms needed to correlate maintenance messages in the case of multiple failures will be complex and will be different for the USHR/P and BSHR architectures. An observation in Reference [18] states that path-level rings may create a larger volume of

156

messages than line-level BSHRs, which could impact a data communications network. Communication with each node is needed to support routine maintenance activities. Knowledge of the ring state at all times is also needed.

Maintenance on two-fiber rings (USHR and BSHR) requires a higher degree of coordinated maintenance activities and an overall ring view. For span maintenance, removing a span from service may be easier on the BSHR architecture than on the USHR/P architecture. However, it is important to mention that routine maintenance on a ring may be infrequent enough that it can be dealt with on an exceptional basis. Rings could probably be supported on an exceptional basis by monitoring line-level alarms. Alarm messages (e.g., LOS, LOF) do exist. In the future, line-level alarm messages may not be sufficient to support the USHR/P architecture. The study [18] also showed that in a two-fiber SHR (USHR or BSHR), the self-healing function of the ring is unavailable during maintenance activity.

The need for automated tools becomes more important when rings are widely deployed in the network. It may be easier to plan and administer bandwidth for USHRs than for BSHRs, but BSHRs are easier to expand and add on new nodes with point-to-point demand, without impacting the bandwidth utilization of the ring.

In summary, each type of ring has a certain degree of operations impact, although the BSHR is more consistent with today's point-to-point methodology. Of the three architectures mentioned, the BSHR/4 architecture may have the least impact on operations because it is most like today's networks (although more study is needed). The BSHR (particularly the BSHR/2 architecture) can become as complicated or as detailed as the path switched USHR because of operation issues associated with bandwidth rearrangement, capacity upgrades, misdirected traffic under multiple failures, and administration of self-healing capacity for low-priority services.

The same general operations considerations would also apply in the case of managing interconnected rings. Interconnected rings are a combination of two or more rings interconnected with dual- serving nodes. There may not be significant operations savings in using one ring architecture rather than another. In a network scenario requiring a multiple number of rings to meet traffic demands, there may be operations savings with a ring architecture that minimizes the total number of individual rings (and ring ADMs) that need to be managed.

4.3.5 Summary

Table 4-3 summarizes the discussions in Sections 4.3.1 through 4.3.4. Figure 4.17 depicts a diagram for SONET ring architecture alternatives and their development status. As discussed in Section 4.3.1, applications determine which type of ring architectures could be used in terms of the network cost. The restoration time for the path protection switching scheme can be as fast as the 1+1 point-to-point system because the path AIS carried by transit STS-1s (or VTs) does not stop at intermediate nodes of the ring. In contrast, the SHRs using line protection switching may take longer to restore demand because the K1/K2 byte has to be terminated and regenerated at each intermediate node. Bellcore Technical Reference TR-TSY-000496, Supplement 1 [14] defines the equipment criteria for USHR using path protection switching, which became available in 1991. The USHR or BSHR using proprietary line protection switching may be available in 1992, and its standard line protection switching is under consideration in the ANSI T1X1.5 standards group (expected to be available in 1992-1993).

Table 4-3. A Relative Comparison among SONET SHR Architectures

SHR architectures	ADM type	Carry demand	Node component costs	SONET K1, K2 changes	System complexity	ADM protection	Restoration speed
BSHR/4	basic ADM+	more	higher	yes	medium	yes	slower
BSHR/2	ADM/TSI*	more/less**	medium	yes	complex	no	slower
USHR/L	basic ADM	less	lower	yes	medium	no	slower
USHR/P	basic ADM	less	lower	no	simple	no	faster

+ Basic ADM: ADM with signal add-drop feature only
* ADM/TSI: ADM with both add-drop and grooming features
** depends on the demand pattern

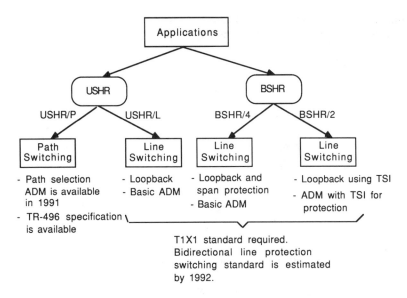

Figure 4.17. SONET ring alternatives and development status.

4.4 PASSIVE PROTECTED SONET BSHR/4 ARCHITECTURE

As discussed in Section 4.3, among SONET SHR alternatives, the BSHR/4 has the largest available capacity and can work with today's OSs with minimum changes (where the BSHR/4 is made up of one working ring and one protection ring). However, the BSHR/4 involved in the analysis in Section 4.3.1 uses a conventional implementation method that makes it relatively expensive compared to other SONET ring architectures. Figure 4.18 depicts three possible conventional methods for implementing the BSHR/4. The first method is to equip two independent ADMs (one for working and one for protection) and a protection switch between two ADMs in each ring node [see Figure 4.18(a)]. The second method is to use a 1+1 ADM (i.e., two ADMs sharing a common unit) in each ring node [see Figure 4.18(b)], which costs less than the first implementation method but has no protection capability when the common unit of the 1+1 ADM fails in each node. The third method is to use one ADM for the working ring and a SONET regenerator as the protection component for each ring node [see Figure 4.18(c)]. The third implementation method may be the least costly of the three alternatives. However, an additional control may be needed for this implementation because the SONET regenerator does not terminate the line overhead, which is a necessary part of the BSHR line protection switching protocol (i.e., to

159

determine the failure messages, such as K1 and K2 bytes, that should be passed through to the next node or terminated and processed).

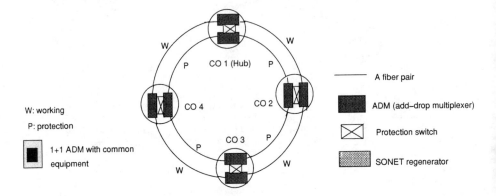

(a) BSHR/4 with Two Independent ADMs at Each Node

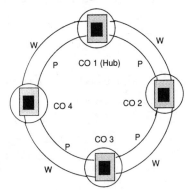

(b) BSHR/4 with a 1+1 ADM at Each Node

Figure 4.18. Three possible conventional implementations for the BSHR/4 architecture.

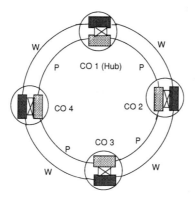

(c) BSHR/4 with One ADM and One Regenerator at Each Node

Figure 4.18. (Continued)

For each conventional implementation method for the BSHR/4, the cost of the protection ring is at least half of the cost of the working ring. In this section, we discuss a new, more cost-effective implementation method for the BSHR/4, called a *passive protected SONET BSHR/4 architecture (BSHR4/PPR)* [19], which significantly reduces the cost of the protection ring while preserving the capacity advantage of the BSHR/4.

The BSHR4/PPR, which uses passive optical components to reduce the protection cost, essentially extends from an optical switching protection system that is primarily designed for point-to-point systems [20]. This architectural extension is natural because the BSHR/4 evolved from today's point-to-point networks. The following sections discuss the BSHR4/PPR network architecture and its SONET control scheme, and then show that the BSHR4/PPR can be implemented using existing optical technology. The economical merit of the BSHR4/PPR, as compared to conventional implementations, is also discussed.

4.4.1 Passive Protected Bidirectional Self-Healing Ring Architecture (BSHR4/PPR)

Figure 4.19 depicts a generic BSHR4/PPR architecture for a ring network including a Hub and its associated COs (CO-1 and CO-2). The hub in this ring configuration provides an access point to the rest of the network for CO-1 and CO-2. A BSHR4/PPR is composed of two parts: a SONET ring for bidirectional working signals and a totally passive protection ring for bidirectional protection signals. One SONET basic ADM is required for each working ring node to carry bidirectional working signals. The passive protection ring is a standby ring dedicated to the working SONET ring; it is composed of optical switches and optical amplifiers if the ring size is large. The passive protection ring is essentially an optical add-drop protection ring, where optical switches of two ends of the failed facility act as optical "add-drop" components, and optical switches of intermediate ring nodes are used as optical "pass-through" components. The optical signal add-drop is controlled by SONET ADMs on the working ring.

Figure 4.20(a) depicts a more detailed BSHR4/PPR functional configuration. The high-speed loopback function is performed via 1×2 and 2×2 optical switches. Optical protection switches eliminate the need for duplicate ADMs. As depicted in Figure 4.20(a), each CO requires one ADM for bidirectional working signals and three pairs of 1×2/2×2 optical switches for protection switching (in case of network component failures). One or more optical amplifiers may be needed for optical signal amplification if the protection ring length exceeds a repeater spacing threshold. Two 1×2 optical switches at each office in the working ring terminate two fibers for transmitting and receiving normal traffic. The protection ring is composed of 2×2 optical switches, a pair in each office, and two fibers terminated at each 2×2 optical switch. To simplify the discussion of SONET ring control operation, Figure 4.20(b), which includes one working fiber system and one protection fiber system with opposite directions, shows only one direction for working signals. The A-AA port pairs shown in Figure 4.20(b) for all switches are used to carry normal working traffic, whereas A-BB and B-AA port pairs are used to carry recovery traffic if the network component fails. For convenience, we refer to the 1×2 optical switches that connect to input and output ports of the ADM as OP_A and OP_B, respectively. OP_C is referred to as the 2×2 optical switch in the protection ring. The OP_C optical switches control signal adding-dropping to/from the protection high-speed optical ring from/to the working ring.

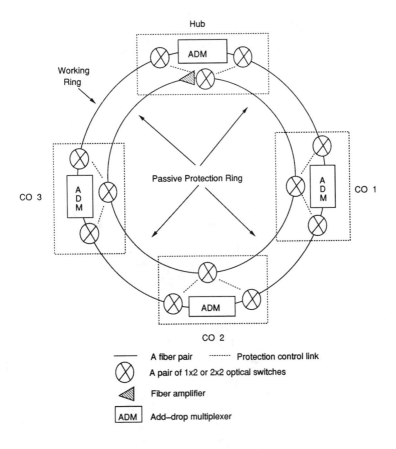

Figure 4.19. BSHR/4 with passive protection ring (BSHR4/PPR).

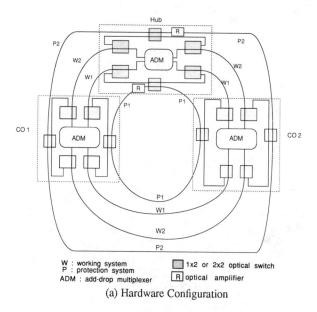

W : working system
P : protection system
ADM : add-drop multiplexer

1x2 or 2x2 optical switch
R optical amplifier

(a) Hardware Configuration

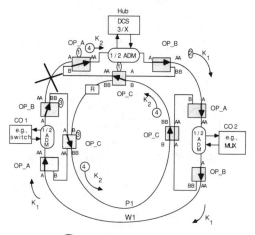

(n) : n-th step of control protocol

(b) Fiber Failure Scenario

Figure 4.20. BSHR4/PPR operations.

164

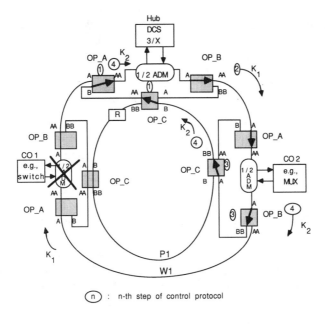

(n) : n-th step of control protocol

(c) Node Failure Scenario

Figure 4.20. (Continued)

Opposite signal directions are used for the working and protection rings. In the normal situation, working signals are circulated around the working ring. All 1×2 optical switch default configurations in the working ring are in state A-AA. The default configurations of all 2×2 optical switches in the protection ring are in the bar state (i.e., A-AA and B-BB). When the working fiber between CO-1 and the Hub fails [see Figure 4.20(b)], the ADM at the Hub detects LOS or other hard failure signals and issues commands to change the configurations of OP_A (from A-AA to B-AA) and OP_C (from A-AA/B-BB to A-BB/B-AA[7]) via internal control circuits. Then the Hub's ADM sends the K1 byte to CO-1 via the working ring in the clockwise direction.

---

7. Connection A-BB is used to move signals from the working system to the protection system, whereas connection B-AA is used to drop signals from the protection system to the working terminal.

165

When the ADM at CO-1 detects the K1 byte, it issues commands to change configurations of OP_B (from A-AA to A-BB) and OP_C (from A-AA/B-BB to A-BB/B-AA) at CO-1 via internal control circuits, and the ADM of CO-1 sends the K2 byte back to the Hub's ADM via the protection ring. The loopback protection switching is completed after the Hub's ADM receives the K2 byte, as shown in Figure 4.20(b). Thus, the channels from CO-2 to the Hub can be rerouted via the following path [see Figure 4.20(b)]:

ADM(CO-2)--> OP_B(CO-2)--> (working fiber W1)--> OP_A(CO-1)--> ADM(CO-1)-->
OP_B(CO-1)--> OP_C(CO-1)--> (protection fiber P1)--> OP_C(CO-2)-->
OP_C(Hub)--> OP_A(Hub)--> ADM(Hub)

The optical switch configurations are held until the failed fiber is repaired. The protection switching control for a cable cut that breaks both the working and protection fibers between CO-1 and the Hub is the same as that for one working fiber failure, as just described.

The protection switching time for the proposed control system in the cable cut scenario can be approximately calculated as follows (assume that the maximum number of nodes in the ring is 16, as is being considered in ANSI T1X1.5):

1. The time to change the optical switch configuration in the head end = 10 ms (for an MOS [20]) or 0.15 µs (for a lithium niobate crystal switch[8] [21])
2. The time for detecting the K1 byte (via the working ring of 16 nodes) in the tail end = 125 µs×2×14 (it takes about two SONET frames in each intermediate node to process and load a new K1 to the next frame) + 3×125 µs (time to detect three consecutive K1 bytes as required by the standards) = 3.875 ms
3. The time to change the optical switch configuration in the tail end = 10 ms (for the MOS) or 0.15 µs (for the lithium niobate crystal switch)
4. The time for detecting the K2 byte (via the passive protection ring) in the head end = 3×125 µs (three consecutive byte rule).

Thus, the protection switching time is approximately 24.3 ms when MOSs are used or 4.25 ms when crystal switches are used; either approach meets the 50-ms protection switching requirement.

---

8. A lithium niobate crystal switch may have an insertion loss of 5 dB, and a commercially available mechanical optical switch has an insertion loss of 0.3 dB.

The SONET line protection switching scheme described above can also be applied to an ADM or CO failure; however, protection switching is performed at two COs that are adjacent to the failed ADM or CO. Figure 4.20(c) depicts a demand-rerouting scheme for a CO-1 failure. In the node failure scenario, a timer, which defaults to twice the 50-ms requirement (i.e., 100 ms), is introduced. When the Hub detects a LOS, it loads a valid K1 byte and sends it to the other end as in the cable cut scenario. The Hub expects to receive the corresponding K2 byte within 50 ms, if CO-1 functions well. If the Hub does not receive the corresponding K2 byte before the timer expires (i.e., within 100 ms), it determines that CO-1 is down and then loads a new valid K1 byte that is destined to the node located immediately before CO-1 (i.e., CO-2 in this example). Based on this protocol, demand restoration for a node failure will take longer than for fiber cable cuts; however, it is acceptable because a CO failure rarely happens.

Other more sophisticated SONET signaling protocols, such as the one reported in Reference [8] (which uses the matching method of the ring node number for a node failure scenario), can also work on the BSHR4/PPR as long as the signaling protocol allows the K1 byte to travel through the working ring when a failure is detected.

4.4.2 Implementation Technologies for the Passive Protection Ring of BSHR4/PPR

The complexity and cost of implementing the passive protection ring of the BSHR4/PPR depend on the ring's total length and on optical switching technologies. If the ring length is short enough that it does not need to regenerate optical signals in the protection ring, then only optical switches are needed. Figure 4.21 depicts a relationship between the number of ring nodes and the maximum passive protection ring length (without optical amplification) for the OC-12, OC-24, and OC-48 BSHR4/PPRs. A ring length less than the maximum length described in Figure 4.21 represents the best implementation scenario for the BSHR4/PPR architecture. Two wavelength windows, 1300-nm and 1550-nm, are shown in the figure because they have been commonly used in today's fiber-optic systems. Appendix A gives detailed assumptions and results that correspond to Figure 4.21. For example, for the OC-48 BSHR4/PPR of six nodes with 1300-nm wavelength, the maximum passive protection ring length without optical signal amplification is about 18 miles, which may be too short to be useful. In this case, eight MOSs are passed through for an end-to-end protection optical signal under the worst case scenario. An MOS is suggested for the initial implementation because it is commercially available and is very reliable [20]. In the future, a crystal switch or other high-speed optical switch may be used to provide "hitless" services if the high insertion loss for these types of optical switches can be improved or tolerated.

167

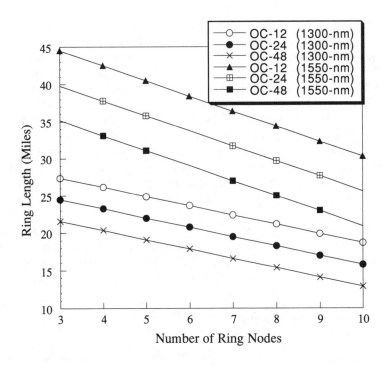

Figure 4.21. Number of ring nodes vs. maximum passive ring length without optical amplification.

If the ring length exceeds the distance threshold, optical signal amplification is needed. A possible and cost-effective approach for providing optical amplification is to use optical amplifiers that amplify optical signals directly without an E/O conversion (like SONET regenerators). It is noted that although the optical amplifier(s) is placed in the ring topology, it does not form an optical signal amplification ring (loop) because the optical signal carried on the protection ring is added and dropped at two different optical switches. Reference [22] discusses recent research in optical amplifiers and their potential applications. Compared to its electronic counterpart (i.e., the regenerator), the fiber amplifier not only has lower equipment costs, but also lower maintenance costs because the fiber amplifier is spliced into the fiber and has less system complexity. Also, the fiber amplification system is essentially independent of line rates, wavelength, and data signal formats. This makes the system attractive in a diverse or rapidly growing environment.

Two possible optical amplifier placements for the protection ring of the BSHR4/PPR are (1) place low-gain amplifiers at each ring node [23], and (2) place high-gain amplifiers at some appropriate locations of the protection ring. If the high-gain amplifiers are used, it is desirable for the optical amplifiers used in the BSHR4/PPR to be able to accept a large variation of input power (from different ring nodes) and to provide a near constant output power (to simplify the receiver design). A high-gain *Optical Limiting Amplifier* (OLA) reported in Reference [24] is one optical amplifier that has this desirable characteristic. The OLA can provide a constant output power despite a large variation of input power due to widely spaced COs along the ring. Thus, the optical receiver[9] of a CO does not have to cope with the extremely large dynamic range of the ring. This facilitates easier network design and implementation by using optical amplifiers in the protection ring. The following sections discuss a protection ring architecture of the BSHR4/PPR and some system analyses.

4.4.2.1 BSHR4/PPR Protection Ring Architecture using OLAs

The OLA is a high-gain fiber amplifier that keeps the output power constant (about +5 dBm) despite variations in the input power over a 30-dB dynamic range (i.e., -30 dBm to 0 dBm). Figure 4.22 depicts a functional diagram for the OLA [25]. As shown in Figure 4.22, an OLA uses two cascaded stages of *Erbium-Doped Fiber Amplifiers* (EDFAs) and two *Optical Bandpass Filters* (OBPF). OBPF#1, which is a grating filter with a 3-dB bandwidth of 1 nm and an insertion loss of 4.5 dB, is used to eliminate most of the *Amplified Spontaneous Emission* (ASE) noise generated in the first-stage amplifier so that the gain of the second stage can be largely provided to the signal. OBPF#2 is an angle-tuned Fabry-Perot filter with a 3-dB bandwidth of 1.5 nm and an insertion loss of 7 dB. The OLA maintains a constant total output power of about 5 dBm, which includes the loss of OBPF#2, for an input power ranging from -30 to 0 dBm [25].

---

9. The typical range of input power that the optical receiver can detect is about 10 dB. Increasing the range of input power variation could make the optical receiver complex and expensive.

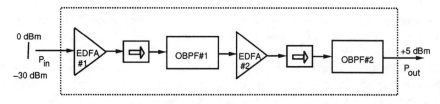

EDFA : erbium–doped fiber amplifier
OBPF : optical bandpass filter

Figure 4.22. Functional diagram of an OLA.

The OLA works on the 1550-nm wavelength window and, thus, is applied only to dispersion-shifted fibers that have zero dispersion near 1550 nm. The major advantage of using OLAs in the BSHR4/PPR protection ring is that the OLA can provide a constant output power despite the large variation of input power, due to widely spaced COs along the ring.

Figure 4.23 depicts a protection ring configuration of the BSHR4/PPR with an OLA that is applied to dispersion-shifted fibers. The commercially available dispersion-shifted fiber may have a power loss of 0.2-0.25 dB per km.

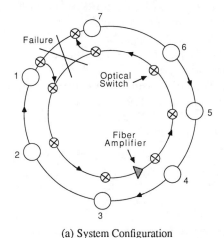

(a) System Configuration

Figure 4.23. A system configuration and protection path for BSHR4/PPR using optical amplifiers.

170

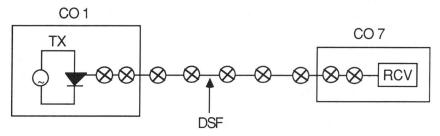

CO 1

TX

CO 7

RCV

DSF

TX: transmitter   RCV: receiver
DSF: dispersion-shift fiber

(b) Protection Path

Figure 4.23.   (Continued)

As depicted in Figure 4.24, the maximum distance that the signal can be carried by a single optical limiting fiber amplification system depends on the location of that fiber amplifier. Assume that the allowable fiber loss is 34 dB[10] for OC-48 systems. If the fiber amplifier is near the transmitter [i.e., 0 dBm input, see Figure 4.24(a)], the power entering the fiber amplifier saturates it and results in a maximum distance of only 136 km (= 34/0.25) (i.e., 85 miles). If the fiber system is separated from the transmitter by a transmission distance equivalent to a 30-dB optical loss [i.e., -30 dBm input, see Figure 4.24(b)], then the maximum distance that an OLA system can carry is about 256 km [(34+30)/0.25)] (i.e., 160 miles). This is the best case for the OLA system, in terms of distance, that optical signals can be carried. Thus, the worst case for our ring application is where the fiber amplifier is placed after the optical switch at one of the COs on the ring [see Figure 4.24(a)]. Figure 4.24(c) depicts a possible placement location for the first OLA on the protection ring of the BSHR4/PPR, which places the first fiber amplifier at the end of the longest link before the optical switch on the ring. The second optical limiting fiber can be placed at a location following the first amplifier in the same signal direction at some distance that allows a total of 30-dB

---

10. Assume that the receiver sensitivity for OC-48 is -29 dBm and the output power of the optical amplifier is +5 dBm. Thus, the allowable fiber loss is 34 dB.

171

power loss as depicted in Figure 4.24(c). The 30-dB power loss is equivalent to a signal traveling a distance of (30 dB-n×0.3 dB)/[(0.25 dB/km)×1.6 km/miles] miles where n is the number of MOSs between two OLAs on the protection ring of the BSHR4/PPR and the power loss of each MOS is assumed to be 0.3 dB. For n=3 (i.e., three optical switches between two fiber amplifiers), a 30-dB power loss means that optical signals can be picked up and amplified at the second OLA, which is about 72 miles from the first optical fiber amplifier. The optimum placement of fiber amplifiers on the protection ring of the BSHR4/PPR remains to be determined.

One unidirectional OLA is estimated to cost about $15K in the 1992-1993 time frame (which is also the time frame for SONET OC-48 ring deployment). The cost for OLAs is expected to drop significantly when these devices become commercially available, which is expected in the 1992-1994 time frame. Of course, promising applications may shorten the time frame for commercial availability. The cost of dispersion-shifted fiber is nearly twice the cost of normal fiber (with a wavelength of 1300 nm). It is, however, still negligible in our ring applications because the equipment (i.e., ADM) cost is the dominant factor of the total fiber transport cost.

4.4.2.2 System Experiment

Recently, the concept of the BSHR4/PPR protection ring using OLAs has been demonstrated experimentally for 1.244-Gbps (OC-24 line rate) transmission [25]. This experiment uses a configuration described in Figure 4.23, and the results have been reported in Reference [25]. This experiment verified that a single OLA can handle a ring size ranging from 160 to 200 km. It has also been estimated that, without considering fiber dispersion penalty, a ring size as large as 1000 km is possible by using 10 concatenated OLAs along the protection ring.

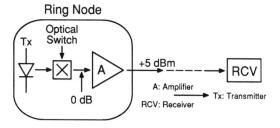

(a) Worst Case for Fiber Amplifier Placement

(b) Best Case for Fiber Amplifier Placement

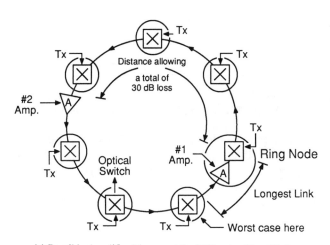

(c) Possible Amplifier Placement for PPR using Two OLAs

Figure 4.24. The impact of OLA location on maximum signal transport distance for BSHR4/PPR.

173

4.4.3 Economic Merit of BSHR4/PPR

The economic merit of the BSHR4/PPR compared to the BSHR/2 architecture has been reported in Reference [19] and is summarized in Table 4-4. The OLA system (as discussed in Section 4.4.2.1) is used here to amplify optical signals, if necessary. This economic study considered a ring of seven nodes with a total ring length of 150 miles[11] and equal link length of 21.4 miles. For this considered ring length, the passive protection ring with dispersion-shifted fiber of 1550-nm wavelength can be implemented using a total of two OLAs in the worst case scenario [19].

In Table 4-4, the parameter p is the cost factor of the cost of the ADM with TSI (used for the BSHR/2) to the cost of the basic ADM (used for the BSHR/4) at the same line rate. The value of p may vary from 1 to 1.5, representing the best and worst cases, respectively, for the BSHR/2. The OC-N per node cost for the BSHR/4 is computed as (q×basic OC-N ADM cost), where q indicates the ratio of the per-node cost of both the working and protection rings to the per-node cost of the working ring for the conventional BSHR/4 implementation. Note that q=2 if two independent basic ADMs are used in each node, and q=1.5 if the 1+1 basic ADM (two ADMs share a common control unit and software) or one basic ADM plus a regenerator is used in each node.

Table 4-4. Cost Savings for Seven-Node BSHR4/PPR Compared to Other BSHRs (Ring Length = 150 Miles)

BSHR STS-1 capacity requirement	OC-N (BSHR/2)	BSHR/2 (%)		OC-N (BSHR/4)	BSHR/4 (%)	
		p=1.5	p=1.0		q=2.0	q=1.5
7-12	24	57%*	36%	12	36%	14%
13-24	48	49%	24%	24	43%	24%

* +57% means that the BSHR4/PPR cost is 57 percent less than the cost of the BSHR/2 with p=1.5

Table 4-4 shows cost savings for the BSHR4/PPR compared to other BSHRs under two practical capacity requirement levels (7-12 STS-1s and 13-24 STS-1s) that can be carried by a single ring of up to OC-48. For example, for the practical capacity requirements (i.e., between seven STS-1s and 24 STS-1s), cost savings for the BSHR4/PPR are 24 to 36 percent compared to the best case of the BSHR/2 (i.e., p=1). As shown in Table 4-4, for most cases of interest, the BSHR4/PPR has a significant

---

11. The ring length of 150 miles is considered a practical upper bound for the ring length in metropolitan areas.

cost advantage over other BSHR implementations because the cost savings is at least 14 percent and may reach 57 percent for some cases. This economic merit of the BSHR4/PPR can be even greater when the number of nodes in the ring increases, providing the total ring length remains unchanged.

4.4.4 General Comparison between BSHR4/PPR and Other BSHRs

Table 4-5 summarizes the general comparison (between the BSHR4/PPR and other BSHRs) that has been discussed in previous sections.

Table 4-5. General Comparison between BSHR4/PPR and other BSHRs

SHR architectures	ADM Type per node	# of required fibers	TSI required for protection	Per-node cost	Protection control complexity	Self-healing capability during ADM maintenance
BSHR/4	2 basic ADMs+	4	no	high	simple	yes
	1 basic 1+1 ADM	4	no	medium	simple	no
	1 ADM plus 1 regenerator	4	no	medium	simple	no
BSHR/2	ADM/TSI*	2	yes	medium/ low**	complex	no
BSHR4/PPR	basic ADM%	4	no	low	simple	no

+ Basic ADM: ADM with signal add-drop feature only (using TSA)
* ADM/TSI: ADM with built-in time slot cross-connect switching matrix
% For the working ring only
** Depending on the vendor pricing strategy

As discussed in previous sections, the BSHR4/PPR architecture is technologically implementable and has a cost advantage over other BSHR implementations. The control for the passive ring of the BSHR4/PPR is simple, but the K1 byte has to travel through the working fibers when a failure is detected. Thus, as long as the future SONET line protection switching standard for the ring application allows that, the BSHR4/PPR may work with other proposed BSHR line switching standards. Like all two-fiber SHRs (USHR or BSHR/2) and all BSHR/4s that do not use two independent ADMs for each node, the self-healing function of the ring is unavailable during ADM maintenance activity. Note that service maintenance for the BSHR4/PPR may be different from that of other SONET ring architectures. SONET ring architectures (except the BSHR4/PPR) can perform in-service maintenance on a 1/2-ADM basis, whereas the BSHR4/PPR may need to perform in-service maintenance on a component

175

basis using the ADM built-in redundant components.[12]

Furthermore, the passive protection ring of the BSHR4/PPR is independent of line rates, wavelength, and signal formats. In other words, if the ring needs to be upgraded to a higher rate in the future, only the working ring of BSHR4/PPR needs to be upgraded. Also, the working ring of the BSHR4/PPR can be implemented using today's asynchronous technology and then upgraded to the SONET working ring when the SONET ring technology becomes available.

4.5 APPLICATION OF SONET SHR ARCHITECTURES IN INTEROFFICE NETWORKS

We have discussed the SONET SHR architectures and their relative differences; the next question for planners is how and where the SONET ring can be deployed in the present fiber-hubbed or point-to-point interoffice networks. Before discussing interoffice networking applications for SONET SHRs, we first discuss some planning and bandwidth management issues.

4.5.1 Planning Considerations and Bandwidth Management for SONET SHRs

This section discusses where the ring structures previously discussed can be economically used in place of the more traditional survivable fiber network architectures using diverse protection routing. Additionally, it is of interest to measure the cost penalty even when rings may not be economical because their inherent advantage of being able to partially recover from a hub building failure may outweigh the difference in cost.

It is important to realize that the point-to-point (or, in this case, the building-to-building) demand requirement in today's telephone networks normally falls into the neighborhood of a few DS1s. That is, the usual demand requirements are in the range of several groups of 24 circuits (DS0s), and most are well below the capacity of a DS3, or equivalently, an STS-1. This signal rate is the primary building block of fiber

---

12. Most commercially available ADMs have built-in redundant components to achieve the equipment availability objective and provide an in-service maintenance capability. For high-speed components (e.g., OC-48 transceivers), 1:1 or 1:M redundancy is usually used; for low-speed components (e.g., STS-3 multiplexers), 1:N component redundancy is usually used. The component redundancy within the ADM is designed to achieve the necessary equipment availability and in-service maintenance, and is independent of the network architecture considered. (Note that when all component redundancy is engineered on a 1:1 or 1+1 basis, it may provide applications for SONET ring architectures.)

transmission equipment because fiber multiplexing equipment normally accepts it as input. To realize an efficient fiber network using present technology, one would concentrate the demands from a particular building onto a single fiber span going to a building selected as a hub building. To economically *groom* or sort the point-to-point demands into parcels large enough to fill a DS3 level demand, we can employ a DCS having the capability to cross-connect at the DS1 signal rate. Accomplishing the grooming function at a central point, the *hub*, is very cost-effective.

SONET ring architectures, on the other hand, may alter this picture in the near future. The framing structure employed in the SONET standards makes it relatively easy to locate and extract groups of VT1.5-level (equivalent to DS1-level) signals within an STS-1 signal. Byte-interleaving of STS-1 bytes is used to form rates higher than STS-1. The idea behind a SONET ring is to multiplex demands inexpensively to the highest line rate possible—a rate that hopefully exceeds the total of all demands to be carried on the ring. The receiving office then removes the data of interest and places the transmit side of the communication in the vacated time slots. The ring architecture thus may replace the hub DCS function by using distributed electronics, if applicable.

A question that naturally arises is whether it is advantageous to use a DCS with a SONET ring to concentrate the demand for a particular office into only a few STS-1 signals. Thus, the offices involved need only ADMs with the TSA feature to access these STS-1s of interest for channel adding and dropping in the same time slots. Because a hub DCS function will most likely be required to support additional offices that cannot be economically served by a ring, the cost of the hub DCS can be shared among many demand parcels, and thus the cost of using a DCS for the ring in question could be low. Should survivability of the network using only one hub DCS be insufficient, a dual-homing arrangement can be employed using two DCSs. Thus, connectivity can be maintained using the ring recovery ability and the second hub. The scheme of using the DCS for SHR bandwidth management is called the *centralized ring bandwidth management* (grooming) scheme.

An alternative approach is to use ADMs with the TSI feature to utilize the transmission bandwidth efficiently in a distributed manner (called the *distributed ring bandwidth management* scheme). Figure 4.25 depicts these two possible SONET bandwidth management schemes. The use of the centralized (DCS grooming) or distributed (ADM/TSI grooming) demand grooming for SONET SHRs depends upon relative economics and the global network planning strategies. The centralized ring grooming scheme using a DCS can best utilize today's fiber-hubbed network architectures, which have a DCS at each major hub. The distributed ring grooming scheme has flexibility for rearranging demands at each node and is not constrained by any particular network architecture. More discussion of SONET bandwidth management schemes can be found in Reference [26].

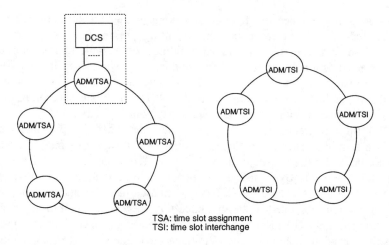

TSA: time slot assignment
TSI: time slot interchange

(a) Centralized Bandwidth Management   (b) Distributed Bandwidth Management

Figure 4.25.   Two possible SONET bandwidth management schemes.

In addition to the above two SONET ring grooming schemes, a non-grooming SONET ring bandwidth management scheme has also been used for applications where high point-to-point demand is involved. In this scheme, point-to-point STS-1s are assigned to each node pair, and DS1 demands for each node pair are placed directly into these dedicated STS-1s. For example, if five DS1s are required between Node 1 and Node 3, an STS-1 will be assigned to this node pair to accommodate five DS1s. This type of ring is sometimes referred to as a *building-to-building ring*.

Ring exhaustion is perhaps the largest problem facing the SONET SHR/ADM architecture. In the SHR using *Time Division Multiplexing* (TDM), all optical components run at the same data rate; should the ring capacity be exceeded, major additions may be needed to carry the excess demand. One method that is easy to visualize, but probably not cost-effective, is simply to build an entirely new ring with ADMs in the same place as the old ring to carry the excess demand. All growth could be placed on the new ring. Another method, probably more attractive from an installed first-cost point of view, is to place the largest demand parcels into a ring that has ADMs only at a few buildings. By reducing the load on the original ring in this manner, the smaller parcels should find enough capacity to be carried with ease. As the larger parcels on the new ring grow, new ADMs can be added, when appropriate, to maintain adequate utilization of installed equipment. Of course, another slightly different approach is to handle the larger demands via a hubbing strategy with terminals in the hub and the largest buildings. Demand from the largest buildings

178

could be handled on the new facilities, thus leaving the original ring free to carry the demand from the other buildings.

The relative economics of these approaches, along with associated algorithms to accomplish network optimization, need to be fully ascertained. As growth is very important to ring strategies, multiple-year scenarios would certainly have to be studied to ascertain the viability of the ring approach. Of particular interest is which buildings should be placed together on a ring. Also important is the question of how one decides whether a ring should be used at all, and if rings are attractive, what operational questions need to be considered for the final analysis.

4.5.2 SONET SHR Application Feasibility Studies

Reference [27] details a feasibility study for using the SONET USHR with path protection switching architecture (USHR/P, see Section 4.2.4) in interoffice network applications. This feasibility study investigated two types of network applications: single-homing SHRs, where one hub is in the SHR, and dual-homing SHRs, where two hubs are considered in the SHR. The purpose of the SHR feasibility study was to investigate conditions that make the SHR architecture attractive compared to its current technology counterparts in the interoffice applications. Because we want to compare both architectures based on the same survivability for cable cuts (i.e., both have 100 percent protection for cable cuts), the counterpart of the SHR in this feasibility study is the hubbing architecture with 1:1/DP.

The ADM cost in this study was based on an ADM equipment model that may be used in the USHR/P architecture (see Figure 4.26). In Figure 4.26,[13] an incoming optical OC-48 is demultiplexed into a maximum of 16 STS-3[14] (155.52-Mbps) channels. Some of the STS-3 channels may directly pass through and terminate at an outgoing STS-48 (2.488-Gbps) multiplexer if the STS-3 channels do not need to be dropped at this CO. These "through" STS-3 channels, together with other STS-3 channels added from this CO, are then multiplexed and converted to an OC-48 optical signal and sent to the next CO in the SHR over an outgoing fiber. If an STS-3 channel must drop at this CO, an STS-3 *multiplexer/demultiplexer* (MUX/DMUX) is used to

---

13. In Figure 4.26, we assume an optical line rate of OC-48 for the SHR. However, any OC-N with $N \geq 3$ can be applied to our model.

14. An intermediate multiplexing stage of STS-3 between STS-48 and STS-1 is used here because it can be built by present cost-effective CMOS technology and has flexibility for supporting both the STS-3c and STS-1 channels.

demultiplex an STS-3 signal to a maximum of three STS-1 (51.84-Mbps) channels and then to convert each STS-1 to the DS3s (44.7 Mbps) via STS-1 interface cards. In our cost model, we assume that the DS3 rate is used to interface with existing equipment. The cost model can be easily modified to incorporate new signal rates if appropriate. The 2:1 selector accepts the STS-1 signals from the primary ring in the normal situation; it is changed to accept the STS-1 signals from the secondary ring should network components fail. The 1:2 splitter generates duplicate STS-1 signals from an added STS-1 to both the primary and the secondary rings. Thus, the protection level (i.e., redundancy level) in our SHR ADM configuration model is the STS-1 signal and above. The minimum SONET electrical signal considered in our ADM model is the STS-1.

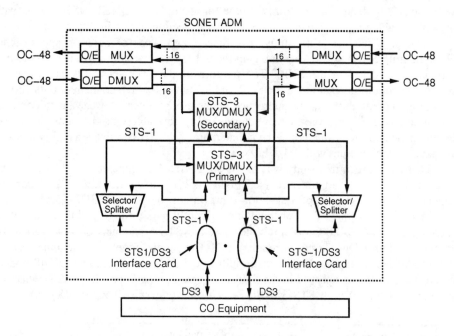

Figure 4.26. An example of ADM equipment model for USHR/P.

The total ADM cost is the cost for all optoelectronic components and startup costs, which include the non-optoelectronic hardware costs due to protection switching, frames, power, microprocessors (control, interface, and monitor), and control software.

180

In this feasibility study, *candidate area survivability* is defined as the percentage of total demand that is still intact when a network component fails. Because the survivability for fiber cable cuts for two competitive architectures (SHR and hubbing with 1:1/DP) is the same (100 percent protection for cable cuts), we compare network survivability for both architectures only for the case of major hub failures. *Hub survivability* is defined as the percentage of total demand that is still intact when the hub building fails. Note that if a DS3 or DS1 is terminated at the hub building, it may be either terminated at the hub DCS or terminated at other equipment within the hub.

For single-homing SHRs, only demands not terminating at the hub or passing through the hub for other offices that are not on the ring will survive when that hub fails. For dual-homing SHRs, the DS3s or DS1s that pass through the hub can be restored via the other hub when that hub fails.

The LATA network used in this study is a metropolitan area LATA network (including 36 nodes and 64 links) that is the same as the one used in Chapter 3. This study considers three line rates: OC-12, OC-24, and OC-48.

Figures 4.27(a) and 4.27(b) depict study areas for single- and dual-homing applications in a metropolitan LATA network. In the study, only special COs are considered candidate nodes to be included in SHRs. There are five study areas for the single-homing application, where each area is made up of a cluster of offices. Two areas [shown in Figure 4.27(b)] were studied for dual-homing applications. Area 1 covers six nodes: two clusters with three nodes for each cluster (served by two hubs). Area 2 covers five nodes: two clusters with four nodes in the one cluster, and the remaining node in the second cluster (this node is also the hub of the second cluster).

Table 4-6 reports feasibility results associated with Figure 4.27 and compares cost and survivability between two competitive architectures [27]. Detailed methods for obtaining these results can be found in Reference [27]. The cost considered here is the transport cost that includes terminal costs, fiber material and placement costs, and the SONET regenerator cost. The cost of the OLTM, used for the 1:1/DP network, is assumed to be 70 percent of one ADM at the same rate.

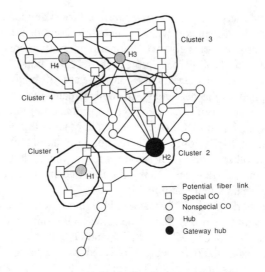

(a) Single Homing

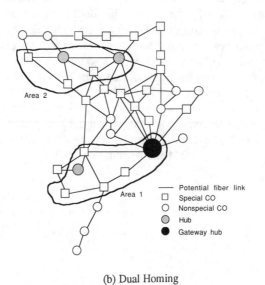

(b) Dual Homing

Figure 4.27. SONET SHR feasibility study areas.

Table 4-6. Cost/Survivability Comparisons between SHR and Hubbing/Diverse Protection Structures

Study areas	Ratio of SHR cost to hub-DP cost	Hub surv. (HS)	
		SHR (%)	Hub* (%)
Cluster 1 (SH)	0.85	8.3	0.0
Cluster 2 (SH)	0.84	9.8	0.0
Cluster 3 (SH)	0.80	15.6	0.0
Cluster 4 (SH)	0.83	5.0	0.0
Inter-hub (SH) subnetwork	0.69	7.7	0.0
Area 1 (DH)	0.57	63.6	21.7
Area 2 (DH)	0.63	72.7	17.8

HS: Hub survivability
SH: Single homing
DH: Dual homing
Hub-DP: Fiber-hubbed network with 1:1/DP

As shown in Table 4-6, the SHR architecture shows a promising role (compared to SH/1:1/DP) in all study areas, especially for the dual-homing application. Cost savings for dual-homing SHRs (compared to DH/1:1/DP networks) are significantly higher than cost savings for single-homing SHRs (compared to SH/1:1/DP networks). This is because the design cost for dual-homing hubbing with diverse protection networks is much higher than that for single-homing hubbing with diverse protection networks. The number of span pairs required for the dual-homing network architecture is almost twice that required for the single-homing architecture.

This study also suggests that the wider the area, the more cost savings the SHR can provide (compared to its hubbing counterpart), as long as a single SHR can support the demand in this area. When the area demand requires multiple SHRs, use of multiple overlapped SHRs is preferable to multiple overlaid SHRs. Two SHRs are defined as *overlaid* if they have the same set of nodes and fiber routes. Two SHRs are defined as *overlapped* if they are not overlaid and have a non-empty common set of nodes and fiber routes.

We also see a better hub survivability for SHRs should the hub building fail. For each candidate area (an intra-cluster subnetwork or the inter-hub subnetwork) in the SH/1:1/DP network, the hub survivability is always zero because all demands from COs (or hubs) in the area pass through or terminate at the hub (or gateway [2]) in the single-homing architecture. Compared to 1:1/DP networks, the gain of hub survivability for dual-homing SHRs is much higher than for single-homing SHRs (41.9 to 54.9 percent compared to 5 to 15 percent in Table 4-6). Thus, the dual-homing SHR architecture is a good choice for maximizing the potential of cost and survivability advantages inherently associated with the SHR architecture.

Some other SONET ring application studies using BSHRs and/or USHRs have also been reported in Reference [28]. The results of these studies suggest that results discussed in Table 4-6 can also be applied to BSHRs. When comparing the USHR and BSHR/4 under the fiber-hubbed network environment, the BSHR/4 can connect more nodes in the ring than the USHR. This is because, for the same set of ring nodes, the total demand that can be carried by a USHR can also be carried by a BSHR/4. The BSHR/4 has more remaining capacity than the USHR, which can be used to connect other small, nearby COs into the ring boundary. Thus, the relative economics between the BSHR/4 and the USHR really depend on the per-node average cost. Given a fiber-hubbed network, it is expected that the number of rings built using the BSHR/4 architecture would be fewer than the number built using the USHR architecture, but with a larger ring size (in terms of the number of ring nodes) for the BSHR/4 compared to the USHR. Thus, determining whether to use the BSHR/4 or the USHR really depends on the total capital savings and the possible operations costs, as well as whether the number of USHRs is significantly higher than the number of BSHR/4s in the fiber-hubbed network.

For areas where the STS-1 demand distribution is highly unbalanced (i.e., only a few demand pairs have significantly higher demand requirements than other demand pairs), the point-to-point systems may be used to carry demands to a level that can be economically carried by the ring and/or hubbing networks. This engineering rule has been supported by a case study reported in Reference [29]. Figure 4.28 shows a result of network costs that is the function of the number of point-to-point systems, assuming the percent fill is 75 percent [29]. The model network used in this case study is part of the metropolitan-area LATA network, which consists of nine COs (one of which is the hub). There are 16 demand pairs with a total of 82 STS-1s in a five-year planning period. Among these demand requirements, one demand pair has a significantly higher demand requirement than other demand pairs. As shown in Figure 4.28, the design using two point-to-point systems in addition to the ring may have a 10-percent cost saving compared to the design without using point-to-point systems.

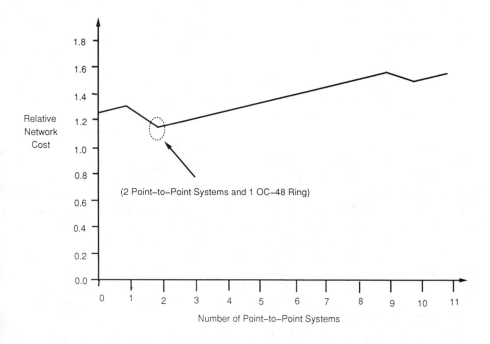

Figure 4.28. Advantage of using point-to-point systems in a demand-unbalanced area.

4.5.3 Interconnection of SONET SHRs

In some applications, geographical limitations or the need for extra bandwidth dictates a network solution to the survivability problem that requires multiple, interconnected rings. When multiple rings are connected together, it may be desirable (from survivability and control simplicity points of view) for the control mechanism of different rings to be independent of each other. Figure 4.29 shows some possible ring interconnection configurations for single-access and dual-access applications. For the single-access application, two rings are interconnected via an access node, which may be a hub in the fiber-hubbed network. The simplest nodal configuration for this interconnection access node is to use two ADMs, one for each ring. Connection is made between two SONET ADMs on the low-speed side. The two USHR/P rings,

185

each having different capacities ($OC-N_1$ and $OC-N_2$), communicate via an OC-M link. The symmetry and, thus, independence of the two rings is obvious from Figure 4.29(a). The second approach is to use a *Broadband DCS* (B-DCS) to interconnect inter-ring demands that are dropped from ADMs, as depicted in Figure 4.29(a-ii). The benefit of this approach is the flexibility of adding and removing rings. Another alternative nodal configuration is to use a DCS (W-DCS or B-DCS) to interconnect multiple rings [see Figure 4.29(a-iii)]. Compared to the previous two configurations, this configuration may reduce ADM costs but imposes more delay for normal services and for emergency restoration. Also, DCS failure in the last configuration may disable the multiple rings that are connected via this DCS (i.e., the multiple rings that are interconnected using the last configuration are not independent of each other).

For the dual-access application, the ring interconnection needs two common access points as depicted in Figure 4.29(b). The possible nodal configurations for each access node are similar to those described for the single-access application. Like the single-access application, connections are made through the access node on the low-speed rate, which is primarily on the path level. In general, the dual-access interconnection is more easily implemented using the path protection switching scheme rather than the line protection switching scheme. For SHRs using line protection switching, individual rings are restored on the line level but signals are passed from one ring to another at the path level. Thus, a mixture of line protection switching and path protection switching is needed for rings using line protection switching if they need to be interconnected and if the dual-access application is required. The above configurations for the interconnection of multiple rings are general structures that can be used in the loop as well as in the interoffice environment.

The interworking schemes for interconnecting two undirectional path protection switching rings and two rings with at least one bidirectional line protection switching ring have been reported in Reference [14] and References [30,31], respectively.

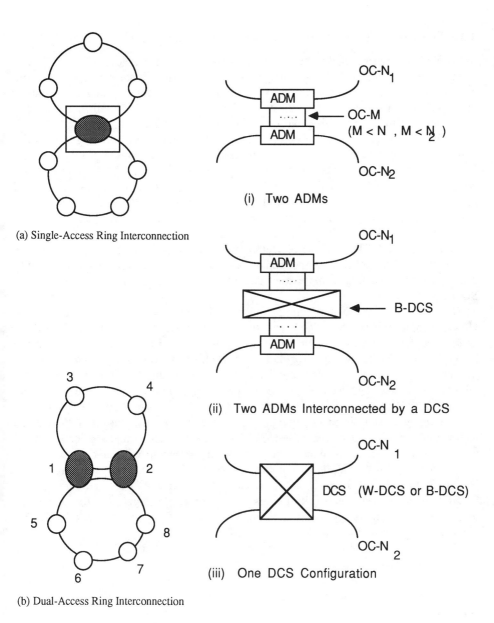

(a) Single-Access Ring Interconnection

(i) Two ADMs

(ii) Two ADMs Interconnected by a DCS

(iii) One DCS Configuration

(b) Dual-Access Ring Interconnection

Figure 4.29. Possible interconnection configurations of SHRs.

4.5.4 Potential Role for OC-192 Technology in SONET Interoffice Networks

SONET SHRs are expected to be deployed in LATA interoffice networks in 1991 due to their cost and survivability advantages over other conventional survivable network architectures. The current OC-48 (2.5-Gbps) ring capacity limit may be insufficient for metropolitan interoffice networking applications. A high interoffice demand requirement with a limited ring capacity (up to OC-48) situation has resulted in small SONET ring sizes (about five to eight nodes). This relatively small ring size means that a LATA interoffice fiber network with rings would be composed of many smaller SONET rings interconnected at hub buildings using either the back-to-back ADM configuration or in conjunction with B-DCSs. A large interconnection network of many SONET rings could result in higher operation and maintenance costs. To overcome such an expensive scenario, using ultra high-speed ADMs such as OC-192 (i.e., 9.6 Gbps) seems a natural approach to maximize potential cost and survivability advantages of SONET rings by connecting more office buildings on a ring. This may reduce the number of interconnections needed at hub buildings and also simplify operations and planning complexities due to a smaller interconnection network of SONET rings.

The OC-192 fiber transmission rate has been demonstrated in research laboratories. However, unless these ultra high-speed systems (such as OC-192 ADMs) prove to be economical for applications, their future is very uncertain. Bellcore recently conducted an application feasibility study for these ultra high-speed ADMs in LATA interoffice networks. The purposes of this application feasibility study were to determine (1) whether using OC-192 technology could lower survivable SONET network design costs compared to using today's line rates of up to OC-48, and if so, (2) whether the cost saving was significant enough to justify development of this ultra high-speed technology for ring applications. Bellcore also wanted to determine what architectural changes should be made to take advantage of this ultra high-speed technology.

To analyze this technology application feasibility, prototype software called STRATEGIC OPTIONS (which will be discussed in Chapter 7) was used to design the SONET interconnection network of rings and to compare design costs for using both today's OC-48 technology and the future OC-192 technology. The results of this application feasibility study are reported in Reference [32].

The network costs considered in this study include costs of SONET ADMs, fiber material, fiber placement, and SONET regenerators. The costs of ADMs of up to OC-48 were provided by an industrial source, and the OC-192 ADM cost was estimated from Bellcore subject matter experts. The model network is a metropolitan LATA network with a total of 53 nodes (COs). The total number of links is 84, where each link is a potential place for fiber route deployment. The total DS1 demand over a five-year planning period is about 6300 DS1s.

The study results suggest that SONET technology with the OC-192 rate provides significant cost savings for SONET interoffice network applications when compared to existing SONET OC-48 ADM technology. Additional savings over the OC-48 technology are up to 24 to 27 percent for the considered case study. The results also suggest that the SONET network clustering strategy should allow for a larger cluster size to take advantage of ultra high-speed ring capacity.

4.6 MULTI-WAVELENGTH PASSIVE INTEROFFICE SHR ARCHITECTURE

The SHR architectures discussed in previous sections are to use high-speed electronics (e.g., SONET OC-48 ADM) to process signals in the area covered by the ring. Alternatively, the demands in that area can be also carried by a *multi-wavelength passive SHR network* (SHR/WDM). The function of the SHR/WDM is essentially the same as the 1:1/DP network with a ring topology [see Figure 4.1(a)], but the multi-wavelength SHR architecture requires fewer fibers and regenerators than the 1:1/DP architecture. One optical amplifier can amplify multiple wavelengths on a fiber simultaneously, whereas one regenerator is needed for each fiber pair in a 1:1/DP network.

4.6.1 Multi-Wavelength Passive SHR Architecture (SHR/WDM)

Figure 4.30(a) illustrates a conceptual example of an SHR/WDM with four offices having a fully interconnected "logical" mesh network [33]. This figure represents an extreme case for the SHR/WDM architecture, which dedicates separate wavelengths to each pair of offices. The SHR/WDM is essentially a passive optical add-drop ring architecture using WDM and filtering technology, and lower speed electronics (compared to SONET high-speed electronics) to terminate and process local optical signals. The self-healing function of the SHR/WDM is similar to that of the USHR/P described in Section 4.2.4 (see Figure 4.11).

In the SHR/WDM [Figure 4.30(a)], each office is equipped with a pair of *Multi-Wavelength Filters* (MWFs), a pair of WDMs terminating a pair of fibers, and three lower-speed terminal multiplexers with one dedicated to each other office. Figure 4.30(b) shows a more detailed nodal configuration and associated optical signal add-drop processing for Office B.

189

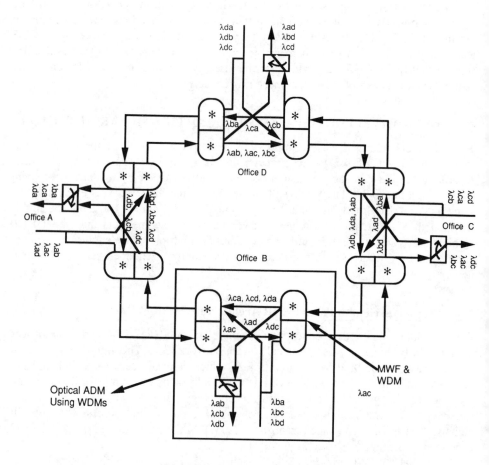

(a) Wavelength Routing in SHR/WDM

Figure 4.30. A multi-wavelength passive SHR architecture.

190

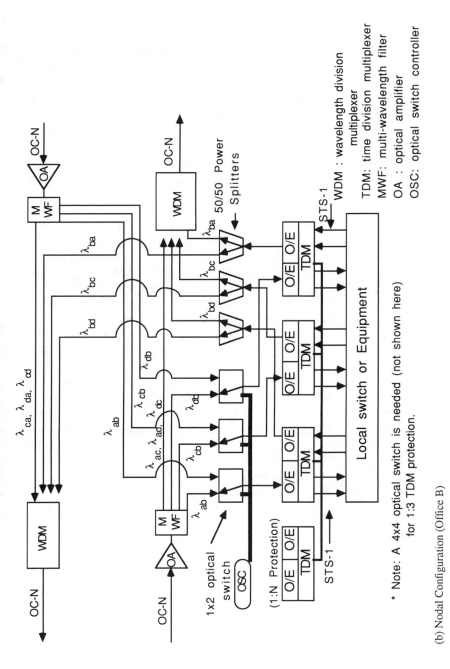

(b) Nodal Configuration (Office B)

Figure 4.30. (Continued)

191

To reduce the required number of wavelengths for the SHR/WDM, we use a fiber pair (one for working signals, another for protection signals), rather than a single fiber. As will be discussed later, the number of available wavelengths is a key factor that limits the number of offices that can be connected by a SHR/WDM.

Before transmission, local switch output is groomed according to its destination, and all signals destined for a given node are time-multiplexed to a single, higher bit rate data stream. These high-speed streams modulate lasers having unique wavelengths corresponding to the destination nodes. The "add" optical signal is duplicated using a 50/50 PS. One optical signal is for the working ring, and the other is for the protection ring. The "add" optical signal is wavelength-multiplexed with the other data streams (which do not drop in this node), and the resulting data stream is transmitted over a fiber to the next office of the working or protection ring. For the receiving end of Office B [see Figure 4.30(b)], an MWF is used to perform the optical signal "drop and continue" function, which drops a set of wavelengths destined to this office [e.g., λ_{ab}, λ_{cb}, λ_{db} in office B of Figure 4.30(b)]. The other portion of the data stream (i.e., λ_{ac}, λ_{ad}, λ_{dc}) goes directly to the outgoing WDM and is multiplexed with data streams added from this office [e.g., λ_{ba}, λ_{bc}, λ_{bd} in Office B of Figure 4.30(b)]. Normally, a 1×2 optical switch (for each receiving terminal) accepts dropped signals from the working ring and then switches to accept signals from the protection ring during working ring failures. The dropped wavelength signal is then detected and time-demultiplexed so the input to the local switch is the original, lower bit rate STS-1 signals.

Optical amplifiers may be needed to amplify optical signals not dropped in this office if the ring length exceeds a certain distance threshold. A total of n-1 terminal multiplexers are needed for each office to accept dropped optical signals from the SHR, where n is the number of offices in the ring. An additional wavelength and terminal multiplexer is needed in this architecture to provide 1:N protection for electronic terminals in each office. Thus, the SHR/WDM provides 1:1 protection for a single fiber cable cut and 1:N protection for lower-speed terminating multiplexer failures.

4.6.2 Passive Optical Technologies for Implementations

Five optical components required for the SHR/WDM implementation are WDMs, MWFs, optical amplifiers, optical PSs, and optical switches. In the SHR/WDM, the wavelength division multiplexers multiplex adding wavelengths with transit wavelengths at each CO. The MWFs perform an optical "drop-continue" function. The optical amplifiers provide optical signal continuity for signal transport, and the PSs and optical switches are used for the network self-healing function.

Two major system parameters that will impact the system design of the SHR/WDM are (1) the number of channels that can be supported, and (2) the power budget.

192

However, the use of broadband optical amplifiers to simultaneously amplify all optical wavelengths can alleviate, to a large degree, the power budget issue. More discussion on optical amplifiers can be found in Reference [22] or Section 4.4.2. Thus, under the described ring architecture, the size of the network is primarily limited by the number of wavelengths available.

Unlike the hubbing (or star) architecture, where the wavelength can be reused so that only N wavelengths are required for N offices [34], the number of wavelengths needed for a SHR/WDM that fully interconnects N nodes is $N(N-1)/2$ for two-way transmission [35]. Based on the state-of-the-art, grating-based WDM technology, the WDM can accommodate as many as 50 wavelengths[15] [36]. Thus, the SHR/WDM ring can support a maximum number of 10 offices, which is the typical size of one office cluster in current fiber-hubbed LATA networks. Generic requirements for WDM components can be found in Bellcore Technical Reference TR-TSY-000901 [37].

The maximum number of wavelengths, N, that can be supported by wavelength filters is determined by the tunable filter technology. It is the ratio of the total tuning range to the minimum channel spacing required to guarantee a minimum crosstalk degradation. For example, 1-GHz filter bandwidth, 5-GHz channel spacing (which may guarantee <0.5-dB crosstalk), and 200-GHz tuning range (equivalent to about 1.5 nm around the wavelength of 1.5 μm, and 1.1 nm around 1.3 μm) imply a system capacity of about 40 channels. Several types of tunable wavelength filters available today are summarized in Reference [38]. Among them, the acousto-optical tunable fiber, which is controlled electronically and can be switched fairly quickly (on the order of μs), and can simultaneously and independently select multiple optical waves from a wavelength-multiplexed optical stream [39]. A typical insertion loss of the acousto-optical tunable fiber is about 5 dB. For an N-node ring, the filter must be able to drop N-1 wavelengths from $N(N-1)/2$ wavelengths. For example, for a 10-node SHR/WDM, the filter must be able to drop nine wavelengths simultaneously from a 50-wavelength-multiplexed optical stream. With future technology acoustic waveguides fabricated on optical waveguides [40], several tens of wavelengths can be selected simultaneously from a wavelength-multiplexed optical stream of about 100 wavelengths.

---

15. Commercially available WDMs may support up to 32 wavelengths for several hundred megabit-per-second systems.

4.6.3 Comparison between SONET SHR and Multi-Wavelength Passive SHR

Table 4-7 lists a general comparison of system characteristics between the SONET SHR and the SHR/WDM. For convenience, a SONET SHR using SONET ADMs is denoted by SHR/ADM. In Table 4-7, the network throughput is assumed to be the same for both the SHR/ADM and the SHR/WDM.

Table 4-7. Comparison between SONET SHR/ADM and SHR/WDM

Attributes	SHR/ADM	SHR/WDM
Signal add-drop unit	STS-1/VT	optical channel up to fiber line
Electronic equipment	ADM	OLTM*
Number of terminals	fewer	more
Transceiver speed+	higher	lower
Mixed transceiver speeds	no	yes
Upgradability	difficult/ expensive	easy
Ring size limitation	ADM capacity	number of available wavelengths
E/O Conversion for pass-through signals	yes	no++
Average transport delay	longer	shorter

* OLTM: Optical line terminating multiplexer
\+ Assuming network throughput is the same
++ Assuming optical amplifiers are used

The major difference between SHR/ADMs and SHR/WDMs is the add-drop signal unit: SHR/ADMs add-drop the electrical STS-1 or VT, whereas SHR/WDMs add-drop local optical channels directly with the signal rate up to the fiber line rate. Compared to the SHR/ADM, the SHR/WDM needs more (but low speed and less expensive) electronics (i.e., OLTMs). Because the SHR/WDM is functionally similar to a 1:1/DP network, OLTMs dedicated to different office pairs (i.e., different wavelengths) may be different depending on the demand requirements for these office pairs. In contrast, the SHR/ADM must operate under a single rate for all ADMs on the ring. Thus, it is easier and less expensive to upgrade the SHR/WDM than the SHR/ADM. In an extreme scenario, the SHR/WDM can be upgraded to carry data of up to n×2.4 Gbps (where n is the number of available wavelengths for the SHR/WDM) for a fiber transmission rate of 2.4 Gbps without adding more fibers.

The primary limiting factor of the ring size for the SHR/WDM is the number of available wavelengths, rather than the electronics (i.e., ADM) capacity for the SHR/ADM. It is expected that the average transport delay for the SHR/WDM device is shorter than that for the SHR/ADM because signals do not need to be processed electrically until they are dropped at the local office.

The relative economic merit between the SHR/ADM and the SHR/WDM depends on the number of nodes on the ring, the demand pattern and requirements, and the relative costs of high-speed ADMs to low-speed OLTMs. For example, assume the demand pattern is the mesh demand pattern (see Section 4.3.1) with three STS-1s required for each node pair, the ring has four nodes, and the cost ratio of an OC-24 ADM to an OC-3 OLTM is 6.6.[16] The SHR/ADM needs four OC-24 ADMs, assuming the SHR/ADM is a undirectional ring, and the SHR/WDM needs 16 OC-3 OLTMs (three working and one protection OC-3 OLTMs for each office). Thus, the SHR/WDM is less expensive than the SHR/ADM (16 versus 26.4) in terms of equipment costs.

On the other hand, let us consider the case where the demand requirement for each node pair is changed from three STS-1s to four STS-1s, and the number of ring nodes is changed from four to five. Assume that only OC-3, OC-12, OC-24, and OC-48 line rates are available for use. In this case, the SHR/ADM needs four OC-48 ADMs, whereas the SHR/WDM needs 25 OC-12 ADMs (four working and one protection OC-12 OLTMs are needed for each office). Also assume that the cost ratio of an OC-48 ADM to an OC-12 OLTM is 3.3. Thus, the SHR/WDM is more expensive than the SHR/ADM (25 versus 13.2) in terms of equipment costs.

4.7 BROADBAND BROADCAST SONET SHR ARCHITECTURE

Figure 4.31(a) depicts a broadband *broadcast* SHR using SONET ADMs [denoted by SHR/ADM(B)], which serves COs A through G [33]. CO-H also receives the broadcast signals from CO-E via a 1:1 APS fiber-optic system. *Broadcast provider 1* (BP1) transmits its broadcast signals into the network via CO-A. Broadcast providers 2 and 3 transmit their signals into the network via CO-E. As many as sixteen 135-Mbps HDTV channels could be served by each SHR/ADM(B) operating at the OC-48 rate. Broadband broadcast signals traverse the network in the clockwise (solid line) and

---

16. Assume that the cost ratio of an OC-24 ADM to an OC-3 ADM is about 5.3 and the cost ratio of an OC-3 OLTM to an OC-3 ADM is 0.8.

195

counterclockwise (dashed line) directions of the two fiber-optic rings that comprise the SHR/ADM(B). In Figure 4.31(b), we depict an STS-3c signal from BP1 being added to channel 1 of an SHR/ADM(B) with a line rate of OC-12. The returning signal is then delivered (usually via port 1) to BP1, which can thereby monitor its quality. In this case, port 1 (P1) is operating in the drop-add mode, and ports 2, 3, and 4 are operating in the drop-continue mode. Broadband broadcast signals carried on STS-3c channels served by these ports are dropped at CO-A, but are also retransmitted to the next CO on the SHR/ADM(B). In near real time, ports 2, 3, and 4 can be reconfigured to serve broadcast providers at CO-A; similarly, port 1 can be converted to the drop-continue mode when needed.

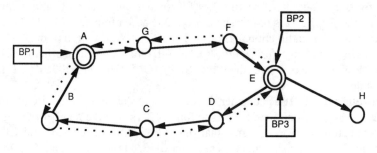

(a) General Configuration

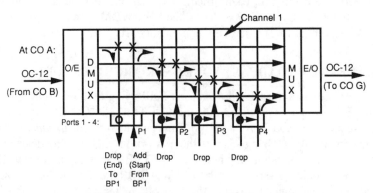

* Depicts the ADM transmitting in the clockwise direction at CO A. Port 1 is operating in the drop/add mode. Ports 2, 3 and 4 are operating in the drop/continue mode.

(b) Example of ADM of SHR (Broadcast) at Office A

Figure 4.31.   Broadband broadcast SHR.

196

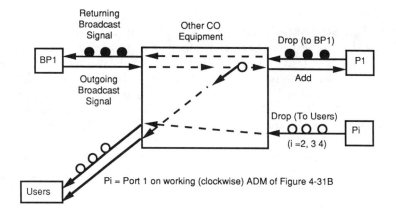

(c) Signal Processing at CO A

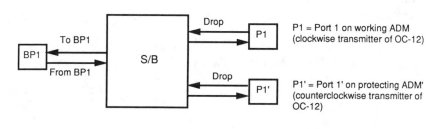

(d) Selector/Broadcaster (S/B)

Figure 4.31. (Continued)

In Figure 4.31(c), we further depict the flow of broadband broadcast signals in the vicinity of CO-A. Note that the signal from BP1 is distributed to local users at the same time it is sent to P1 of the ADM in Figure 4.31(b). This is done because a cable cut on the SHR/ADM(B) can prevent BP1's signal from returning to CO-A. Drop-continue signals from ports 2, 3, and 4 are also distributed to local broadcast users.

In Figures 4.31(b) and 4.31(c), we depict only the ADM and ports of the clockwise direction of the SHR/ADM(B). In Figure 4.31(d), we depict port 1 from the clockwise ADM and its corresponding port 1' from the counterclockwise ADM. The *Selector/Broadcaster* (S/B) function in Figure 4.31(d) is also an integral part of the ADM at CO-A (and of all other ADMs on the SHR). It receives the broadcast signal from BP1 and "broadcasts" it to ports 1 and 1', and also receives the returning signals from ports 1 and 1'. It normally selects the port 1 signal, but selects the port 1' signal when the port 1 signal is bad, e.g., due to an ADM failure affecting the clockwise portion of the SHR/ADM(B).

Figure 4.32 depicts a possible generic equipment model for an ADM with a drop-continue feature. This model is similar to the one depicted in Figure 4.26 except that local signals are duplicated before they are dropped, and these duplicated signals are transmitted to the next node on the ring.

Consider an alternative broadcast structure where broadcast providers transmit their broadcast signals to a single CO in an area. From that CO, the signals are broadcast in a tree-like architecture to other COs as depicted in Figure 4.33. The advantages of the SHR/ADM(B) over the single-node broadcast structure include

1. Higher survivability in the face of fiber cable cuts, ADM failures, and CO failures
2. More economical access to the broadcast network for broadcast providers (because each ADM provides a potential broadcast access point to broadcast service providers)
3. More reassurance (via a returned signal) to broadcast providers that their video programs are reaching customers in the area.

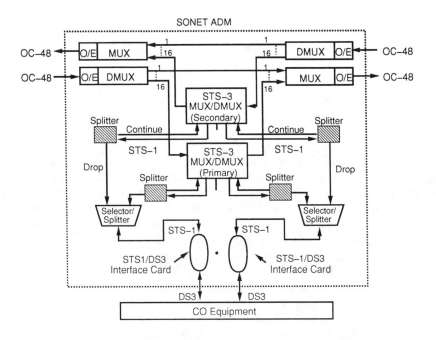

Figure 4.32. A possible equipment model for ADM with drop-continue feature.

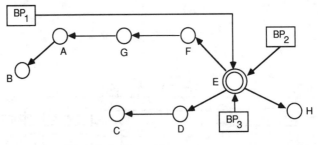

BP: broadcast provider

Figure 4.33. Broadband broadcast alternative.

Possible disadvantages of the SHR/ADM(B) are a somewhat higher capital cost and more complex control of broadcast programming changes. In particular, more fiber is needed to form the SHR/ADM(B) than the broadcast network of Figure 4.33. Also, changing the BP using a particular STS-3c channel may require changes at two ADMs' on the SHR, whereas the single broadcast source alternative requires changes at only one location (CO-E). However, the advantages cited previously far outweigh these possible disadvantages. In addition, many customers will use video programming simultaneously. A failure that disrupts video programs could therefore generate many trouble reports. The SHR/ADM(B), unlike the tree-like broadcast structure, will prevent interruption of video programming due to interoffice cable cuts,[17] and so on. Hence, fewer trouble reports will be received, the expense of processing these trouble reports will be saved, and customer satisfaction will be higher.

4.8 RADIO SYSTEM APPLICATIONS USING SONET SHR ARCHITECTURES

SONET SHRs can be implemented only if physical fiber connectivity exists for a ring topology. However, this may not be the case for small metropolitan or suburban areas that usually have a tree-like network topology. The creation of the second route for

---

17. Cable cuts in the distribution plant (connecting broadcast users to their local CO) will interrupt video programming to affected customers. The broadcast signal from a local BP may also be prevented from reaching the SHR/ADM(B).

diversity protection may be too expensive to justify the investment in small metropolitan and suburban areas. Also, in some areas, it may not be possible to build fiber-ring connectivity because of geographical limitations, such as rivers, mountains, reservoirs, and so on. On the other hand, many digital radio systems have been used in these areas either as normal transport systems or as backup systems. Thus, using radio systems to supplement insufficient fiber connectivity becomes a natural solution for providing service assurance at affordable costs in areas where the fiber connectivity may be too expensive or impossible to build. For convenience, the SONET SHR using radio systems as part of the ring architecture is referred to as a *SONET SHR/Radio*.

4.8.1 SONET/Radio SHR Architectures

Figure 4.34 depicts a generic OC-12 SHR/Radio architecture with three fiber spans and one radio span. The radio span helps provide the ring connectivity.[18]

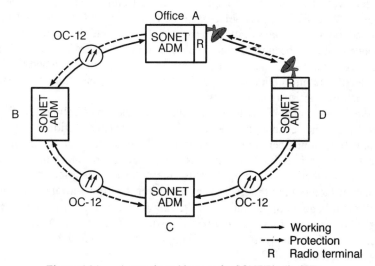

Figure 4.34. A generic architecture for SONET/radio SHR.

---

18. The line protection switching scheme protects the ring at the line level, whereas the path protection switching scheme protects the ring at the end-to-end path level (i.e., STS-1 or VT channels). See Section 4.2 or Reference [5] for more details.

Like the fiber-based SONET SHR, the SONET SHR/Radio can be implemented using either the unidirectional or the bidirectional ring architecture, depending on the traffic pattern and demand requirements. For the unidirectional SHR/Radio architecture, the SONET radio span acts like an optical link and is transparent to the ring operation, as is the fiber-based USHR. In this USHR implementation, the path protection control scheme may be more desirable than the line protection control scheme not only because path protection switching is simpler than line protection switching (see Section 4.3.1), but also because path protection switching makes the ring independent of the radio span. In the normal situation, signals added to the ring are duplicated and sent to the destination via two rings with opposite direction. When channels pass through the radio span, the radio terminal converts the STS-1 format to a high-rate format for radio transmission. The radio signal is then converted back to the STS-1 format at the receiving radio terminal of the radio span, and is processed by the adjacent SONET ADM at the same CO. At the receiving end, the SONET ADM always receives two identical signals with different delays and performance parameters such as BER. A 2:1 selector in the receiving ADM always selects the best signal based on protection switching criteria. Thus, even if the radio span suffers performance problems because of a hostile environment (e.g., rain), the receiving ADM can always select a good signal from the fiber span of the ring, provided the fiber route is not under the failure condition. If a fiber cable is cut, the radio span can provide backup capabilities of sufficient quality.

For the bidirectional SHR/Radio architecture, the self-healing function can be provided using either channel redundancy (i.e., two-fiber bidirectional ring) or facility redundancy (i.e., four-fiber bidirectional ring). In this architecture, it may be more desirable to always use the radio span as the protection resource in order to reduce frequency for protection switching, because the radio span is affected more easily by environmental factors than the fiber span. However, this route arrangement may result in an inefficient bandwidth allocation for the ring (i.e., capacities of some fiber links on the ring will exhaust more quickly). The tradeoff between efficient bandwidth allocation and minimal protection switching frequency should be considered when planning a bidirectional SHR/Radio.

4.8.2 Radio Span Design for SONET SHR/Radio

Figure 4.35 depicts the functional diagram for a radio span in the SONET SHR/Radio. The SONET radio span is composed of two SONET-compatible radio terminals that transmit microwave signals bidirectionally. If the span length exceeds a critical spacing threshold, the radio span also includes SONET digital radio regenerators.

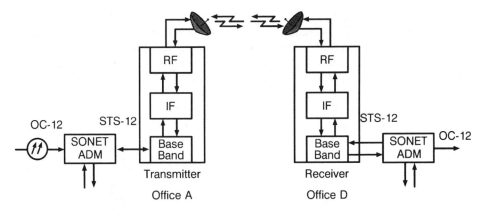

Figure 4.35. Functional diagram for a radio span in SONET SHR/radio.

The SONET-compatible radio terminal is *Line Terminating Equipment* (LTE) that includes a SONET-compatible radio modem and a pair of transmitters/receivers. It interfaces to SONET ADM equipment at STS-N (in this case, STS-12). The SONET ADM, which is co-located with the SONET radio terminal, is required to access section and line overhead bytes. The radio terminal can also interface to an asynchronous transmission rate using a bit rate adapter. More details on SONET radio systems can be found in Reference [41].

To form a SONET SHR/Radio architecture, a coordinated scheme is needed to incorporate different technology characteristics into a single, universal technology and to ensure that total ring network performance is not degraded. This coordination includes not only technology differences, but also SONET formats that both technologies can use. Optical systems may carry very high-speed data rates (at least 2.4 Gbps) with excellent quality (BER $\leq 10^{-10}$) and free of interference from rain, electrical noise, and so forth. In contrast, radio systems that may be less expensive than fiber systems have limited capacity for transmitting data up to several hundred megabits per second (e.g., OC-12) due to spectrum allocation and bandwidth limitations. Radio systems may encounter interference from other electromagnetic sources, such as terrain scattering. Furthermore, radio wave propagation at 11 GHz and above is also limited by rain attenuation.

To reach a radio system availability objective, space diversity, frequency diversity, and angle diversity are often used to reduce the outage time caused by multipath fading [42]. However, the design of the SONET radio span in the SONET SHR/Radio could differ from the present point-to-point protected radio system. The self-healing design of the SHR/Radio may eliminate the protection radio system in the radio span because a combination of the working radio system (in the radio span) and the diversely routed fiber systems (used as the protection resource of the working radio system) may satisfy the stringent availability requirement for the point-to-point radio span [43].

Figure 4.36 depicts a relative comparison between a normal protected radio span and the radio span in the SONET SHR/Radio architecture. As shown in Figure 4.36, the traditional protected radio span using space diversity protection must duplicate radio terminals at two ends. However, this radio terminal duplication may not be necessary for a radio span within the SHR/Radio, as explained below.

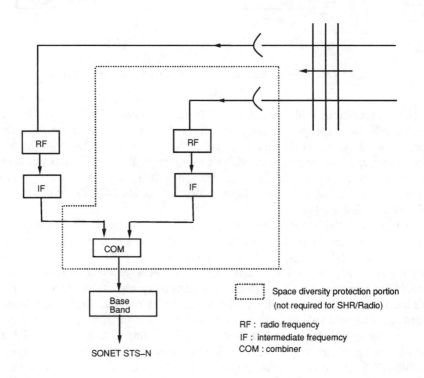

Figure 4.36. Comparison of a traditional protected radio span and a radio span within SHR/radio.

In general, the availability requirement for intraLATA systems is 99.98 percent [44] for a 250-mile, two-way channel. This requirement may be applied to both fiber-optic and radio systems. (Regarding 0.02 percent unavailability, for example, 75 percent is allocated to transmission media failure, and 25 percent is allocated to

204

equipment failure [44].) For illustration, assume that the probability of the network using all the fiber path or the fiber/radio path is the same. Furthermore, assume that failures in radio systems and failures in fiber-optic systems are independent. Should we allocate 0.015 percent transmission unavailability to an SHR system when each channel can have at most 1.2 percent[19] unavailability (if the reliability of both paths is equal and independent)? The 1.2 percent unavailability may exceed the forecast for fiber-optic transmission systems because these systems have better transmission BER performance[20] than radio systems. Therefore, significant portions of the 1.2 percent unavailability can be allocated to the radio systems, which consequently allows them to meet the availability objective without further protection techniques. Because radio systems are very sensitive to multipath fading, the loose availability requirement may simplify the radio system design and lower the radio span cost in the SONET SHR/Radio. In other words, the space diversity or frequency diversity for radio systems may not have to be installed for the radio-protected fiber systems. In addition, some of the unused fiber unavailability in one route may be allocated to the fiber/radio systems in another route, further reducing their costs, while still achieving the overall availability of the ring network.

4.9 SUMMARY AND REMARKS

SONET technology and associated flexible, high-speed ADM technology have made SONET SHR implementation practical and economical for interoffice network applications. SONET OC-48 SHRs are scheduled to be commercially available and deployed during 1992-1993. The commonly used point-to-point systems are very efficient and cost-effective when the two points involve a demand requirement close to the utilized system's capacity. In contrast, SONET SHRs offer cost-effective solutions by sharing equipment and fiber facilities among several offices. In particular, the SONET SHR offers cost-effective and survivable services to the dual-access (or dual-homing) survivability application, which cannot be implemented economically using traditional approaches.

---

19. The square root of 0.015 percent is 1.2 percent. This is based on the assumption that the failure of a fiber-optic system (event A) and the failure of a SONET radio system (event B) are two independent events (i.e., if A and B are independent, then P(A,B)=P(A)P(B), where P is the probability of an event).

20. It should be noted that most of the transmission unavailability of a fiber-optic system is associated with catastrophic failures, such as an accidental cable cut. Such outages persist for much longer periods of time (hours) than outages caused by radio propagation anomalies (seconds).

Two types of SONET ring architectures have been discussed according to the channel routing in the normal circumstances: USHR and BSHR. In general, the USHR is cost-effective in areas where the STS-1 demand is highly centralized; the BSHR is cost-effective in areas where the STS-1 demand is more decentralized. Each type of ring can use line switching or path switching as its protection switching scheme. The line protection switching ring uses the SONET line-level signal to trigger protection switching and restores all channels on the failed facility at the same time. In contrast, the path protection switching ring uses the SONET path-layer signal to trigger protection switching and restores end-to-end STS-1 or VT channels as fast as the point-to-point SONET system. The path protection switching ring uses the existing SONET signaling protocol. The line protection switching ring needs a different set of SONET standards [45] because present standards are primarily designed for point-to-point systems using line protection switching.

New technologies that have been developed recently are expected to play a key role in SONET applications. These new technologies include optical switching, optical amplification, wavelength division multiplexing, and ultra high-speed transmission (e.g., OC-192 systems). This chapter discussed using these new technologies to design cost-effective SONET SHR network applications.

Optical switches and optical amplifiers have been used to implement a passive protection ring for a four-fiber BSHR. The resulting architecture reduces significant protection costs for the four-fiber BSHR implementation and retains the advantages of conventional four-fiber BSHR implementations, such as the simple OS requirement and highest-available capacity among SONET ring alternatives. The passive protection ring is transparent to line rates, fiber wavelength, and signaling formats. The passive protected four-fiber ring can evolve from today's point-to-point systems, conventional four-fiber BSHR architectures, or from any two-fiber ring architectures (including the two-fiber BSHR and the USHR) with a minimum capital investment and twice the capacity of any two-fiber ring.

Ultra high-speed transmission systems, such as OC-192 systems, have been shown not only to maximize potential cost savings for SONET ring technology, but also to simplify the interconnection structure of SONET rings, thus reducing operations costs. As an alternative to the high-speed SONET ADM rings, multi-wavelength rings using low-speed SONET terminal multiplexers may provide an economical solution to applications where the 1:1 point-to-point diverse protection strategy may be best applied.

In addition to interoffice networking applications, broadband broadcast service is another valuable application of the SONET SHR architecture. The broadband broadcast SHR architecture would provide highly survivable interoffice facilities to serve HDTV and other high bandwidth broadcast services.

SONET SHRs can be implemented only if physical fiber connectivity exists for a ring topology. However, this may not be the case for small metropolitan or suburban areas where creating the second route for diversity protection may be impossible, due to geographical limitations (e.g., a river) or too expensive to justify. Using SONET radio systems to supplement insufficient fiber-ring connectivity is a natural solution for providing service assurance at affordable costs for these areas. Due to the ring architecture's self-healing design, the radio span design within the SONET ring may be less expensive than the traditional protected point-to-point radio span design. Thus, the protection radio system may not be needed in the radio span of the SONET ring architecture.

Based on our discussion of engineering concerns for SHRs, we conclude that more global planning methods are needed to determine the optimal mix of SHR and traditional fiber-hubbed architectures in the interoffice networks of the future. Chapter 7 will discuss one such SONET planning method.

Finally, due to the service survivability requirement and different restoration performance for different restoration schemes, the target SONET network architecture is likely to contain a variety of survivable SONET network architectures because each network architecture has an inherent application area. These survivable network architectures include hubbing/diverse routing, point-to-point/diverse routing, SHRs, and dynamically reconfigurable network architectures. The hubbing (or point-to-point)/diverse routing architecture and rings determine physical survivable network topologies. The reconfigurable network architectures are built on a topology based on a mix of hubbing, point-to-point, and ring architectures. The reconfigurable network architectures include SONET/STM, using SONET DCSs and SONET/ATM using ATM/DCSs or a hybrid approach, which we will discuss in Chapters 5 and 9, respectively.

REFERENCES

[1] Langmeyer, P., "Considerations for SONET Deployment," *Proceedings of National Fiber Optics Engineers Conference (NFOEC'90),* April 1990, pp. 1.2.1-1.2.9.

[2] Wu, T-H., Kolar, D. J., and Cardwell, R. H., "Survivable Network Architectures for Broadband Fiber Optic Networks: Model and Performance Comparisons," *IEEE Journal of Lightwave Technology,* Vol. 6., No. 11, November 1988, pp. 1698-1709.

[3] TR-TSY-000496, *SONET Add-Drop Multiplex Equipment (SONET ADM) Generic Requirements and Objectives,* Bellcore, Issue 2, September 1989.

[4] Wu, T-H., and Lau, R. C., "A Class of Self-Healing Ring Architectures for SONET Network Applications," *Proceedings of IEEE GLOBECOM'90,* December 1990, San Diego, CA, pp. 403.2.1-403.2.8.

[5] Sosnosky, J., and Wu, T-H., "Definitions of Ring Architectures," T1X1.5/90-179, September 1990.

[6] Hawker, I., et al., "Self Healing Fiber Optic Rings for SONET Networks," British Telecom Contribution to T1M1/T1X1 Ad Hoc Committee, October 1988.

[7] Hasegawa, S., et al., "Protection Switching in a SONET Ring Architecture," NEC Contribution to T1X1.5/89-053, April 12/July 20, 1989.

[8] AT&T Network Systems, "SONET Line Protection Switched Ring APS Protocol," T1X1.5/91-026, February 6, 1991.

[9] Baroni, J., et al. "Further Considerations for a SONET Line Switched Ring Protocol," AT&T Network Systems, T1X1.5/91-042, April 15, 1991.

[10] *American Standard for Telecommunications - Digital Hierarchy - Optical Interface Rates and Formats Specification,* T1.105/90-025R2, September 1990.

[11] Northern Telecom Inc., "SONET Rings: Proposal for K Byte Definition," T1X1.5/90-124(R1), 1990.

[12] Boehm, R., et al, "Overhead Usage for Protection Switching in a Self-Healing SONET Ring," Fujitsu Contribution to T1X1.5/90-066, April 20, 1990.

[13] Lau, R., "An Architecture for a SONET Self-Healing Ring," Bellcore Contribution to T1X1.5 Standards Project, May 10, 1989.

[14] TR-TSY-000496, *SONET Add-Drop Multiplex Equipment (SONET ADM) Generic Criteria: A Unidirectional, Dual-Fed, Path Protection Switched, Self-Healing Ring Implementation,* Bellcore, Issue 2, September 1989, Supplement 1, September 1991.

[15] Conlisk, J. K., "How Fragile Is Your Network?" *Telephony,* October 31, 1988, pp. 27-35.

[16] Alexander, A., "The Critical Umbilical," *Telephony,* May 1989, pp. 62-70.

[17] Howells, D. L., "High Capacity Light Wave Technology Comes to Age," *AT&T Technologies,* Vol. 3, No. 4, 1988.

[18] Sosnosky, J., Wu, T-H., and Alt, D. L., "A Study of Economics, Operations and Applications of SONET Self-Healing Ring Architectures," *Proceedings of IEEE GLOBECOM,* December 1991, pp. 57.3.1-57.3.7.

[19] Wu, T-H., and Way, W. I., "A Novel Passive Protected SONET Self-Healing Ring Architecture," *Proceedings of IEEE Military Communications Conferences (MILCOM'91),* November 1991, pp. 38.5.1-38.5.7.

[20] Wu, T-H., and Habiby, S., "Strategies and Technologies for Planning a Cost-Effective Survivable Fiber Network Architecture Using Optical Switches," *Journal of Lightwave Technology,* Vol. 8, No. 2, February 1990, pp. 152-159.

[21] Edinger, D., Duthie, P., and Prabhakara, G. R., "A New Answer to Fiber Protection," *Telephony,* April 9, 1990, pp. 53-55.

[22] Nakagawa, K., and Shimada, S., "Optical Amplifiers in Future Optical Communication Systems," *IEEE Lightwave Communication System,* Vol. 1, No. 4, November 1990, pp. 57-62.

[23] Goldstein, E.L., et al., "Use of Quasidistributed Optical Amplification in SONET Self-Healing Inter-exchange Networks," Technical Digest, Conference on Optical Fiber Communication (OFC), San Jose, CA, February 1992, pp. ThL2.

[24] Way, W., et al, "A High Gain Limiting Erbium-Doped Fiber Amplifier with Over 30 dB of Dynamic Ranges," OFC'91/ThM5, San Diego, CA, February 1991.

[25] Way, W. I., Wu, T-H., et al, "Applications of Optical Amplifiers in a Self-Healing Ring Network," *Conference Digests of European Conference on Optical Communications (ECOC'91)*, Paris, September 1991.

[26] Yan, S., To, M., and Oxner, S., "Bandwidth Management in a SONET Transport Network," *Proceedings of IEEE GLOBECOM'90*, San Diego, CA, December 1990, pp. 705B.2.1-705B.2.7.

[27] Wu, T-H., and Burrowes, M., "Feasibility Study of a High-Speed SONET Self-Healing Ring Architecture in Future Interoffice Fiber Networks," *IEEE Communications Magazine*, Vol. 28, No. 11, November 1990, pp. 33-42.

[28] Flanagan, T., "Fiber Network Survivability," *IEEE Communications Magazine*, Vol. 28, No. 6, June 1990, pp. 46-53.

[29] Wu, T-H., Cardwell, R. H., and Boyden, M., "A Multi-Period Architectural Selection and Optimum Capacity Allocation Model for Future SONET Interoffice Networks," *IEEE Transaction on Reliability*, Vol. 40, No. 4, October 1991, pp. 417-427.

[30] Kremer, W., "Ring Interworking with a Bidirectional Ring," AT&T Network Systems, T1X1.5/91-043, April 15, 1991.

[31] Copley, G., and Betts, M., "Further Considerations on 2-Fiber Bidirectional Line Switched OC-48 Rings," T1X1.5-019, February 1991.

[32] Wasem, O. J., Wu, T-H., and Cardwell, R. H., "OC-96/OC-192 Technology Impact on SONET Self-Healing Rings," Technical Digest, Conference on Optical Fiber Communication (OFC), San Jose, CA, February 1992, pp. ThL4.

[33] Wu, T-H., Kolar, D. J., and Cardwell, R. H., "High-Speed Self-Healing Ring Architectures for Future Interoffice Networks," IEEE GLOBECOM'89, November 1989, pp. 23.1.1-23.1.7.

[34] Wagner, S. S., and Kobrinski, H., "WDM Applications in Broadband Telecommunication Networks," *IEEE Communications Magazine*, March 1989, pp. 22-30.

[35] Hill, G. R., "A Wavelength Routing Approach to Optical Communications Networks," *Proceedings of IEEE INFOCOM*, April 1988, Section 4B.1.1-4B.1.9.

[36] Laude, J. P., and Lerner, J. M., "Wavelength Division Multiplexing/Demultiplexing (WDM) Using Diffraction Gratings," *Proceedings of SPIE*, 503, 1984, pp. 22-28.

[37] TR-TSY-000901, *Generic Requirements for WDM (Wavelength-Division-Multiplexing) Components*, Bellcore, Issue 1, August 1989.

[38] Kobrinski, H., and Cheung, K-W., "Wavelength-Tunable Optical Filters: Applications and Technologies," *IEEE Communications Magazine*, October 1989, pp. 53-63.

[39] Cheung, K. W., Smith, D. A., Baran, J. E., and Heffner, B. L., "Multiple Channel Operation of an Integrated Acousto-Optic Tunable Filter," *Electronics Letter,* Vol. 25, 1989, pp. 375-376.

[40] Hinkov, V. P., Opitz, R., and Sohler, W., "Collinear Acousto-Optical TM-TE Mode Conversion in Proton Exchanged Ti:LiNbO$_3$ Waveguide Structures," *IEEE Journal of Lightwave Technology*, Vol. LT-6, 1988, pp. 903-908.

[41] Shafi, M., Davey, L., and Smith, W., "The Impact of Synchronous Digital Hierarchy on Digital Microwave Radio: A View from Australia," *IEEE Communications Magazine*, May 1990, pp. 16-20.

[42] Lin, S. H., Lee, T. C., and Gardina, M. F., "Diversity Protection for Digital Radio - Summary of Ten-Year Experiments and Studies," *IEEE Communications Magazine,* Vol. 26, No. 2, February 1988, pp. 51-64.

[43] Lee, T. C., and Wu, T-H., "Feasibility of Deploying SONET Compatible Digital Radio Systems to a SONET Self-Healing Ring Architecture," National Radio and Engineers Radio Conference, Schaumurg, IL, May 1991.

[44] TR-NWT-000499, *Transport Systems Generic Requirements (TSGR): Common Requirements,* Bellcore, Issue 4, November 1991.

[45] TA-NWT-001230, *SONET Bidirectional Line Switched Ring Equipment Generic Criteria,* Bellcore, Issue 1, October 1991.

CHAPTER 5

Reconfigurable DCS Networks

5.1 OPPORTUNITIES AND TECHNOLOGIES FOR RECONFIGURABLE DCS NETWORKS

5.1.1 Physical and Logical Layer Protection

For a fiber facility network topology, network survivability can be provided using two network protection methods:[1] physical facility protection and logical channel protection. The physical facility protection method uses a dedicated standby fiber facility for protection. In contrast, the logical channel protection method reserves a portion of capacity within the working facilities for protection. As shown in Table 5-1, the tradeoffs between these two protection methods are required protection capacity, restoration time, and required system complexity.

Table 5-1. Physical Facility Protection and Logical Channel Protection

Attributes	Physical facility protection	Logical channel protection
Required spare capacity	more	less*
Restoration time	faster	slower
System complexity	simple	complex
Network size	smaller	larger**
Network reliability	higher	lower

* due to a higher degree of spare capacity sharing
** due to economies of scale for protection

---

1. These protection methods also work for other non-fiber networks.

211

Survivable network architectures that implement the physical facility protection method include SHRs and APS systems. Survivable network architectures that implement the logical channel protection method include BSHR/2 and reconfigurable DCS networks. The DCS is a type of switch that inputs high-rate digital signals and internally cross-connects low-level tributary channels.

Network designs based on physical facility and logical channel protection methods are sometimes referred to as *physical* and *logical layer network designs,* respectively. Figure 5.1 depicts examples of a physical layer network design [see Figure 5.1(b)] and a logical layer network design [see Figure 5.1(c)] for a given fiber facility topology [see Figure 5.1(a)]. In the physical layer network design depicted in Figure 5.1(b), SHRs and point-to-point spans with DP are interconnected by either back-to-back ADMs or intelligent DCSs. The logical layer network design shown in Figure 5.1(c) is composed of intelligent DCSs and controllers.

The basic differences between the physical and logical layer networks are network design and protection resources. The physical layer network design is a sectionalized design and uses separate fiber facilities and/or equipment as protection resources. In contrast, the logical layer network design is more global and uses spare capacity within the working facilities as protection resources. In general, the logical layer network (e.g., DCS network) uses spare capacity more efficiently than the physical network (e.g., ring) at the expense of increased control complexity. Section 5.4 discusses the tradeoffs between these two network designs in more detail.

5.1.2 DCS Functional Architecture

Figure 5.2 depicts an asynchronous DCS 3/1 functional diagram, which includes electrical signal terminations (i.e., DS3 terminations), a channel (DS1 in this case) cross-connect matrix, and a switching controller. The switching controller manages the cross-connect matrix configuration via the TSI switching method (see Chapter 2 or 4). The DCS also has operations interfaces to the central OSs and local operator consoles, which allow the operator to control the DCS reconfiguration manually for maintenance and other activities.

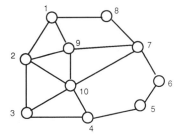

(a) Facility Network Topology

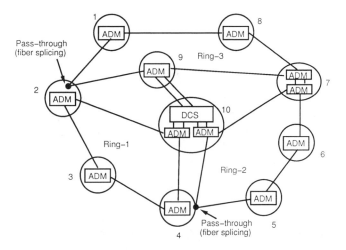

(b) Physical Network Layout Consisting of 3 Rings

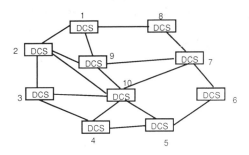

(c) Logical Network Layout Using DCSs

Figure 5.1.   Examples of physical and logical layer network designs.

213

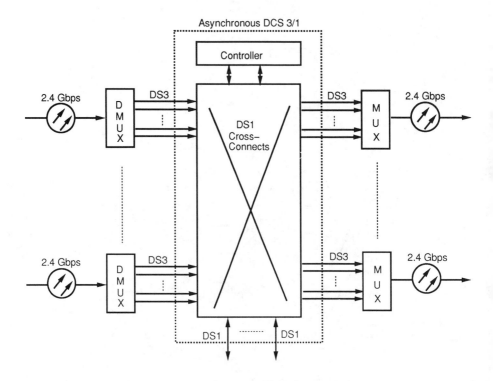

Figure 5.2. Asynchronous DCS 3/1 functional diagram.

The basic operation of a DCS 3/1 is as follows. When a high-speed signal (i.e., DS3) is terminated at the DCS, it is first demultiplexed to a set of tributary channels (e.g., DS1s). These tributary channels are then cross-connected to appropriate output ports for either service or facility grooming. The tributary channels that are switched to the same output port are multiplexed together and converted to a high-speed, time division multiplexed stream (i.e., DS3). This stream is then sent to the appropriate destination over the fiber facility. The cross-connect matrix, which performs channel switching, is managed by a switching controller that changes the switching matrix, if necessary. (Section 2.6 contains more details on the DCS architecture and its operation.)

Asynchronous DCSs are categorized based on the signal level being cross-connected. A W-DCS switching fabric interfaces at DS3s and cross-connects tributary channels at the DS1 and/or DS0 rates (e.g., asynchronous DCS 3/1). A B-DCS switching fabric interfaces at the DS3 rate and cross-connects channels at the DS3 rate (e.g., asynchronous DCS 3/3).

5.1.3 Benefits of SONET DCSs

SONET DCSs perform the same functions as asynchronous DCSs but in a more efficient and cost-effective way. Compared to the asynchronous DCS (Figure 5.2), the SONET DCS (Figure 5.3) integrates multiple functionalities of today's asynchronous networks (e.g., O/E conversion, add-drop capabilities, and the cross-connect function) into a single system. Thus, it eliminates back-to-back multiplexing and reduces internal overhead processing. It also reduces the need for intermediate electrical distribution frames and labor-intensive jumpers between frames.

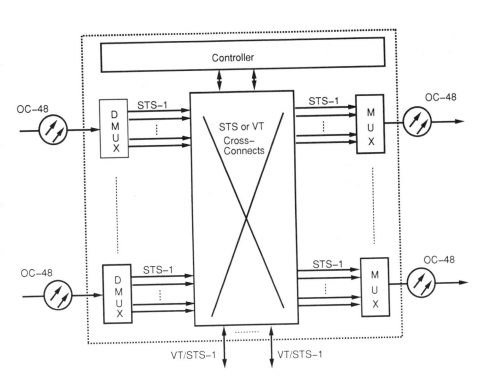

Figure 5.3. SONET DCS functional diagram.

215

As shown in Figure 5.3, the SONET DCS directly terminates fiber facilities, and the asynchronous DCS terminates electrical signals, such as DS3s for the DCS 3/3. Like asynchronous DCSs, SONET DCSs are categorized based on the signal level being cross-connected. A W-DCS switching fabric interfaces at the DS3 or STS-1 rate and cross-connects tributary channels at the VT1.5/DS1 or VT group rate. A B-DCS switching fabric interfaces at DS3, STS-1, and STS-Nc rates and cross-connects channels at the DS3, STS-1, and STS-Nc rates.

Table 5-2 summarizes a relative comparison between the asynchronous DCS network and the SONET DCS network. The SONET DCS directly terminates at fiber facilities, and the asynchronous DCS terminates at electrical signals up to the DS3 rate. In addition, using DCCs as EOCs in the SONET transport overhead allows subnetworks to be monitored and controlled without the need for an external control network to the remote locations. Using the SONET overhead to perform the necessary DCS control functions may also eliminate the need to use vendor-specific proprietary technology. Furthermore, the SONET B-DCS has the STS-1 or DS3 grooming capability, but the asynchronous B-DCS does not because it terminates OC-N facilities and can cross-connect STS-1s or DS3s. In general, the SONET DCS network has higher network reliability than the asynchronous DCS network because it integrates more functional components. Thus, fewer components are passed through for an end-to-end optical channel.

Table 5-2. Asynchronous DCS and SONET DCS Networks

Attribute	Asynchronous DCS network	SONET DCS network
DCS line termination	electrical (up to DS3)	optical (OC-N)
Grooming devices	W-DCS	W-DCS and B-DCS
Performance monitoring	external	internal
Network reliability	lower	higher*

* fewer components need to be passed through for an end-to-end optical channel

5.1.4 Reconfigurable DCS Network Applications

The DCS network's reconfiguration capability is provided through the DCS switching matrix, which is controlled by a network-wide centralized or distributed control system. This flexible network reconfiguration capability provides the following benefits for DCS network applications:

1. Reduced operations costs
   — Processes service orders, provides provisioning and system administration

216

2. Enhanced revenue
   — Provides new flexible bandwidth-switched DS1 and DS3 services
3. Better utilized network capacity
   — Reduces equipment needed
   — Responds efficiently to demand uncertainty
4. Enhanced network survivability
   — Reduces system capacity requirement (needed for restoration)
   — Responds more economically and efficiently to multiple failure cases.

To provide the above benefits, the DCS network requires a more complex control system than the SHR. It may be more vulnerable to software failures than other survivable network architectures, such as SHRs. Software failures in the DCS switched network, which is essentially a software-oriented system, may affect more services on a larger scale than the SHR and the diversely protected point-to-point system because they are more difficult and take much longer than hardware failures to detect and repair. Section 5.4 discusses the impact of software failures in more detail.

Two major applications of reconfigurable DCS networks are flexible network control and efficient network restoration. However, the different purposes of these applications may complicate DCS network design. The flexible network control application allows the DCS network to provide fast provisioning and efficiently utilize network resources. For this application, the primary concern is the long-term growth of the network (according to the forecast of working demands) as well as the best use of capacity (i.e., best shared use of capacity from multiple demand pairs given an uncertain forecast). The network reconfiguration time may not be a primary concern in this application. On the other hand, the primary goal of the network restoration application is to restore as many services as possible within a very short period of time. Thus, the DCS network designed only for flexible network control may not be appropriate for network restoration, and vice versa. For convenience, the DCS network designed for flexible network control is called a *DCS bandwidth management network,* whereas the DCS network designed for network restoration is called a *DCS self-healing network.* Table 5-3 summarizes the major differences between these two networks.

217

Table 5-3. Comparison between DCS Bandwidth Management and DCS Self-Healing Networks

Attributes	DCS bandwidth management network	DCS Self-healing network
Goal	utilize resources and provide flexible control	restore affected demands as quickly as possible
Failures concerned	frequent/limited failures (e.g., DS1/DS3 failures)	catastrophic failures (e.g., cable cut)
Control architecture preferred	centralized	distributed

As shown in Table 5-3, the DCS self-healing network competes directly with SONET rings and DP networks for the same type of failures (catastrophic failures). Sections 5.4 and 5.5 discuss this architectural competition and its relative merits.

5.1.5 DCS Bandwidth Management Networks

The flexible network control application is likely to be the first target application for reconfigurable DCS networks because it is directly related to revenue and operations cost reductions. Additionally, its operations are similar to today's voice circuit-switched networks. Thus, the self-healing capacity for network restoration is likely to be built on DCS bandwidth management networks designed for flexible network control. According to this scenario, reconfigurable DCS networks are likely to be implemented in two phases. The first phase is to build reconfigurable DCS networks for fast provisioning and efficient network utilization. The second phase is to add the self-healing capability to the first-phase DCS network. The DCS self-healing feature may need extra spare capacity for restoration, in addition to the spare capacity reserved for growth in the DCS bandwidth management network. Because this chapter focuses on network restoration, this section only briefly discusses DCS bandwidth management networks (designed for flexible network control). DCS self-healing networks will be discussed in the remaining sections.

5.1.5.1 Flexible Network Control

Business needs may be the primary motivation for planning and deploying the DCS switched network. Demand for DS1 service is growing at an explosive pace. Northern Business Information/Datapro has projected that both DS1 and DS3 service revenues will increase 350 percent by the end of 1993, compared with revenues at the end of

218

1988 [1]. The competition for these services is intensifying, and, as a result, rates for DS1 and DS3 services are decreasing. Network and service providers must expand the bandwidth management capabilities of their networks to provide fast provisioning and to allow customers to flexibly control their networks in this rapidly growing market.

Figure 5.4 shows an example of time-of-day flexible trunk group control. In Figure 5.4, CO-B is primarily a business office, and CO-A and CO-C are primarily residential offices. During the day, two DS1s carry trunk groups between CO-A and CO-B; two DS1s carry trunk groups between CO-B and CO-C; and one DS1 carries trunk groups between CO-A and CO-C. At night, the B-C and A-B demand is reduced, and A-C demand is increased. Therefore, the DCS can change the switch configuration at night to cross-connect one of the DS1s between CO-A and CO-B with one of the DS1s between CO-B and CO-C to create an additional DS1 between CO-A and CO-C without installing a new DS1 facility.

5.1.5.2 Centralized DCS Bandwidth Management Network Architectures

DCS bandwidth management networks, which have the capability to efficiently use network resources, can be made available through the use of a centralized network management system. In this case, DCSs receive commands from the centralized network management system to reconfigure their DCS switching matrixes, if needed. In general, this centralized control system is used for bandwidth management and routing control for more frequently occurring congestion and electronic failures. The centralized routing capability in the DCS bandwidth management network is designed to optimize network throughput; it is not designed to provide network restoration from catastrophic failures. However, this restoration capability can be added to the DCS bandwidth management network to build a dynamic and robust reconfigurable DCS network. (It can be provided in a distributed or centralized manner.) Due to the slow reconfiguration times attributable to centralized processing, the central reconfigurable DCS network architecture may not be appropriate for network restoration if the restoration time is a primary concern.

The *SONET Switched Bandwidth Network* (SSBN) [2] and DACSCAN [3] are two examples of the centralized DCS bandwidth management network architecture designed primarily for flexible network control and efficient bandwidth management. SSBN uses SONET capabilities with a centralized OS to automatically and quickly provision bandwidth. It uses status-based routing methods to efficiently utilize network resources.

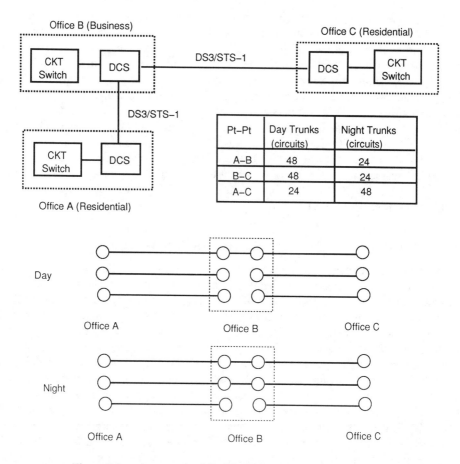

Figure 5.4. An example of flexible DCS switched network control.

Figure 5.5 depicts a functional architecture for SSBN. The SONET NEs used as the bandwidth-switched devices in SSBN have a remotely programmable TSI capability, as do DCSs or ADMs equipped with TSI. The DCC of the embedded SONET overhead serves as the signaling medium for the SONET DCS switched network. The bandwidth is managed from end-to-end via a centralized OS, functionally called the *bandwidth manager*. The bandwidth manager provides routing logic via dynamic routing methods. The SSBN concept expands the digital bandwidth provisioning process to encompass efficient and automatic demand routing principles, similar to the automatic routing of calls in the public circuit-switched voice network.

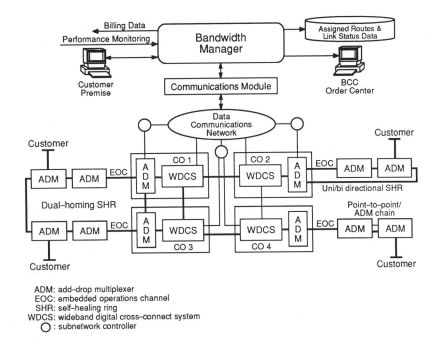

Figure 5.5. SSBN network control architecture.

5.1.5.3 Centralized Reconfiguration Algorithms for DCS Bandwidth Management Networks

DCS network reconfiguration is achieved by rerouting demands. Routing is defined as the procedure of selecting a path through the DCS to serve the demand. As with today's centralized circuit-switched voice networks, routing can be implemented via a hierarchical or non-hierarchical scheme. In hierarchical routing, all DCSs in offices are organized into a tree hierarchy. When a demand arrives, an attempt is made to route it on the direct link, i.e., the link whose endpoints coincide with the demand pair. If this link is either non-existent or full, then the routing algorithm searches alternate routes via offices in the next hierarchical position (e.g., hubs). Hub routing, which is discussed in Reference [4], is an example of hierarchical routing.

Non-hierarchical routing can be static or dynamic. In static routing, a priority list of routes between various demand pairs is maintained and updated at regular planning intervals. Dynamic routing methods can be either (1) time dependent, where the priority list of routes changes at prespecified times (e.g., time-of-day, day-of-week), or (2) state dependent, where the route is determined by the state of the network at the

221

time of the request for service. Examples of the time-dependent and state-dependent routing algorithms are DNHR [5] and *Dynamical Routing-5* (DR-5) [6], respectively.

For non-hierarchical routing schemes, the state-dependent routing algorithm usually provides better resource utilization than the time-dependent routing algorithm because the time factor is usually implicitly incorporated into the state factor. For the state-dependent routing algorithm, the route for any particular demand pair is determined by the state of the network when the demand arrives. A weight function is associated with each link, where the weight function is a function of the existing demand on the link and reflects the expected number of rejected future demands if the current demand is routed on that link.

The state-dependent routing algorithm reported in Reference [6], which was proposed for circuit-switched networks for voice traffic, can be modified for efficient DS1 routing [7]. The major modification involves changing the weighting function of the link to reflect DS1 routing rather than circuit routing. Another possible modification of the state-dependent routing method involves a rerouting method [7]. This is not a major issue in the voice network because the holding time for demands is short compared to changes in the network status and in the demand arrival rates. However, traditional private DS1 lines tend to have long holding times. Therefore, the network status, as well as the arrival rates, can change substantially over the duration of the demand. Although the routing heuristic is usually designed to take into account future traffic effects, it can be potentially enhanced by rerouting long duration demands at periodic intervals. The rerouting algorithm applies the same routing method to the candidate demand as is applied to new connect requests. If the demand is rejected by this method, it is left in place; otherwise, it is rerouted to the new path.

5.1.6 DCS Self-Healing Networks

The DCS network is called a self-healing network if it can automatically restore affected demands when the fiber facility, equipment, or office fails. SONET DCSs provide a network restoration capability via alternate routing of DS1s and VT1.5s for SONET W-DCSs, and DS3s, STS-1s, and STS-Nc's for SONET B-DCSs. Priority restoration through DS1 and VT1.5 path rearrangement may be implemented only if limited spare capacity is available in the network.

Figure 5.6 depicts a simplified diagram showing how an STS-1 is restored in a DCS self-healing network. In this example, an STS-1 is sent from CO-A to CO-D via CO-C under normal conditions. Let us assume that the fiber link between CO-C and CO-D is broken, i.e., link failure. If centralized routing control is assumed, the failure is reported to the central controller, and the new route A->B->D is computed and downloaded to CO-A, CO-B, and CO-D to change DCS switching matrixes at CO-A, CO-B, and CO-D. The routing decision can be reached either by a sequence of

222

preplanned paths or by a dynamic rerouting decision.

If distributed switching control with path restoration (which will be discussed in Section 5.2.2) is used, the originating DCS at CO-A starts to send control messages to search for alternate paths, after receiving a path AIS from CO-D, by communicating with other nodes, including the DCS at CO-D. If an alternate path exists (in this example, A->B->D), the DCS at CO-D responds with an acknowledgment to CO-A (via the same path) for successful connection, after receiving a valid restoration message from CO-A, and configurations of DCSs at Nodes B and D are changed. Once CO-A receives the acknowledgment message from CO-D, it reconfigures its DCS switching matrix and then sends a confirmation message along with the particular STS-1 channel to CO-D via the DCS at CO-B. The confirmation message is used for bidirectional protection switching.

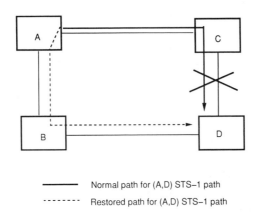

—————— Normal path for (A,D) STS–1 path

-------- Restored path for (A,D) STS–1 path

Figure 5.6.   An example of DCS network restoration.

5.1.7 DCS Switched Services vs. Real-Time DS1/DS3 Switched Services

A switched service provided by the reconfigurable DCS network should not be confused with a fast-switched type of service such as *Switched DS1/Switched Fractional DS1* (SWF-DS1) [8]. Table 5-4 provides a relative comparison between these two types of service. SWF-DS1-like services, which have been offered by major U.S. long-distance carriers, are real-time switched services. They have a maximum level of DS1 or DS3 based on high-rate switching in extended circuit switch fabrics or integrated circuit switch/DCS systems. The switching for a reconfigurable DCS network requires a slow setup capability communicated through an OS, rather than a quick setup capability provided through distributed switches.

Table 5-4. DCS Switched Services and Real-Time DS1 Switched Services

Service Characteristics	Services	
	DCS switched services	*Real-time switched DS1 services*
Applications	leased-line DS1s/DS3s	SWF-DS1
Setup time	minutes	seconds
Connect time	hours, months, years	minutes, hours
Types of failures concern	catastrophic failures	frequently occurred non-catastrophic failures
Network reconfiguration rate	moderate	frequent

A potential SWF-DS1 application is video teleconferencing using DS1s. Such an application has characteristics similar to *Plain Old Telephone Service* (POTS) (i.e., human-to-human communication) and, therefore, requires fast setup and relatively short holding times (i.e., short when compared to current, private DS1 line-holding times). The short holding times are particularly important due to the expensive DS1 connection costs. On the other hand, the leased-line DS1/DS3 services supported by the DCS switched network have a longer connection time (hours to years) and require a longer setup time (minutes).

SWF-DS1 service provides a customer network survivability application that addresses the speedy setup of an alternate communications path and provides services similar to the original connection when that connection fails. Thus, this customer network survivability application is limited to providing protection from frequently occurring network problems, as opposed to providing protection from catastrophic problems, such as fiber cable cuts.

On the other hand, like the APS system discussed in Chapter 3, the DCS switched network operates under a moderate network reconfiguration rate, i.e., the DCS switched network configuration remains the same until the network component failure has been detected and the protection switching process is initiated. The time required for service restoration depends on the protection control architecture used and the failure type. The protection switching time for the DCS switched network may range from seconds to several hours.

5.2 A CLASS OF SONET DCS SELF-HEALING NETWORKS

The DCS network restoration methods typically respond to a network failure as follows [9]:

224

1. Detection - Detection of the failure by the appropriate OS, controller, or NE
2. Control message propagation - Broadcast of the failure to other NEs (if necessary) and instigation of the restoral process
3. Route selection - Selection of an alternate route for each affected demand, either through data loop-up or a dynamic routing algorithm
4. Rerouting - Cross-connection of existing demands to the alternate routes
5. Return to normal - Detection of failure repair and return to the fully operational state of the network.

For each of the above steps, there are a variety of implementation techniques. DCS self-healing networks can be classified as follows based on these different implementation techniques:

1. Self-healing control scheme (centralized versus distributed)
2. Rerouting path planning (preplanned versus dynamic)
3. Signal restoration level (line restoration versus path restoration).

Sections 5.2.1 through 5.2.3 discuss these classifications in more detail, and Figure 5.7 depicts a class of SONET DCS self-healing network architectures.

5.2.1 Self-Healing Control Architectures (Centralized vs. Distributed)

As depicted in Figure 5.8, DCS self-healing control architectures can be divided into two classes: centralized and distributed.

The centralized self-healing control architecture operates from a central controller, as depicted in Figure 5.8(a). Information about all network nodes, connectivity maps, and spare facilities is stored in the database of the central controller. In Figure 5.8(a), the normal route between locations A and B is via a link between DCS#1 and DCS#2. Once the central controller is informed of the link failure between DCS#1 and DCS#2 (via separate control links), it then searches for the best rerouting path based on the most current network status.[2] The controller then sends commands to remote DCS#1, DCS#2, and DCS#3, prompting them to change their switching matrixes and reroute demand between locations A and B (via a link between DCS#1 and DCS#3 and a link between DCS#3 and DCS#2). Examples of centralized DCS self-healing networks include those reported in References [2,3,10].

---

2. The ability to choose the optimum alternate routes depends on the capacity and connectivity information that is available on the network.

225

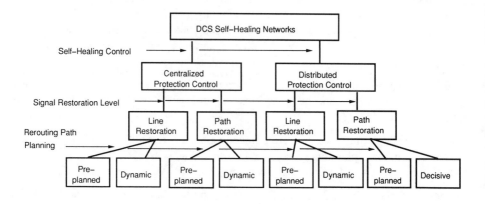

Figure 5.7. A class of SONET DCS self-healing network architectures.

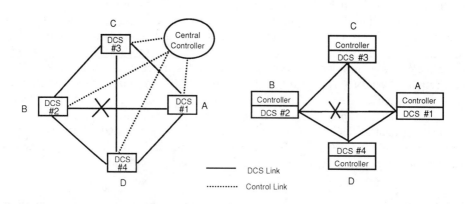

(a) Centralized DCS Protection   (b) Distributed DCS Protection
Figure 5.8. DCS self-healing control architectures.

For the distributed self-healing control architecture, one external controller is needed for each DCS, as depicted in Figure 5.8(b). However, in the future, it may be possible to integrate the function of the external controller into the DCS. Each DCS stores local information that includes working and spare capacity associated with each link terminating at that DCS. In Figure 5.8(b), when a link failure between DCS#1 and DCS#2 is detected, one of the affected DCSs begins searching for available paths to reroute demand using local information stored at each DCS. In this architecture, no central control is used. Examples of distributed DCS self-healing networks include those reported in References [11-19]. Table 5-5 summarizes a relative comparison between the centralized and distributed self-healing control architectures.

Because the centralized self-healing control scheme is controlled by a central controller and does not require communication among DCSs, concerns about intervendor compatibility and system complexity are minimized. The centralized scheme may or may not require the use of DCC bandwidth for restoration. If the DCC channel is used, that channel needs to be protected for survivability purposes. The centralized self-healing control scheme can be easily incorporated into the centralized OS (designed for fast provisioning and flexible network control), which may make it easier to coordinate capacity planning and restoration. The survivability strategy can only function well if (1) the planning algorithms provide adequate spare capacity throughout the network, and (2) the routing algorithms access spare capacity in a controlled and efficient manner. Predictability of the algorithms is necessary to achieve this level of coordination between planning and restoration. This predictability factor tends to favor centralized self-healing control because control and predictability of routing may be difficult to achieve in a distributed setting. Disadvantages of the centralized scheme include higher administration overhead, higher system vulnerability, and lower restoration speed (on the order of minutes or longer).

On the other hand, the distributed self-healing control scheme restores demands faster,[3] lowers system vulnerability, and lowers administration costs. However, it does so at the expense of system unpredictability, a complex control system, greater standardization efforts, and, possibly, less efficient use of spare capacity.

---

3. Most distributed self-healing control schemes set 1 to 3 seconds as the target time for finding alternate paths. Note that this time does not include cross-connect time.

Table 5-5. Comparison between Centralized and Distributed Self-Healing Control Systems

Attributes	Centralized control	Distributed control
Vendor independence and interoperability	easy+	difficult*
System complexity	simple	complex
Use of DCC bandwidth	yes/no%	yes/no++
Demand restoration probability	higher	lower
Return-to-normal	easy	difficult%%
Standards needed	messages only†	algorithm and messages
Coordination between capacity planning and real-time restoration	easy	difficult
Restoration time	slower (minutes or longer)	faster (seconds)
System vulnerability	higher	lower**
Administration overhead	higher	lower
Dependence of a centralized OS	yes	no/yes%%
Spare capacity and diverse-route planning	needed	needed

+   Intelligence resides only at the central node.
*   Unless standards related to DCS methods have been standardized
%   It may use a conventional packet-switched network as the control network.
++   Some proposed distributed self-healing networks do not use the DCC for message communications (e.g., [11]).
%%   It may need to use the centralized OS to return the network to the normal condition after failed network components are repaired.
†   The message issue is under study in the T1 (SONET and OAM&P) and CCITT standards groups.
**   Assume that software in the distributed scheme is properly designed and functions as expected.

From a survivability point of view, the centralized DCS self-healing network is more vulnerable than the distributed DCS network because of its centralized structure. However, there are methods to deal with this problem through replication. From a reliability point of view, the distributed DCS network may be less reliable than the centralized DCS network because the distributed DCS network is operated under an unpredictable environment that makes efficient and correct operator intervention more difficult than the centralized DCS network.

Determining which DCS self-healing control architecture to use is a complex task that depends on applications and many factors shown in Table 5-5. However, these factors can be summarized in one fundamental planning question: do we want to take advantage of the existing centralized operations environment for restoration or are we primarily concerned with the restoration time? If the primary concern for restoration is the restoration time, distributed self-healing control is a preferred choice; otherwise, the centralized self-healing control scheme may be a better choice.

5.2.2 Signal Restoration Level (Path vs. Line Restoration)

Like the protection switching schemes used for SONET SHRs, demand restoration can be considered at two different levels: path and line. Path restoration restores the end-to-end logical channel, and line restoration restores all affected demands carried by a failed facility [i.e., a replacement route(s) is found for a failed link]. Figure 5.9 depicts examples of line and path restoration techniques. In these examples, the link between Nodes 2 and 3 carries two STS-1s [STS-1 (1,6) and STS-1 (4,6)]. If that link fails and the line restoration method is used [see Figure 5.9(a)], route 2-3 is replaced by route 2-5-3, and all channels use the new route when they pass from Node 2 to Node 3. On the other hand, if the path restoration method is used [see Figure 5.9(b)], each channel [i.e., STS-1 (1,6) and STS-1 (4,6)] affected by the link failure selects a new route for restoration. For example, STS-1 channel (4,6) may select new route 4-5-6, and STS-1 (1,6) may select route 1-4-5-6.

Table 5-6 shows a relative comparison between the line and path restoration methods. Path restoration uses spare capacity more efficiently than line restoration. However, line restoration requires a simpler routing decision. Thus, line restoration is expected to be faster than path restoration in terms of restoration time.

Table 5-6. Comparing Line Restoration and Path Restoration

Attribute	Line restoration	Path restoration
Restoration speed	faster	slower
Rerouting decision complexity	simple	complex
Efficient use of spare capacity	worse	better

229

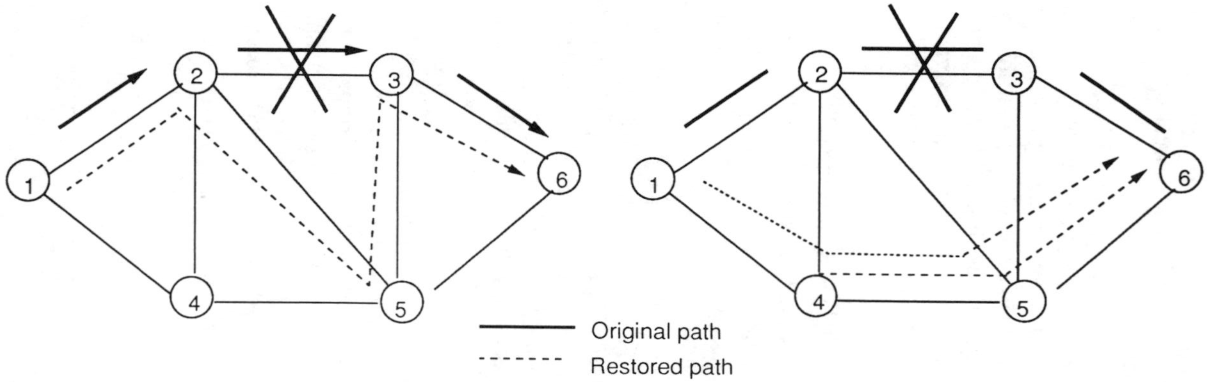

STS-1 path	Original path	Restored path
(1,6)	1-2-3-6	1-2-5-3-6
(4,6)	4-2-3-6	4-2-5-3-6

(a) Line Restoration

STS-1 path	Original path	Restored path
(1,6)	1-2-3-6	1-4-5-6
(4,6)	4-2-3-6	4-5-6

(b) Path Restoration

Original path
Restored path

Figure 5.9. Demand restoration methods.

Path restoration requires circuit endpoints to be identified and notified. To calculate the alternate route beginning at the endpoints, the node detecting the failure must be able to notify the circuit endpoints of the need for recovery. This signaling scheme may be difficult to implement in today's asynchronous networks because the in-service DS3 does not have signaling overhead of sufficient bandwidth. This constraint can be lifted when the SONET network is introduced. However, line restoration may result in inefficient rerouting for low-connected networks (e.g., rings) because the custodial nodes have no path-level knowledge. Figure 5.6 depicts an example of inefficient traffic rerouting for line restoration. In Figure 5.6, a fiber cable on the link between Nodes C and D is cut, and all STS-1 demand that is carried over link C-D is rerouted. If the DCS at C or at D controls the new route and has no knowledge about the demand's origin and termination, the A-C-D demand may be rerouted via path A-C-A-B-D. This type of inefficient rerouting could quickly use up the available spare capacity in the network.

5.2.3 Rerouting Methods (Preplanned vs. Dynamic)

In DCS self-healing networks, rerouting paths for demand restoration can be identified using either a preplanned or a dynamic routing method. The preplanned method requires that each local DCS network controller carry all, or a majority of, rerouting information related to the network restructuring for all preplanned failure scenarios. In contrast, the dynamic routing method requires each DCS network controller to store only necessary local information, and the rerouting decision is made according to the network status (e.g., configuration, available spare capacity, and so on) at the time of network component failures. Table 5-7 summarizes a relative comparison between these two methods. In general, compared to the dynamic rerouting method, the preplanned method requires higher memory and may have more difficulty adapting to rapid network changes. However, the preplanned method has faster restoration capability and lower system complexity. The preplanned rerouting method may also have lower network reliability than the dynamic rerouting method because it is difficult to implement all possible failure scenarios in the preplanned method.

Table 5-7. Comparing the Preplanned and Dynamic Rerouting Methods

Attribute	Preplanned	Dynamic
System complexity	lower	higher
Network adaptation	difficult	easy
Restoration speed	faster	slower
System reliability	lower	higher
Memory requirement	higher	lower

5.2.4 DCS Self-Healing Network Proposals

Table 5-8 summarizes examples of DCS self-healing networks and associated classifications. Because hardware and software systems for DCS self-healing networks are still in the development stage, most reported DCS self-healing protocols are simply proposals. As shown in Table 5-8, most proposed DCS self-healing protocols implement the distributed self-healing control scheme, and demands are restored on a link or path basis using the dynamic rerouting method.

Table 5-8. Examples of DCS Self-Healing Networks

Network proposal	Self-healing control	Rerouting planning	Restoration technique
AT&T (DACSCAN [3])	centralized	dynamic	path
ATRC*[11]	distributed	dynamic	line
Bellcore (SSBN [2])	centralized	dynamic	line/path
Bellcore (FITNESS [12])	distributed	dynamic	line
Bellcore (NETSPAR [13])	distributed	preplanned	path
Fujitsu [14]	distributed	dynamic	line
NEC (NETRATS [15])	distributed	dynamic	line/path
Rockwell [16]	distributed	dynamic	line
NTT** [19]	distributed	dynamic	line

* ATRC: Alberta Telecommunications Research Center (Canada)
** NTT: Nippon Telegraph and Telephone

5.3 DCS SELF-HEALING NETWORK ARCHITECTURES

This section reviews architecture selection criteria, design, and self-healing architectures.

5.3.1 Architecture Selection Criteria

A class of SONET DCS self-healing network architectures may be evaluated based on the relative weight of the following parameters:

- Restoration time: the time a DCS network takes to restore affected demands once the network failure is detected.
- Restoration ratio: the ratio of the number of demands restored to the number of demands affected by network component failures.
- Coordination with the DCS bandwidth management network for fast provisioning and flexible network control.
- Memory requirement for network restoration.
- DCS hardware/software modifications required to add the network restoration capability.
- Cost ratio: the ratio of the cost of adding the self-healing feature to the network cost without the self-healing feature.

5.3.2 DCS Self-Healing Network Design Characterization

The goal of the DCS network restoration method is to restore as many affected demands as possible and as quickly as possible. Thus, the rerouting path-finding problem in the DCS restoration method is essentially a combination of two methods: finding maximum available spare capacity and finding an appropriate set of k shortest paths for each affected node pair. The problem of finding the maximum available spare capacity is sometimes called the *maximum flow problem*. These two methods are very similar in terms of rerouting paths found in the restoration process. Typically, if the primary criterion is to restore affected demands as soon as possible, the k-shortest path method may be used; otherwise, the maximum flow method may be used. The method used depends upon the specific DCS self-healing protocol. For example, self-healing protocols discussed in References [11,14,15] essentially use the k-shortest path method, whereas a self-healing protocol discussed in Reference [12] uses the maximum flow method (also see Table 5-10). Note that the shortest path criterion may be the minimum number of nodes between the affected node pair, or other measures, depending upon the definition of the link function on the path. The locations of the affected node pair depend on the restoration method used. For path restoration, the affected node pair is the two endpoints of each affected demand pair. For line restoration, the affected node pair is the two endpoints of the affected facility.

Figure 5.10 depicts an example that may characterize DCS restoration using the k-shortest path method for a single fiber cable cut. For example, if the line restoration method is used in Figure 5.10, the system is to find two paths 1-3-5 and 1-3-4-5, which restore three STS-1s and six STS-1s, respectively, for a total of nine affected STS-1s in link (1,5). It may also be possible to reroute demands from paths 1-3-5 (three STS-1s) and 1-2-3-4-5 (six STS-1s), depending upon the rerouting criterion, where these two paths are link-disjoint paths.

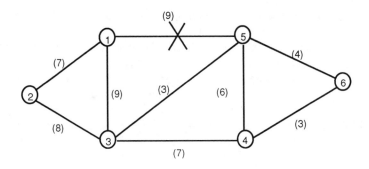

(x) = available spare STS–1s

Possible restoration paths for affected 9 STS–1 paths (1,5):

(1) (i) 1–3–5 restores 3 STS–1s, and (ii) 1–3–4–5 restores 6 STS–1s.

(2) (i) 1–3–5 restores 3 STS–1s, and (ii) 1–2–3–4–5 restores 6 STS–1s.

Figure 5.10.   Problem description of DCS restoration methods.

Thus, according to the simple example depicted in Figure 5.10, the DCS restoration problem using the k-shortest path approach can be characterized as follows: *find a set of k paths whose total spare capacity is at least the amount of the affected demands, such that all affected demand pairs can be restored completely, as soon as possible.*

Mathematically speaking, the above problem can be explained as a problem of finding a minimum set of k shortest paths whose total spare capacity is not less than the affected demands. These k shortest paths may or may not be link-disjoint paths, depending upon whether or not the balance of the spare capacity assignment is a concern. Compared to the k-non-link-disjoint path approach, the k-shortest link-disjoint path approach has a more balanced spare capacity assignment over the network, but requires a much higher network connectivity.

5.3.3 Centralized DCS Self-Healing Networks

Figure 5.11 depicts a centralized DCS self-healing operation. The centralized DCS self-healing system requires a central controller that stores the global network information, including working and available spare capacity available on each network link. The major function of the central controller is to identify the "best" paths for demand rerouting [after it receives the failure report from some DCS(s)] and to send commands to affected DCSs to change their switching matrix configurations. The central controller for restoration may be incorporated into the centralized OS to facilitate the coordination of spare capacity planning and network restoration.

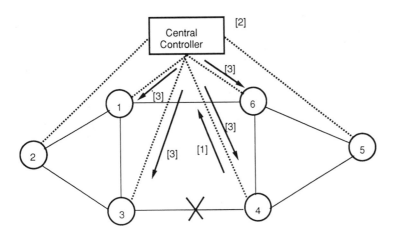

[1] Affected node (e.g., Node 4) sends the help message to the central controller.

[2] The central control processes the incoming messages and computes the restoration paths.

[3] The central controller sends commands to affected DCSs to change their DCS switching matrixes based on the restoration paths just computed.

Figure 5.11. An example of centralized DCS self-healing operation.

235

As depicted in Figure 5.11, when one or more DCSs detect a failure through SONET maintenance signals, the affected DCS(s) sends a restoration message to the centralized controller via separate data links. When the central controller receives the restoration message, it updates the network configuration to reflect this change and then identifies new routing paths for the channels affected by this failure. After the central controller has the rerouting information, it then sends commands to affected DCSs to change their switching matrixes, and channel rerouting is completed.

In the centralized self-healing control architecture, the algorithms used for dynamic bandwidth management (designed for flexible network control, see Section 5.1.5) can be used for network reconfiguration (and thus for restoration) by adding one feature — handling catastrophic failures. A facility or CO failure should trigger an immediate interruption to execute the network reconfiguration algorithm.

For the centralized DCS self-healing architecture, the line restoration method may be more efficient and cost-effective than the path restoration method because it requires fewer algorithm recomputations and fewer DCS reconfigurations (and thus less rerouting). Also, the global optimization of spare capacity utilization can still be preserved because of the method's centralized characteristics. Finally, the centralized DCS self-healing architecture can take advantage of existing centralized OSs for restoration, but with slow restoration time.

To ensure a reliable centralized DCS control system, dual central controllers are usually used. These two identical central controllers are placed in different locations, but only one central controller is in control at any given time. Both controllers are online and synchronized so that the standby controller can instantly assume the restoration control function. Both controllers are connected to all DCS systems via redundant control networks.

5.3.4 Distributed DCS Self-Healing Networks

5.3.4.1 Description of Architecture

In the distributed self-healing architecture, each node includes one DCS and one controller. The controller may be integrated into the DCS, depending upon the vendor's equipment. Each local network controller stores only the local information, including spare capacity for links terminating at that node. Network reconfiguration for restoration is based only on the local information stored in the DCS controller. Examples of distributed self-healing networks include those reported in References [11-16, 19].

A distributed self-healing architecture requires a significant portion of the network to participate in the decision to make necessary network reconfigurations. The architecture considers every node as an equivalent intelligent node. Thus, it can function properly only when all these nodes execute the distributed self-healing protocol properly.

5.3.4.2 Message Flooding for Restoration

Because the distributed DCS restoration method does not have a global network view when network components fail, it must rely on communication between nodes to obtain information to set restoration paths. The flooding technique used in packet-switched data networks is a common technique used in all proposed DCS restoration methods. However, the reasons for using this flooding technique in packet-switched data networks and DCS switched networks are different due to requirement differences. Table 5-9 summarizes these differences.

Table 5-9. Using Flooding in Packet-Switched and DCS Switched Networks

Factors	Flooding in packet-switched networks	Flooding in DCS self-healing networks
Purpose	update routing tables at each node	find alternate paths
Decision timing	real time	near real time for restoration
Periodic message flooding	necessary	not necessary

For packet-switched data networks, routing decisions are made based on the network status at the time the call request is placed. Routing decisions must be made in real time on a per-packet basis. Thus, each node requires a routing table that stores the necessary network connectivity information. The table uses an adaptive distributed routing algorithm to make the real-time routing decision, which reflects the available network resources at the time of the new call request. Thus, the routing table has to maintain the most updated network resource information. The flooding method is a common and simple technique used to periodically convey network management messages to update the routing table. In this case, the flooding technique is used only to update the network configuration and resource information stored in each node. The routing decision is made by other routing algorithms.

On the other hand, the demand routing for DCS self-healing networks may remain constant for a relatively long period of time, but will react to unexpected network interruptions (e.g., fiber cable cuts) as quickly as possible. The flooding method used here identifies a set of alternate routes that have enough spare capacity to restore affected demands.

5.3.4.3 Distributed DCS Self-Healing Protocols

Most proposed distributed DCS self-healing protocols have two major features: (1) multiple messages (flooding) for route search, and (2) a single message for route selection. Figure 5.12 depicts a general model for DCS line restoration under a single-link failure scenario.

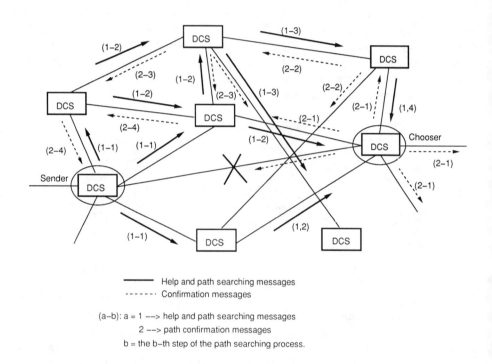

Figure 5.12. A distributed DCS line restoration scenario.

In Figure 5.12, when two adjacent nodes detect a communication failure, one becomes the Sender and the other becomes the Chooser. (All other nodes that participate in the restoration process are called tandem nodes.) The Sender then begins transmitting restoration messages in all possible directions (except the Chooser direction). To restrict the number of restoration messages, some algorithms require transmission only through the network spare capacity. Tandem nodes update the DCS configuration map,[4] based on the received messages, and rebroadcast the received messages to other adjacent nodes based on the particular flooding algorithm used. To implement time constraints on the algorithm execution, message flooding is restricted by some parameters, such as hop count. When the message reaches the Chooser, a rerouting path for restoration is identified, and a confirmation is sent in the opposite direction. The confirmation is conveyed back to the Sender via the path just identified. The DCS switching matrixes at tandem nodes in this path are changed according to instructions stored in the confirmation message. After the Sender receives the confirmation message, it changes its DCS switching matrix to cross-connect the affected demands from the failed facility to newly identified alternate routes. The restoration process is then completed.

A typical message used in the restoration process includes information about

- Bandwidth required for restoration
- Available spare capacity of the route
- Sender/Chooser identifications (IDs)
- Link ID of the restoration message
- IDs of nodes passed through by the restoration message
- Hop count indicating the distance from the Sender or other parameters based on the considered flooding method.

The purpose of the message exchange is to identify the "best" alternate path(s) between the Sender and the Chooser, based on some performance criteria (e.g., minimum hop count), and to instruct affected DCSs to make cross-connections to restore affected demands.

Figure 5.13 depicts an example that shows how to identify alternate paths (using the flooding technique) by using the three-phase procedure described previously for a single-link failure scenario.

---

4. The map is located in the DCS controller to provide necessary cross-connections for network recovery or reconfiguration.

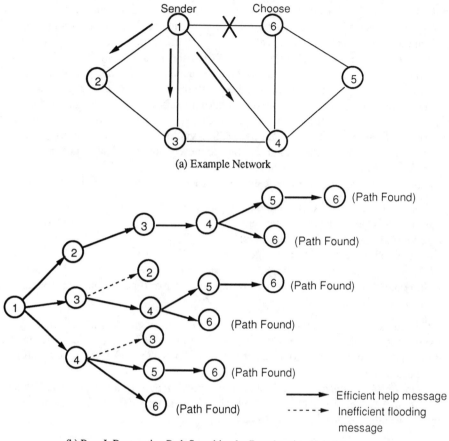

(a) Example Network

(b) Pase I: Restoration Path Searching by Broadcasting Help Messages

Possible Paths	Path Selection
(1) 1–2–3–4–5–6	Choose first k paths found that
(2) 1–2–3–4–6	can restore total affected demands
(3) 1–3–4–5–6	carried by the failed link [e.g.,
(4) 1–3–4–6	Link (1,6)]
(5) 1–4–5–6	
(6) 1–4–6	

(c) Phase II: Path Selection

Figure 5.13. Identifying alternate paths for single-link failure restoration.

240

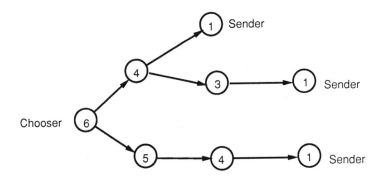

(d) Phase III: Confirmation and DCS Reconfigurations

Figure 5.13. (Continued)

According to the example shown in Figure 5.13, the following questions need to be answered, depending on the algorithm implementation:

- How is the restoration message flooded to its neighborhood nodes?
- How does the node decide which received messages not to process?
- What information does each message need to carry?
- How can the path-finding process be accomplished as quickly as possible?

Messages can be transmitted among DCSs through message channels (e.g., line or section DCCs [12-15]) or physical channels [11,16] (e.g., K1 and K2 bytes). Message communication using physical channels is usually faster than communication using message channels. However, most SONET physical channels have been standardized and have particular applications. Thus, it is likely that the physical channel used for DCS restoration requires a new format structure (e.g., some SONET growth bytes like Z1 or Z2), which needs to be standardized for a multivendor environment. In contrast, message channel protocols for SONET section DCC have been standardized and may be ready to be implemented as soon as some other issues related to DCCs (e.g., DCC protection) have been standardized.

Note that the DCS self-healing protocol proposed in Reference [11] (using physical channels as communication channels among DCSs) is similar to the BSHR/2 architecture (see Chapter 4) in terms of (1) overhead bytes used for restoration and (2) TSI used for restoration. The differences between these two architectures include self-healing protocols and the applicable network topology (mesh for the DCS self-healing protocol and ring for the BSHR/2).

The node-failure case is little different from the link-failure case, which requires multichooser designation [14]. Multiple failure restoration is even more complicated. It can require, for example, multiple Sender-Chooser pairs and a technique to avoid contention, but the major flooding principles do not change. During a multi-link failure, restoration messages coming from different nodes might contend for spare capacity on the same link. If capacity is assigned to arriving messages in turn, the first message reserves the capacity. Even if the reserved capacity is used later for an alternate path, it is not released and, therefore, cannot be assigned to another restoration message. This usually results in a decreased restoration ratio.

The problem just described could be alleviated by a simple message-canceling technique described in Reference [14]. Spare capacity is assigned to restoration messages on a first-come, first-served basis. Assignment is canceled when the message cannot go forward due to hop limits or lack of spare capacity. During message flooding, cancel messages are sent to inform a node that a restoration message, which reserves spare capacity on a specific link, did not reach its destination. The reserved capacity of this link can then be released for other restoration messages. Restoration messages are canceled immediately after reception if (1) they are identical to messages already received, (2) the hop limit is reached, or (3) there is no more capacity at that node. In these cases, the unused capacity can be assigned to other restoration messages.

The DCS self-healing protocols discussed in References [11-15] restore demands on a undirectional basis. For bidirectional restoration, a double-search algorithm was proposed in Reference [19] that assigns two Sender nodes and two Chooser nodes, and process rerouting path finding in parallel in both communications directions. Details of this double-search self-healing protocol can be found in Reference [19].

5.3.4.4 Distributed DCS Self-Healing Network Comparison

Table 5-10 summarizes a comparison among proposed distributed DCS self-healing protocols. Some protocols, such as the one proposed in Reference [15], may be able to restore a mixture of STS-1 and STS-3c rates. Note that the scheme designed for STS-1, or for a bundle of STS-1s, may not be used to restore STS-3c signals because of signal integrity of the STS-3c and different payload sizes between three STS-1s and one STS-3c. The payload of an STS-3c (i.e., 2340 bytes) is larger than the payloads of three consecutive STS-1s (i.e., 2322 bytes) because an STS-3c needs only one path overhead, whereas three consecutive STS-1 frames need three path overheads.

Table 5-10. Comparing Some Proposed Distributed DCS Self-Healing Networks

Network proposal	Routes found per cycle##	Message channel	Route selection criteria	Message broadcast rule	Restoration units
ATRC [11]	multiple routes/ 1 cycle	physical* channel	FIFO	1 message/ 1 spare STS-1/3c	single rate**
Bellcore (FITNESS [12])	1 route/ 1 cycle (repeated)	DCC	maximum bandwidth available+	1 message/ link	bundle of STS-1s
Fujitsu (DIA-FM [14])	1 route/ 1 cycle (repeated)	DCC	FIFO	1 message/ link	bundle of STS-1s
NEC (NETRATS [15])	multiple routes/ 1 cycle	DCC	FIFO	1 message/ link or failed path#	mixed rates++

FIFO: First in first out
* Online memory in physical channels such as K1, K2 bytes or other growth bytes (not DCCs)
** Restore a single rate (DS3, STS-1, or STS-3c)
+ It needs an explicit time-out feature [12].
++ Restore a mixture of STS-1 and STS-3c signals
# One message per link for line restoration; one message per failed STS-1 or STS-3c for path restoration
A cycle is a unidirectional process that begins with the sending of the restoration message from the Sender node and ends after receiving that message at the Chooser node.

5.4 DCS NETWORK PLANNING

This section discusses planning considerations and economic analysis for DCS restoration.

5.4.1 DCS Self-Healing Network Planning Considerations

As discussed in Section 5.1, the required spare capacity for the DCS self-healing network is less than that required for the SHR or the DP network due to a higher degree of spare capacity sharing. However, to achieve such a high degree of spare capacity sharing, the DCS self-healing network requires higher network connectivity (e.g., mesh) and higher equipment (i.e., DCSs) costs than the SHR, which uses ADMs.

243

Because the DCS self-healing network directly competes with the ring and the DP network for the same type of failures, and it is expected that the SONET ring and hubbing/point-to-point with DP network architectures are likely to be deployed before the DCS self-healing network, a key question for determining the potential role of the DCS self-healing network in an integrated network restoration system is: can the spare capacity savings for the DCS self-healing network justify an increased network connectivity requirement and higher equipment costs, compared to SHRs and DP networks? Equivalently, this planning question asks what conditions would make DCS self-healing networks economically attractive, compared to SHRs and DP systems. The answer to this question will help design a cost-effective network restoration system that uses the DCS self-healing feature to complement SHRs and DP networks.

Before analyzing this planning question, it would be helpful to understand qualitative tradeoffs among the DP, SHR, and DCS self-healing networks. These architectural tradeoffs may help derive a model for the considered question. Table 5-11 shows a relative comparison among the DP, SHR, and DCS self-healing networks.

Table 5-11. Comparing the DP, SHR, and DCS Self-Healing Networks

Attributes	DP	SHR/ADM	DCS
Spare capacity needed	most	moderate	least
Per-node cost	moderate (OLTM/APS)	lowest (ADM)	highest (DCS)
Fiber counts	highest	lowest	moderate
Connectivity needed	lowest	moderate	most (mesh)
Restoration time	fastest	fastest	slowest
Mixed line rates	yes	no	yes
Software complexity	least	moderate	most
Service impact due to software failure	least	least	significant
Planning/operations complexity	least	moderate/least*	most
Network size	2 points	regional	global

* Assume that the USHR with path protection switching is used (see Chapter 4).

Chapter 4 compared the DP network and the SHR in detail. Thus, this section focuses on comparisons between the SHR and the DCS self-healing network. Compared to the SHR, the DCS self-healing network requires less protection capacity but has a longer restoration time and requires a more complex planning and operations system. The restoration time for the DCS self-healing network may range from seconds to minutes, whereas the restoration time for the SHR is within 50 ms.

The spare capacity savings for the DCS self-healing network are primarily due to its sophisticated network control system. This sophisticated system provides tremendous advantages when it functions properly but causes multiple problems when a software failure occurs. This is evident from the AT&T software failure that occurred in 1990 [20]. The primary lessons learned from that failure are as follows:

- It is very difficult (and sometimes impossible) to simulate all possible events before software installation.
- It is very difficult to detect and repair a software failure.
- A multivendor environment could help isolate the software failure (due to software bugs) in one area where the products are from the same vendor.
- The network management system may be organized as a hierarchical, regional system to isolate software failures.

The SHR is a sectionalized network due to its capacity constraints, whereas the size of the DCS self-healing network can be bigger than that of the SHR. In general, a sectionalized network has higher system reliability (due to software failures), but less flexibility in spare capacity utilization, than a global network. Thus, to avoid bringing the entire DCS network down when a software failure occurs, sectionalization may need to be incorporated to improve the DCS network's reliability. However, when the DCS network is sectionalized, it loses flexibility for efficient bandwidth allocation and also may retain higher costs than the SHR for fiber placement, fiber material, and equipment. These reliability and capacity concerns suggest that some sectionalized, high-demand (with high embedded fiber connectivity) areas, such as the hub-to-hub subnetwork in the fiber-hubbed network, may be good deployment areas for the DCS self-healing network.[5] Section 5.4.2 discusses a quantitative analysis supporting this scenario.

5.4.2 Planning ADM Rings and DCS Self-Healing Networks

As discussed earlier, the DCS self-healing network has a higher degree of spare capacity sharing than the SHR, but it requires higher network connectivity and equipment costs. If a ring topology is considered, the DCS self-healing network may degenerate to a BSHR/2 (see Chapter 4) that, from a planning point of view, has higher equipment (DCS) costs than a BSHR/2 using ADMs. In this low-connectivity case, the

---

5. The BSHR/4 (see Chapter 4) may also be a good candidate for these areas if the number of hubs is very small.

245

DCS self-healing network cannot take full advantage of its sophisticated control system to save significant spare capacity. The SHR using ADMs may therefore be more cost-effective in this case.

For a higher-connectivity scenario (i.e., a mesh topology), a sophisticated network planning tool may be desirable to see whether or not the spare capacity savings for the DCS self-healing network outweigh the increased equipment (DCS) and connectivity costs. This analysis is performed on a network-by-network basis. However, it may be possible to understand relationships among economical factors of the DCS self-healing network without involving a complex network design tool. This concept, which is sometimes called *trend analysis*, has been successfully used to analyze different application conditions for SONET ring alternatives [21]. Trend analysis can formulate economical factors into a single measurement that represents the total network cost. Appendix B describes a model that may help provide some degree of insight into relationships among the economic factors of the DCS self-healing network. Understanding these relationships may help determine when implementing the DCS network is economically attractive (compared to rings and DP networks). This section summarizes only the important results of the analysis described in Appendix B. Details of the model can be found in Appendix B.

In the analysis discussed in Appendix B, a mesh demand pattern with a constant demand requirement is assumed for simplicity. The mesh demand pattern is a pattern where any two nodes have STS-1 demands between them. Appendix B analyzes factors that may be crucial to determining the economic merit of the DCS network relative to the ring alternative. These factors include network size, demand level, network connectivity, spare capacity savings, and equipment costs.

In summary, a large network with high demands and embedded high connectivity ensures the economic merit of the DCS self-healing network. Spare capacity savings alone may not be a significant factor in contributing to the overall economic merit of the DCS network (i.e., if combined with small size, low connectivity – like a ring topology – and relatively low demand). Reference [22] reports a similar result. Based on these analyses, a reasonably good scenario for the potential DCS self-healing network deployment is to deploy DCSs in a large network that has a high-demand requirement and an embedded, highly connected topology.

Based on the above discussions, an integrated restoration system may be derived to provide a cost-effective network survivability solution to existing fiber-hubbed networks. For fiber-hubbed networks, there are two hierarchies for office buildings. Several offices are grouped together as a cluster of offices and served by a central node, called a *hub*. The hub-to-hub subnetwork usually has a high degree of connectivity and carries high demands that are aggregated from smaller COs. In the cluster of a fiber-hubbed network, some office pairs can be installed with APS systems if these pairs have very high demand requirements that would exhaust SHRs quickly. The SHRs can

246

be used to interconnect smaller offices (within the cluster) and one or two hubs because demand and connectivity may not be high enough to justify adding DCSs at the smaller COs. If ring connectivity exists, these rings can be interconnected via a back-to-back ADM configuration and/or DCSs at hubs. Note that the hub used for ring interconnection serves as both an interconnection point and a demand concentration point (as it does in the hubbing architecture).

For highly connected hub-to-hub subnetworks, the DCSs already installed in each hub can be upgraded to provide the cost-effective DCS self-healing capability to reroute affected demands around the failure areas. The DCS-based network forms a "core" network in this architecture. Figure 5.14 depicts this integrated restoration system for the fiber-hubbed network.

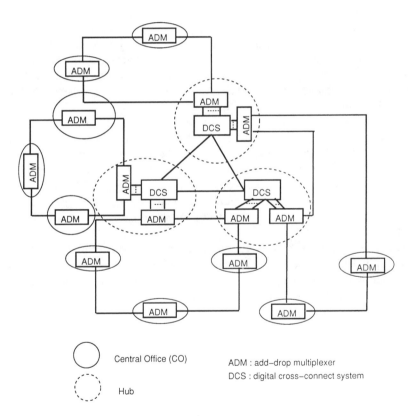

Figure 5.14. An integrated restoration system in a fiber-hubbed network.

247

5.5 INTEGRATED SONET RESTORATION SYSTEMS

5.5.1 An Integrated, Hierarchical Network Restoration Model

For most practical environments, network providers may find that their networks may not match the best characteristics of any single restoration alternative. Thus, a mixture of restoration alternatives, or a hybrid approach, is usually required. This hybrid approach enables network providers to match tariff levels with restoration performance in a cost-effective manner. For example, crucial and priority services (e.g., 911 services) can always be routed on rings or DP switching systems, and other services might use the DCS's alternative routing for restoration.

To provide cost-effective network restoration solutions for different levels of service protection requirements, providers may choose to share physical transport facilities. By provisioning restoration on a per-channel (STS-1 or VT) basis, virtual networks with different restoration techniques (i.e., ring, route diversity, or mesh) can share the same physical transport facilities. In such cases, facility sharing reduces network survivability costs but requires a more complex, integrated OS. This integrated OS views different restoration alternatives as a unit, whereas today's OSs view the restoration alternatives individually.

Restoration tradeoffs among different restoration architectures discussed in Section 5.4 may naturally result in a hierarchical restoration model for service protection, as depicted in Figure 5.15. Reference [16] reports a similar hierarchical restoration model. The hierarchy is structured (top to bottom) as follows: ring/route diverse protection, DCS-based mesh restoration, and operator control. The manual operator control is initiated after the above automatic network self-healing schemes cannot restore services completely, or when sufficient bandwidth is unavailable for DCS restoration.

The *classes of service routing* concept used in packet-switched data networks [23] may be applied to the integrated network restoration system. The service population can be partitioned into classes of services with different parameters based on the service restoration time requirement and costs that customers are willing to pay.

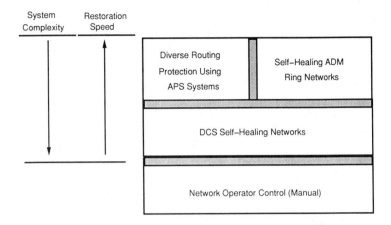

Figure 5.15. A hierarchical network restoration model.

Higher-priority demands can always be carried by the ring or 1:1 point-to-point DP system,[6] whereas lower-priority demands can be carried by the DCS self-healing network.

5.5.2 Integrated Restoration Systems with Spare Capacity Sharing

To reduce the spare capacity needed for the integrated network restoration system, the concept of sharing spare capacity among different restoration schemes has been considered. In general, the 1:1/DP system and the ring do not share spare capacity because they need to protect high-priority demand. Because the DCS self-healing network is essentially a logical layer network, which usually builds on the physical layer network (a network composed of rings and DP systems), sharing spare capacity of the physical network with the DCS network is possible. The DCS network may share spare capacity on the ring and/or DP systems.

Spare capacity sharing reduces the transport cost for protection, but at the expense of a more complex control system. This is because different restoration methods use

---

6. The point-to-point DP system may be used to carry demands between two nodes that have a very high demand requirement.

249

different equipment (e.g., 1:1 APS systems use OLTMs, rings use ADMs, and DCS self-healing networks use DCSs) and operate under different modes (e.g., 1:1 APS and rings have revertive and non-revertive modes, whereas currently proposed DCS self-healing schemes use only the non-revertive mode).

Two spare capacity sharing schemes are possible in an integrated self-healing system: spare facility sharing and spare channel sharing. The first approach is to share the protection facility of the point-to-point system and/or the SHRs with the DCS self-healing mesh network. In this scheme, the demands carried by the DCS mesh network can be restored via the protection facility of the point-to-point system or the ring that is in the normal operations mode. If the point-to-point system or the ring is in the restoration mode, then its protection facility cannot be used by the DCS mesh network for restoration. In this scheme, 1:1 APS, the ring, and the DCS network must operate in the revertive mode to leave the protection facility available for use whenever needed. If the DCS network operates in the non-revertive mode, it will occupy the spare facility of the ring or DP system until the next DCS network reconfiguration (i.e., next failure). During this period, if a network component of the ring or DP system fails, the ring or DP system may not restore services because that spare facility has been occupied by the demand carried on the DCS self-healing network. Therefore, the DCS self-healing scheme must operate in the revertive mode (after failed network components are repaired) to release the spare facility of the ring/DP system that it occupies during restoration.

This revertive operation mode presents a technical challenge for the DCS self-healing design, especially for the distributed DCS self-healing network, because it operates in an unpredictable environment. One method to implement the DCS network revertive mode uses the centralized OS to return the DCS network status back to the normal mode after failed components have been repaired. However, this approach interrupts ongoing services because the process of returning to normal mode may take at least a few minutes to complete. It should be noted that facility sharing between a USHR/P and a DCS network does not work due to the USHR's special self-healing system design (see Chapter 4 for more details).

The second spare capacity sharing approach is to share the "extra" spare channels in the protection facility of the DP system and/or the ring. Let x and y be the working channels and the line rate of the ring, respectively. The "extra" spare channels in the protection facility, which are available for use by the DCS self-healing network, are $y-x$, where x spare channels in the protection facility must be reserved for 1:1 protection for the ring or DP system. Thus, the DCS self-healing system uses only a portion of the ring protection facility, which is never used by the working ring for its 1:1 protection. Reference [24] describes an example of this spare capacity sharing approach.

Compared to the first spare capacity sharing approach, the second approach does not require the DCS network to operate in a revertive mode (i.e., it does not interfere with the ring operations in both the normal and failure situations). However, a more complex control system is needed because the spare capacity sharing is performed on the path level (e.g., end-to-end STS-1 paths) rather than on the line level (e.g., an entire protection facility), as in the first approach. The first approach may require less spare capacity than the second approach for the DCS self-healing network. However, it may lose some services carried on the DCS network if the shared protection facilities of the DP systems and/or rings are not available for use when the DCS network initiates restoration. Chapter 7 discusses a network design method based on spare channel sharing.

5.6 SUMMARY AND REMARKS

Intelligent capabilities of the SONET DCS offer new network capabilities for fast provisioning and flexible customer network control, as well as for efficient network restoration. The use of EOCs within the SONET structure makes the SONET DCS even more powerful in terms of network management. Thus, SONET DCSs can be used to enhance services, which increases revenue, utilize network capacities efficiently, and enhance service survivability. The hardware and software systems for DCS self-healing networks are still in the development stage, although many self-healing protocols have been proposed.

The DCS self-healing networks can be implemented using either a centralized or a distributed self-healing control architecture. The centralized DCS self-healing architecture takes advantage of the centralized OS for fast provisioning and flexible network control. The distributed DCS self-healing control architecture has a restoration speed advantage over its centralized counterpart but requires a more complex distributed control system. The selection of an appropriate DCS self-healing network architecture depends on many technical and non-technical factors. From cost and performance points of view, the centralized architecture is likely to be used for fast service provisioning and flexible network control, which follows the traditional centralized circuit-switched voice network philosophy. The distributed architecture is likely to be used for network restoration against catastrophic failures, which follows the traditional packet-switched data network philosophy.

An integrated network restoration strategy is needed for most networks, because no single restoration alternative can provide all needed restoration characteristics. Integrating different network restoration methods and meeting network requirements in a cost-effective manner are big challenges for network planners and engineers. The resulting integrated strategy will define the target SONET network architecture for their networks. Due to the high degree of planning and engineering complexity involved in

251

an integrated restoration strategy, CAD tools that consider all survivable network alternatives may be necessary to ensure that the SONET target network is being defined and built in a cost-effective manner. (Chapter 7 will discuss some methodologies that may be used to plan and design the SONET networks using rings and DP switching systems.)

The role of passive optical components in future DCS self-healing networks is still under study. Just as passive optical components may be applied to automatic DP switching systems (as described in Chapter 3) and SONET SHRs (as described in Chapter 4), it may be possible to use optical DCSs to reduce protection costs of DCS self-healing networks by reducing the number of spare terminations in SONET DCSs. The future of DCS network deployment is also likely to compete with the new ATM/DCS technology, which employs the virtual path concept. The ATM/DCS network uses SONET as a physical transport medium and combines merits of SONET DCS and flexible, fast packet technologies to provide efficient and cost-effective services. These services may include services that are supported by the SONET DCS network. Chapter 9 will discuss the perspectives and challenges for the ATM/DCS network technology.

REFERENCES

[1] Datapro Reports on Telecommunications, "Private Line Facilities," TC25-003, Datapro Research Group, December 1990.

[2] Doverspike, R. D., and Pack, C. D., "Using SONET for Dynamic Bandwidth Management in Local Exchange Networks," *Proceedings of 5th International Network Planning Symposium,* Kobe, Japan, June 1992.

[3] Doherty, D., Hutcheson, W., and Raychaudhuri, K., "High Capacity Digital Network Management and Control," *Proceedings of IEEE GLOBECOM'90,* San Diego, CA, December 1990, pp. 301.3.1-301.3.5.

[4] Wu, T-H., Kolar, D. J., and Cardwell, R. H., "Survivable Network Architectures for Broadband Fiber Optic Networks: Model and Performance Comparisons," *IEEE Journal of Lightwave Technology,* Vol. 6, No. 11, November 1988, pp. 1698-1709.

[5] Ash, G. R., Cardwell, R. H., and Murray, R. P., "Design and Optimization of Networks with Dynamic Routing," *Bell System Technical Journal,* Vol. 60, No. 8, October 1981.

[6] Krishnan, K. R., and Ott, T. J., "Forward-Looking Routing: A New State-Dependent Routing Scheme," *Proceedings of the 12th International Teletraffic Congress,* Torino, Italy, June 1988.

[7] Doverspike, R. D., and Jha, V., "Comparison of Routing Methods for DCS-Switched Network," Special Telecommunications Issue of *Interfaces,* to be published.

[8] TA-NWT-001068, *Generic Requirements for the Switched DS1/Switched Fractional DS1 Service Capability from a Non-ISDN Interface (SWF-DS1/Non-ISDN),* Issue 2, Bellcore, April 1991.

[9]   Doverspike, R. D., "Restoration Switching of Technical Report on Network Survivability Performance," T1Q1.2/91-038, July 1991.

[10] Yamada, J., and Inoue, A., "Intelligent Path Assignment Control for Network Survivability and Fairness," *Proceedings of IEEE ICC'91*, Denver, CO, June 1991, pp. 22.3.1-22.3.5.

[11] Grover, W. D., Venables, B. D., MacGregor, M. H., and Sandham, J. H., "Development and Performance Assessment of a Distributed Asynchronous Protocol for Real-Time Network Restoration," *IEEE Journal on Selected Areas in Communications*, Vol. 9, No. 1, January 1991, pp. 112-125.

[12] Yang, C. H., and Hasegawa, S., "FITNESS: A Failure Immunization Technology for Network Service Survivability," *Conference Records of IEEE GLOBECOM'88*, Ft. Lauderdale, FL, December 1988, pp. 1549-1554.

[13] Coan, B. A., Vecchi, M. P., and Wu, L. T., "A Distributed Protocol to Improve the Survivability of Trunk Networks," *Proceedings of the XIII International Switching Symposium*, May 1990, pp. 173-179.

[14] Komine, H., et al, "A Distributed Restoration Algorithm for Multi-Link and Node Failures of Transport Networks," *Proceedings of IEEE GLOBECOM'90*, San Diego, CA, December 1990, pp. 403.4.1-403.4.5.

[15] Sakauchi, H., et al, "A Self-Healing Network with an Economical Spare-Channel Assignment," *Proceedings of IEEE GLOBECOM'90*, San Diego, CA, December 1990, pp. 403.1.1-403.1.6.

[16] Ellefson, F., "Migration of Fault Tolerant Networks," *Conference Records of IEEE GLOBECOM'90*, San Diego, CA, December 1990, pp. 301.4.1-301.4.7.

[17] Pekarske, B., "1.5 Second Restoration Using DS3 Cross-Connects," *Trends in Network Restoration Symposium*, Edmonton, Canada, May 24-25, 1990. (Also, "Restoration in a Flash - Using DS3 Cross-Connects," *TELEPHONY*, September 1990, pp. 35-40.)

[18] Chujo, T., Komine, H., Miyazaki, K., Ogura, T., and Soelima, T., "The Design and Simulation of an Intelligent Transport Network with Distributed Control," *Network Operations Management Symposium*, San Diego, CA, February 1990.

[19] Fujii, H., Hara, T., and Yoshikai, N., "Characteristics of Double Search Self-Healing Algorithm for SDH Networks," IEICE CS91-48, 1991.

[20] "AT&T Services Problem Incites Network Security Problems," *Telephone Engineer and Management*, February 15, 1990.

[21] Wu, T-H., and Lau, R. C., "A Class of Self-Healing Ring Architectures for SONET Network Applications," *IEEE GLOBECOM'90*, San Diego, CA, December 1990, pp. 403.2.1-403.2.8.

[22] Tsai, E. I., Coan, B. A., Kerner, M., and Vecchi, M. P., "A Comparison of Strategies for Survivable Network Design: Reconfigurable and Conventional Approaches," *Proceedings of IEEE GLOBECOM'90*, San Diego, CA, December 1990, pp. 301.1.1-301.1.7.

[23] Shaneyfelt, B., and Wu, T-H., "Statistical Analysis of Network Traffic for Class of Service in Packet Switched Data Networks," *IEEE Journal on Selected Areas in Communications*, Vol. 6, No. 4, May 1988, pp. 766-777.

[24] Okanoue, Y., Sakauchi, H., and Haseqawa, S., "Design and Control Issues of Integrated Self-Healing Networks in SONET," *Proceedings of IEEE GLOBECOM'91*, Phoenix, AZ, December 1991, pp. 22.1.1-22.1.6.

CHAPTER 6

Survivable Fiber-Hubbed Network Design

6.1 SURVIVABLE FIBER NETWORK DESIGN CONCEPTS

As discussed in Chapter 1, hubbing may be the most economical fiber network architecture to utilize fiber's high capacity, but it is the least attractive network architecture from a survivability point of view. Automatic diverse routing protection and dual homing are two major network protection strategies used to provide service protection for fiber-hubbed networks. The diverse routing protection strategy provides protection against fiber cable cuts, while dual homing provides protection against major CO hub failures. For the diverse routing protection strategy, 1:1 or 1:N protection may be used depending upon economics and the survivability requirement. For example, today's gigabit-per-second fiber systems are usually protected on a 1:1 basis because the consequences of losing services from these high-speed systems are simply too costly from a business point of view.

Each protection strategy has its level of protection and cost. For example, dual homing provides service continuity for some services when one of the major CO hubs fails. However, the cost for providing such dual-CO protection may be very expensive, and, therefore, it may be implemented only in special cases. Thus, a challenge of planning network survivability is how to choose affordable protection strategies that meet the network or customer survivability requirements. This is the core issue that will be addressed in this chapter and in Chapter 7.

To plan an acceptable network protection strategy for fiber-hubbed networks in terms of cost and survivability requirements, network design tools are needed due to the complexity of choosing appropriate protection strategies for practical fiber networks. Network design tools for planning survivable fiber-hubbed networks are available in some industrial research laboratories. These include Bellcore's *Network Planning System* (NPS) [1], Bellcore's FIBER OPTIONS program [2], Northern Telecom's NetMate Fiber Center Planners [3], and so forth. Each design system has different target application areas and users. For example, the NPS is a complex software system that has an embedded network database. It helps planners and engineers plan an appropriate network transition toward a target network according to today's network environment. On the other hand, FIBER OPTIONS is a simple PC-software program that searches for a target network architecture and helps to set general applications guidelines by analyzing cost and survivability tradeoffs associated with different types of protection strategies. Because the goal of this chapter is to focus on architectural analysis of different survivability strategies, the network design model discussed throughout this chapter primarily follows the one used in FIBER OPTIONS [2]. However, other more appropriate design methods are also discussed wherever applicable.

6.1.1 The Fiber-Hubbed Network Engineering Concept

The following characteristics of fiber have made traditional telephone network design, which uses the mesh topology to optimize link bandwidth and repeater spacing parameters, inappropriate for fiber network design:

- Very high bandwidth
- Long repeater spacing
- Relatively low fiber facility costs
- High fiber-terminating equipment costs.

The economies of scale inherent in high fiber capacity and expensive terminating equipment dictate that demand should be aggregated whenever possible. This results in a hubbing network architecture similar to today's airline and U.S. mail routing systems.

The hubbing concept is essentially one of centralized demand routing. To aggregate demand from a CO so fiber is well utilized, all the demand from the CO is multiplexed on a fiber *span* having terminals in the CO and the hub, as depicted in Figure 6.1. Each fiber *span* is composed of all the terminal electronics and fibers (working and protection) and may pass through[1] one or more intermediate CO on the way to its home hub. At the hub, point-to-point DS1 demands large enough to justify one or more DS3s are cross-connected on a DS3 basis to the multiplex span destined for the proper CO, i.e., not demultiplexed via the hub DCS 3/1. Demand insufficient to fill a DS3 economically is placed into a DS3 with demands for other destinations and sent to the hub DCS 3/1, which rearranges DS1s within DS3s and places each point-to-point demand on a DS3 going to the proper destination. This concept of routing all demand to the hub building is called *hub routing,* which has proven its ability to give near-optimal routing results in terms of economics for intraLATA fiber networks [4].

---

1. Fiber is spliced at intermediate COs.

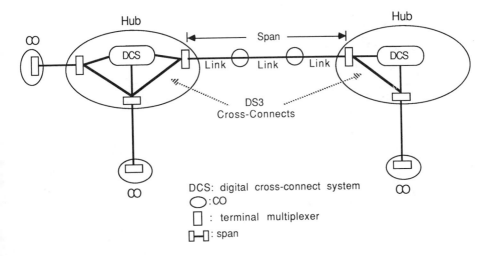

Figure 6.1. Multiplex span and demand aggregation for facility hubbing.

To describe the fiber network design concepts and methods, we assume that a three-level hubbing architecture is used as an example: CO, hubs, and gateways [5,6]. The gateway is also a hub. A group of COs served by the same hub forms a *cluster,* and a group of clusters served by the same gateway hub forms a *sector.* Gateways are fully connected to each other by fiber spans. Figure 6.2 shows such a three-level, fiber-hubbed network architecture.

A fiber-hubbed network can be represented by a span layout, which is a set of fiber spans. Two types of fiber spans are generally used in fiber network design: the *point-to-point span* and the *hubbing span.* A point-to-point span terminates at two nodes belonging to the same facility hierarchy. An example of the point-to-point span is a fiber span that terminates at two non-hub COs [e.g., Span (6,7) in Figure 6.3]. A hubbing span is any span that is not a point-to-point span. An example of a hubbing span is a span that terminates at a CO and its hub [e.g., Span (3,4) in Figure 6.3].

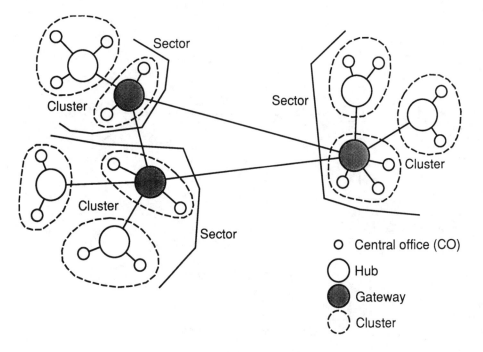

Figure 6.2. Three-level fiber-hubbing network.

A hubbing span always terminates at a DCS of the hub (or gateway), and a point-to-point span bypasses the DCS at the hub. A span layout is called a *centralized span layout* if all its spans are hubbing spans, whereas it is called a *non-centralized span layout* if its spans include at least one point-to-point span. Using the hubbing span or the point-to-point span in a fiber network depends on economics and network survivability requirements.

Figure 6.3 depicts a possible fiber span layout and its demand distribution. The network consists of nine COs with two Hubs, CO-3 and CO-5, where CO-1, CO-2, and CO-4 are assigned to Hub 1 (CO-3), and CO-6, CO-7, and CO-8 are assigned to Hub 2 (CO-5). The example is a non-centralized span layout that includes five spans. Among them, spans #1, #2, and #5 are hubbing spans, and spans #3 and #4 are point-to-point spans, where span #3 is a hub-to-hub span. For demand pair (4, 8), five DS3s are routed over spans as follows: span (3, 4)-span (3, 5)-span (5, 8). Span (3, 5) also carries eight DS3s between CO-3 and CO-5 and four DS3s from pair (3, 8), which results in a total of 17 DS3s including five DS3s from demand pair (4, 8). Note that it is also possible to have demands for pair (5, 8) carried by one point-to-point span, span (7, 8), and a hubbing span, span (5, 7) (not shown in Figure 6.3). The decision to use a hubbing span [i.e., span (5, 8)] or one point-to-point span and one hubbing span depends upon relative economics and survivability considerations.

Demand pair	DS3 requirement
(1,3)	2
(1,4)	3
(4,8)	5
(3,8)	4
(6,7)	6
(3,5)	8

Span#	(s,d)	Span path	DS3s
1	(1,3)	1-2-3	5
2	(3,4)	3-4	8
3	(3,5)	3-9-5	17
4	(6,7)	6-5-7	6
5	(5,8)	5-7-8	9

s = source, d= destination

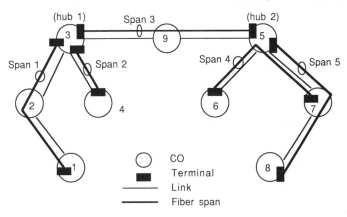

Figure 6.3. A fiber span layout.

6.1.2 Survivable Fiber Network Design

The increasing deployment of interoffice optical fiber transmission systems with large cross-sections supported on a few strands of fiber and the trend toward a fiber-hubbed network architecture have increased concern about the survivability of intraLATA fiber communications networks. As used here, network survivability denotes the network's capability to recover from a single network failure. The hubbing architecture is the network architecture of choice from an economic point of view, but it is vulnerable to

259

facility and office failures (e.g., fiber cable cuts and major hub failures). If a cable between a CO and its hub is cut, that CO will be isolated from other network portions. If a major hub fails, all surrounding COs served by this hub are affected.

Two approaches to designing a survivable fiber network architecture are (1) to design a network with a built-in protection capability so the demand can be switched to a protection system if a network component fails, and (2) to use intelligent NEs to reroute demand through spare transmission paths if a network component fails. The first approach is referred to as *physical layer network design,* which incorporates conventional survivable network architectures (such as APS and dual homing) and SHR architectures. The network using the second approach is referred to as a *logical layer network design,* which uses intelligent DCSs to restore demands if network components fail, provided that the spare capacities exist. Table 1-1 (see Chapter 1) discussed tradeoffs between the physical layer network design and the logical layer network design. This chapter focuses on physical layer network design using conventional restoration techniques, i.e., the physical design incorporating APS and dual homing. These design methods form a component of a SONET network design model (see Chapter 7) that uses a combination of diverse routing protection architectures, SHR architectures, and DCS self-healing architectures.

This chapter defines the fiber network design problem and then discusses a survivable fiber-hubbed network design model and a multi-period capacity assignment model.

6.2 FIBER NETWORK DESIGN PROBLEM

6.2.1 Problem Formulation

The problem of fiber network design can be formulated as follows:

Given: 1. Hubbing network architecture
2. Candidate survivable architecture
3. Link locations and corresponding link distance list
4. End-to-end circuit (64-kbps) demands
5. Cost associated with each network component
6. A list of candidate fiber line rates.

Objective: Build a network that minimizes total network costs.

Subject to: 1. Route all end-to-end demand requirements.
              2. Satisfy any network survivability requirement.

The proposed model does not consider the performance constraints of delay and call-blocking probability because the considered fiber facility network is defined for deterministic forecast capacity requirements, e.g., for the public switched trunk network. The candidate survivable network architecture is any one of the architectures described in Table 3-1 (see Chapter 3) that includes 1:N with and without DP and dual homing. The network cost in the model is the sum of the total facility and equipment costs, including the DCS termination cost, the working and protection terminal cost, the automatic protection switching cost, the fiber material cost, the fiber placement cost, and the regenerator cost. For simplicity, we assume all demand originates at the DS0 (circuit) level because the considered fiber network is a facility network.

The survivability requirement depends on applications that use two approaches. Given several candidate survivable network architectures, the first approach is to compute a minimum cost design and associated network survivability according to the given network connectivity constraint. Each architecture is compared in terms of costs and survivability. These cost and survivability factors can be considered in a single measure, as will be discussed in Section 6.3.3.3. The network connectivity constraint is sometimes referred to as the *qualitative survivability constraint*.

The second approach, given the network survivability level that the network should achieve, is to compare design costs of different survivable network architectures to achieve that level of network survivability. The network survivability constraint associated with this approach should be a quantitative constraint such as the average number of lost calls (if restricted to the voice network) or the percentage of circuits lost due to a link failure.

Network design with the quantitative constraint ensures that the network has a required network survivability level. Network design with a qualitative survivability constraint may ensure only that the network meets the connectivity requirement needed for implementing the candidate survivable network architecture, such as the DP or the dual-homing architecture, and may not necessarily ensure that the desired network survivability level is attained. Network design with the quantitative survivability requirement is more complex than design with the qualitative survivability requirement but can be derived from repeatedly solving network design problems with the network connectivity constraint.

6.2.2 Fiber Network Survivability Measurements

Network survivability is measured based on the type of networks considered. For non-switched networks, such as SONET SHRs, the appropriate survivability measurement would be the percentage of circuits lost due to a network component failure. Non-switched survivability measurements can also be evaluated in a worst or average case, depending on the applications. The worst case allows engineers and planners to locate the most vulnerable segment of the network and is typically used for strategic planning. The average survivability measurement is usually used when the failure probability of each network component and the average time to restore service due to this failure are known. This chapter uses the non-switched network survivability in the worst case because the networks addressed here are non-switched facility networks.

For switched fiber networks, the measurement would be the average network blocking due to a failure, or the average number of lost calls due to a failure, depending upon the types of switched fiber networks considered. In circuit-switched voice networks, existing calls are lost upon failure. These lost calls are expected to retry, and therefore, the measurement is lost calls over time. However, for high-bandwidth switched networks, such as SONET switched networks using DCSs (see Chapter 5), the holding time is long (compared to the outage), and retry may be more difficult; thus, existing demands are rerouted. Existing demands will be lost if there is no sufficient spare capacity to reroute these demands. Thus, a measurement for this type of network would be the number of lost demands due to a failure. The second type of switched network is considered in Chapter 7.

Service survivability is measured similarly to network survivability but is viewed from a user perspective. Acceptable network survivability does not imply an acceptable user's service survivability. For example, if only a single path connects a customer and the serving CO and that single path fails, the customer will lose all services even if the network is protected on a 1:1 basis (i.e., 100 percent survivability).

Figure 6.4 depicts an example of how to evaluate service survivability and its economic impact. For example, assume d_{u1}=6 DS1s, d_{u2}=3 DS1s, and d_{u3}=4 DS1s, where d_{ui} is the DS1 demand requirement between user C_u and user C_i. The normal routing path for demand pair (C_u, C_i) is also depicted in Figure 6.4. For example, DS1 demands between pair (C_u, C_1) are routed through path C_u-CO-1-CO-3-CO-2-C_1. If the link between CO-2 and CO-3 fails, it disconnects the path between customers C_u and C1. We assume that three DS1s can be restored as when a 1:2/DP system is engineered for the fiber span between CO-2 and CO-3. Thus, the service survivability from customer C_u's point of view is 77 percent when that link fails because, the total number of DS1s that are still intact from customer C_u's point of view is 10 (3+3+4) compared to a total of 13 DS1s.

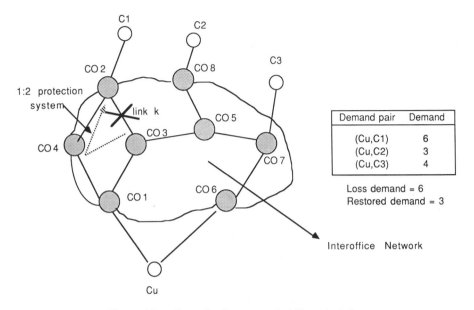

Demand pair	Demand
(Cu,C1)	6
(Cu,C2)	3
(Cu,C3)	4

Loss demand = 6
Restored demand = 3

Interoffice Network

Figure 6.4. Example of a user survivability calculation.

6.3 SINGLE-PERIOD SURVIVABLE FIBER-HUBBED NETWORK DESIGN

The survivable fiber-hubbed network design problem described in Section 6.2 is complex and is unlikely to be solved in a globally optimized manner because the problem involves demand routing, multiplexing, connectivity, and many cost factors that need to be optimized. Reference [4] proposes a model for optimizing routing and multiplexing simultaneously for fiber networks that uses an integer programming approach. However, this model is practical only for small networks. To balance the computational complexity and solution optimality for large fiber network design, the fiber network design problem needs to be divided into two subproblems [5]: network topology selection and network span layout. The topology selection relates to the connectivity requirement, while the multiplexed network span layout relates to the demand requirement. Figure 6.5 shows two subproblem formulations. The methods described in Reference [5] have been tested using many intraLATA fiber networks, and the results indicated that this problem-splitting technique is reasonable. Note that the methodologies described in this section are for a single-year planning period. Section 6.4 discusses some methodologies for multi-period network capacity planning.

263

Given: 1. Hubbing network architecture
        2. Candidate survivable architecture
        3. Link locations and distances
        4. Costs for fiber material, placement,
           and regenerator

Objective: Design a network topology such that
            the total cost of fiber, fiber placement,
            and regenerators is minimized.

Subject to: Network connectivity requirement.

(a) Network Topology Design Problem

Given: 1. Hubbing network architecture
        2. Candidate survivable architecture
        3. Network topology [from Problem (a)]
        4. End–to–end circuit and DS3 demands
        5. Costs for fiber material, placement,
           regenerator and terminating equipment

Object: Design a multiplexing network layout
         such that the total cost of fiber material,
         fiber placement, regenerators and
         terminating equipment is minimized.

Subject to: End–to–end DS3 demand requirement.

(b) Multiplexing Network Layout Problem

Figure 6.5.   Network design subproblem formulations.

The network design model depicted in Figure 6.5 first selects a feasible network topology that meets the network connectivity constraint for the candidate survivable network architecture. It then builds a minimum-cost multiplex span layout that meets the demand requirement and the demand restoration technique specified by the candidate survivable architecture. This process then perturbs the current network topology and repeats the same multiplex span layout procedure until no further improvement is possible or until some stopping criteria are met. A good, initial feasible topology selection can reduce many unnecessary iterations that would eventually reduce the computational complexity of the design model. This results in an alternative approach that designs a minimum-cost network topology and a minimum-

264

cost span layout based on the selected topology. Additionally, this approach involves no topology perturbation once the topology is finalized in the topology design phase.

Figure 6.6 is a flowchart of a single-period network design model. In this model, the topology design module generates a minimum-cost network connectivity that meets the connectivity requirement for the candidate survivable network architecture. The cost is the total cost of fiber materials, fiber placement, and regenerators, because these three cost factors affect the topology selection. Topology selection serves as a starting point for the network span layout to simplify the design complexity.

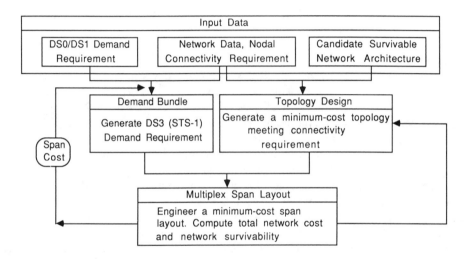

Figure 6.6. A flowchart of a network design method.

Based on the selected topology, the network span layout problem can be stated as follows: How do we generate a minimum-cost multiplexing network layout that determines the amount of fiber facilities and equipment needed to meet demand and protection requirements for the candidate survivable network architecture? The costs in the network span layout are the transport costs, which include costs of terminating multiplexers, fiber material and splicing, fiber placement, and regenerators. This design phase includes demand multiplexing and routing. Because the basic signal transport unit in today's fiber transport systems uses the DS3 signal level (or STS-1 in the SONET network), it is necessary to bundle the end-to-end circuit or DS1 requirements to the end-to-end DS3 (or STS-1) requirements to better utilize DS3 capacities before engineering the networks.

In Sections 6.3.1 through 6.3.4, we discuss the demand-bundling problem (Section 6.3.1), which maximizes the DS3 or STS-1 utilization. We then discuss two major design modules: network topology selection (Section 6.3.2) and multiplexing network span layout (Section 6.3.3). These two design modules are constrained by a qualitative survivability requirement (i.e., network connectivity requirement). We then briefly review some design methods for the quantitative survivability constraint (Section 6.3.4).

6.3.1 Demand-Bundling Algorithms

The demand-bundling algorithm combines point-to-point circuits into appropriate DS3-level demands (or STS-1s in the SONET networks). This process is needed because, otherwise, each DS1 or DS0 demand pair would route over a unique DS3, thus resulting in low-filled DS3 systems and high network costs. The DS3 (or STS-1) signal level is commonly used as input to fiber systems in today's interoffice fiber networks (or SONET networks). For convenience, this section discusses DS3 interfaces only. However, the design method can readily be extended to SONET interoffice networks.

Two demand-bundling algorithms have been proposed in References [5,7]. The first [5] is a single-period demand-bundling algorithm that uses a demand-bundling concept similar to the one used in designing hierarchical trunking networks [8], which assigns traffic (in erlangs) to circuits. The second algorithm [7] is a more sophisticated, multi-period demand-bundling algorithm that is performed in two phases. Phase 1 solves a single-period static model, which first relaxes capacity constraints by rerouting all demands over alternate paths and then enforces the constraints by solving a 0-1 multiconstraint knapsack problem to retain the demands with the most cost-effective routes. Phase 2 performs a more detailed, multi-period optimization on each rerouting path chosen in Phase 1.

For simplicity, we will discuss the single-period demand-bundling concept from Reference [5]. More information on the multi-period demand-bundling algorithm can be found in Reference [7]. As mentioned previously, the single-period algorithm [5] is similar to the algorithm used in designing hierarchical trunking networks [8]. In trunking network design, the fundamental problem involves routing calls (in erlangs or Hundred Call Seconds) over prioritized paths through switches. If the primary path is full, the alternate path is used. The *Economic Hundred Call Seconds* (ECCS) method uses a hierarchical scheme similar to hubbing to prioritize the paths and size the trunk groups in such a way as to minimize trunk and trunk termination costs. The resulting trunk capacity requirements become DS0-level circuit demand requirements for the bundling algorithm.

The demand-bundling algorithm, which routes circuits over DS3s, uses a similar hierarchical approach to evaluate the economics of routing circuits directly between two COs or using an intermediate hub DCS to further aggregate the demand. The DCS discussed here is a DCS 3/1 that terminates DS3s and cross-connects DS1s; it can be a SONET W-DCS when the SONET network is considered. A SONET W-DCS terminates fibers and cross-connects VT signals (see Chapter 2).

Routing involves two types of paths: direct and indirect. A DS1 that routes over a direct path does not demultiplex or rearrange signals at intermediate offices, i.e., it routes them over a single DS3 system from origination to termination. This type of DS3 is called a direct DS3. The indirect path consists of two or more DS3s formed by demultiplexing or rearranging signals at intermediate hub locations. The DS3s may be classified by their role in direct or indirect paths. Indirect DS3s can be further classified as follows:

- CO-to-home hub DS3s, which originate from local CO demand and terminate on the home hub DCS of that CO
- CO-to-foreign hub DS3s, which originate in a CO and terminate on a hub DCS other than the home hub of that CO
- Hub-to-hub DS3s, which terminate on a hub DCS 3/1 at both hubs.

In the considered bundling algorithm, the number of circuits is first converted to DS1s by dividing the given number by 24 (the number of circuits in a DS1) and rounding up. Once the end-to-end DS1 requirement is obtained, the bundling algorithm follows the ordered steps depicted in Figure 6.7 to build a set of DS3 demand requirements. The algorithm first builds direct DS3s for each CO, if possible, and then processes the next higher facility hierarchy. It completes its process by a path from the highest gateway to the home hub and back to the CO.

Once the number of DS1s is obtained, the economics of forming a direct CO-to-CO DS3 is calculated, or a user-specified threshold (called DS3 direct trigger point) is applied. If the number of DS1s is greater than the DS3 direct trigger point or, alternately, if the cost of the direct DS3 is deemed less expensive than the cost of the alternate hub DCS route, a direct DS3 system is formed. If there are more circuits than one DS3 can carry, the process is repeated until the number of remaining DS1s does not exceed the criterion for direct routing. In Figure 6.8, the DS3 direct trigger point is assumed to be 20 DS1s. Thus, the demand requirement of 26 DS1s between CO-1 and CO-2 is enough to build a direct DS3 system between CO-1 and CO-2.

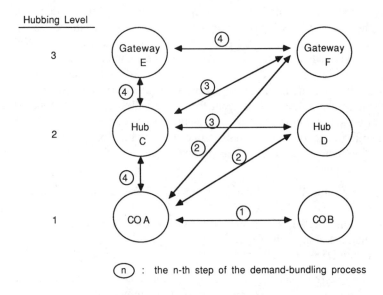

Hubbing Level

⟨n⟩ : the n-th step of the demand-bundling process

Figure 6.7. DS1 routing hierarchy.

For the unrouted DS1s, two options may exist: dual or single bundling. For dual bundling, DS1s are assigned equally between two hub DCSs if the two COs considered are in different clusters. For single bundling, all unrouted DS1s are routed to the home hub DCS of the lowest numbered CO. The planner determines which option to use. Figure 6.8 depicts an example of how to assign indirect DS3 systems for an eight-node network. The DS1 parcel for each node is obtained by using the dual-bundle method. For example, a total of 19 DS1s between CO-1 and CO-6 (they both belong to different clusters) is divided into two parts: nine DS1s are assigned to Hub A's DCS (in CO-1) and 10 DS1s are assigned to Hub B's DCS (in CO-2), where Hub B is the home hub of CO-6. The resulting DS1 parcel for each hub's DCS is also shown in Figure 6.8.

After all the CO-to-CO circuit demand pairs have been processed, DS3s must be formed between buildings and other hub DCSs. In this step, the bundling algorithm forms DS3s that carry DS1s and their associated circuits in the most direct route possible through the minimum number of hub DCSs. This process includes paths from a building to a foreign hub and paths from that foreign hub to the endpoint of the DS1.

For example, consider forming an indirect DS3 between CO-1 and Hub C's DCS (i.e., CO-5). The total available demand for this particular DS3 is 20 DS1s, which is the sum of the parcel between CO-1 and CO-5 and the parcel between CO-1 and CO-6 (see parcel list C).

The indirect DS3-forming process continues by following the hierarchical approach depicted in Figure 6.7. The final result is shown in Figure 6.8. Note that, according to this bundling algorithm, a DS1 may appear in different DS3s. For example, in Figure 6.8, some DS1s between CO-1 and CO-5 are routed through two indirect DS3 systems: one DS3 is between CO-1 and Hub D's DCS (in CO-7), and the other is between Hub D's DCS and CO-5.

6.3.1.1 Cost Calculations

As mentioned previously, if the fiber transport and DCS termination costs are provided, the bundling process can calculate the costs of making a direct DS3, as well as the cost of the alternate route that uses hub DCSs to improve fill. The transport cost of the direct route is calculated as the actual cost of the DS3 section involved (the sum of all the DS3s along the routing path). The transport cost of the alternate route is calculated assuming that using the additional DCS will result in 100 percent fill. The cost of multiplexing to the DS3 level from the DS1 level is also included, and once again the cost of the direct route is calculated using actual demand; the cost of the alternate route is calculated assuming 100 percent fill.

Figure 6.9 depicts an example of the above procedure. Here the decision to be made is whether to establish a direct DS3 from CO-A to CO-B or to use the alternate route from CO-A to the hub DCS and then to CO-B. The costs are as shown; note that the DS1 basis as the fill is assumed to be 100 percent. The term M13 refers to a multiplexer from the DS1 signal rate to the DS3 signal rate. In the example in Figure 6.9, we assume that the M13 common unit cost is $7650; the DCS 3/1 termination cost is $1325 per DS3; and the cost of an M13 plug, which accommodates four DS1s, is $540.

The transport cost for the hub route is the sum of the transmission cost for the hub route, the DCS termination cost, and the M13 termination cost. For example,

DCS termination cost/DS1	$(1325/28) \times 2 = 94.60$
M13 termination cost/DS1	$[(7650/28)+(540/4)] \times 2 = 816.40$
Transmission cost/DS1	$250+800 = 1050.00$
	1961.00 per DSI

For the direct DS3 route, the common cost is $24,300 [i.e., $9000+(7650 \times 2)$], and the per-plug cost is $1080 (i.e., 540×2). The numbers in Figure 6.9 indicate that the direct DS3 would be formed if the number of DS1s exceeded 14; otherwise, the hub DCS path would be used.

269

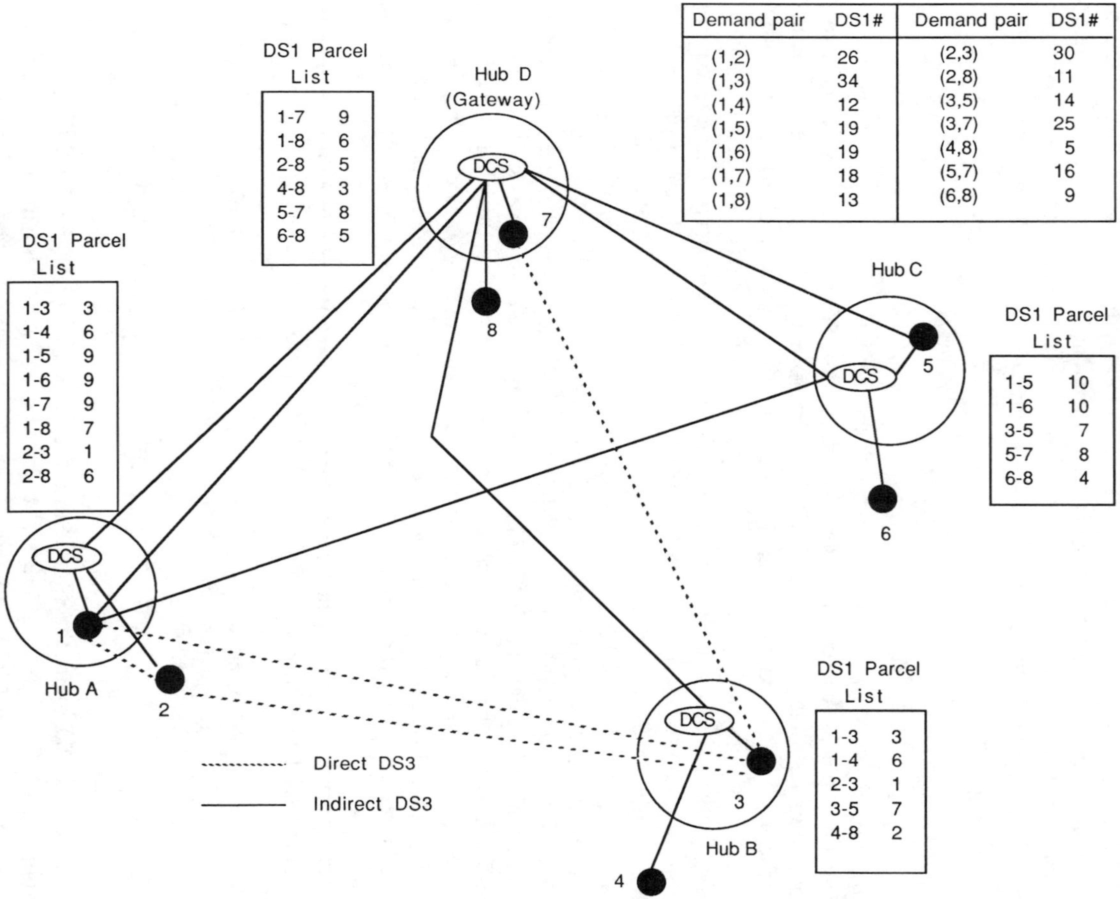

Demand pair	DS1#	Demand pair	DS1#
(1,2)	26	(2,3)	30
(1,3)	34	(2,8)	11
(1,4)	12	(3,5)	14
(1,5)	19	(3,7)	25
(1,6)	19	(4,8)	5
(1,7)	18	(5,7)	16
(1,8)	13	(6,8)	9

DS1 Parcel List (Hub C)

1-5	10
1-6	10
3-5	7
5-7	8
6-8	4

DS1 Parcel List (Hub B)

1-3	3
1-4	6
2-3	1
3-5	7
4-8	2

DS1 Parcel List (Hub D Gateway)

1-7	9
1-8	6
2-8	5
4-8	3
5-7	8
6-8	5

DS1 Parcel List (Hub A)

1-3	3
1-4	6
1-5	9
1-6	9
1-7	9
1-8	7
2-3	1
2-8	6

.... Direct DS3

———— Indirect DS3

Figure 6.8. DS3-forming example.

270

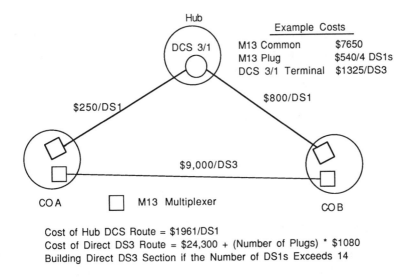

Cost of Hub DCS Route = $1961/DS1
Cost of Direct DS3 Route = $24,300 + (Number of Plugs) * $1080
Building Direct DS3 Section if the Number of DS1s Exceeds 14

Figure 6.9. Threshold example for cost calculation option of demand-bundling process.

6.3.2 Two-Connected Network Topology Design

Given the CO locations and connectivity requirements and a list of potential fiber links between COs, the topology design is intended to select these links to be equipped with fiber and, thus, provide a minimum-cost network that meets the network connectivity requirement. Each CO is either a special CO, which has a connectivity of two, or an ordinary CO, which has a connectivity of one. The particular network architecture selected by the user determines the network connectivity requirement. Table 6-1 depicts a relationship between the network architecture and the network connectivity requirement.

As shown in Table 6-1, the 1:N/DP architecture requires a two-connected topology and two link-disjoint paths between a special CO and a hub. The dual-homing architecture requires a two-connected topology and two node-disjoint paths from the special CO to its home hub and to its foreign hub. A two-connected network is a network where any two nodes in the network have at least two disjoint paths (either link-disjoint or node-disjoint). The two-connectivity, rather than k-connectivity (k>2), requirement is considered here because it is unlikely that two or more network components (link or node) will fail simultaneously for interoffice fiber networks.

271

Table 6-1. Relationship between Network Architectures and Connectivity Requirements

Network architecture	Topology connectivity requirement	Disjoint path requirement
1:N	single-connected topology (tree)	none
1:N/DP	two-connected topology	link-disjoint
Dual Homing	two-connected topology	node-disjoint*

1:N/DP = 1:N with diverse protection
* Node-disjoint paths are also link-disjoint paths.

The cost factors affecting topology selection are fiber placement costs, fiber material and splicing costs, and regenerator costs, which are functions of distance. The fiber placement cost is the cost of fiber installation per mile multiplied by the total route mileage in the topology.

The heuristics commonly used in minimum-cost, two-connected network topology design include initial feasible network design and local topology perturbation heuristics, which may reduce costs of an existing network design while preserving a feasible network. This approach has commonly been used in computer network design [9]. It has also been used successfully for a number of problems including the traveling salesman and graph-partitioning problems [10]. This section describes algorithms that are primarily taken from References [5,11]. The cost factor considered in the initial feasible network design is the fiber placement cost (i.e., to minimize the total route mileage); the cost factor considered in the topology perturbation stage includes regenerator, fiber placement, and fiber material and splicing costs.

6.3.2.1 Initial Feasible Topology Design

An *ear composition* method is commonly used to build a feasible two-connected network. That is, a cycle is first found on a subset of the vertices to form a partial solution, and then a path, called an *ear,* is repeatedly added to the solution until all nodes are two-connected. These paths start at one node on the present solution, pass through nodes not yet on the solution, and end back at a node already on the solution. Once all the nodes with connectivity of two are included in the two-connected part of the solution, the remaining nodes with connectivity of one are linked by spanning trees. A spanning tree of a graph is a subgraph containing all the nodes of that subgraph and some collection of links chosen so that there is exactly one path between each pair of nodes. The efficiency and correctness of heuristics, which build a feasible two-

272

connected network topology, depend on whether the model network being considered is dense or sparse. Because most fiber networks are sparse (due to the use of the hubbing architecture), this section discusses only heuristics for sparse networks. Design heuristics for dense networks can be found in Reference [11]. Note that when no survivability constraints are considered (i.e., each node has a connectivity of one), a network that minimizes the total route mileage will be a minimum spanning tree.

Before describing the network topology design algorithm, called 2-CON, we explain some notations to improve the understanding of the algorithm. An undirected graph G=(V,E) is given, where V and E represent the sets of CO locations and links, respectively. Associated with each link (u,v) between CO u and v is a non-negative distance d_{uv}. Each CO or hub v has an associated connectivity type r_v. The connectivity constraint requires that there be at least r_v disjoint paths between a CO and its cluster hub and at least min $\{r_v, r_w\}$ disjoint paths between hub buildings v and w. The following discussion describes a heuristic based on a *greedy ear* composition method to build a two-connected network. For convenience, a special node in the 2-CON algorithm is defined to be a node with connectivity of two.

The 2-CON Algorithm (Building a Two-Connected Network)

1. Create a cycle

   a. Select a special node v, find a shortest path P_{vz} for each other special node z, and also compute the length l_{vz} of P_{vz} (using the *Shortest Path First* [SPF] algorithm described in Appendix C).

   b. Choose a special node w such that $l_{vw} = \max \{l_{vz}: z$ is a special node$\}$.

   c. Let u be the special node next to w on path P_{vw}. Create a cycle C=$\{(u,w)\}$ ∪ $\{$a shortest path P_{uw} that does not use link $(u,w)\}$.

   d. If no such cycle C exists, the problem is infeasible and the process stops.

2. Add an ear to form a two-connected network

   a. Select a special node z (not yet in the solution) whose shortest path P to the partial solution (to a special node y) is longest among all special nodes not yet included.

   b. Find another shortest path Q from z to the partial solution that does not use any links of P and terminates on the partial solution at a node w other than y.

   c. The combination of P and Q forms an ear, which is added to the partial solution.

   d. If no such ear exists, the problem is infeasible and the process stops.

To implement the 2-CON algorithm, algorithms for computing two shortest link-disjoint or node-disjoint paths for a node pair are needed. Some of these algorithms can be found in Appendix C.

Example Using the 2-CON Algorithm

Figure 6.10 depicts an example of the model network and a resulting feasible network for using the 2-CON algorithm. A subnetwork taken from the model network, which is depicted in Figure 6.10(b), is used to explain this algorithm. This sample network includes 12 nodes. Among them, 10 nodes (say u) are special nodes with connectivity of two (i.e., $r_u=2$) (Nodes 1, 2, 3, 4, 5, 6, 7, 9, 10, and 11), and two nodes are non-special nodes with connectivity of one (nodes 8 and 12). An initial cycle is then constructed following the 2-CON algorithm described above. Starting from Node 1 (i.e., v), Node 4 (i.e., w) is selected because its shortest path (say P), with length 14.6 from Node 1 to Node 4, is longest among all special nodes, where P = 1-2-3-4. Because Node 3 (i.e., u) is the node next to Node 4, a shortest path is found between Nodes 3 and 4 that does not use link (3,4), say $P_{3,4}$, where $P_{3,4}$ = 3-7-6-5-4. Thus, an initial cycle C=3-7-6-5-4-3 is defined, as shown in Figure 6.10(c).

The special nodes that are not yet included in the solution are Nodes 1, 2, 9, 10, and 11. The second step is to greedily add short ears to the solution until all special nodes are contained in this two-connected network solution. First, Node 10 (i.e., z) is selected because its shortest path to a node (i.e., Node 7) on the cycle closest to Node 10, say P, is longest among Nodes 1, 2, 9, 10, and 11, where P= 7-8-9-10 (Node 7 is node y in the algorithm). We now find another shortest path Q from Node 10 to Node 3 (i.e., w) as follows: Q= 10-11-1-2-3, which does not use any link in path P. Thus, an ear is created by combining P and Q: 7-8-9-10-11-1-2-3. Adding this ear to the cycle created in the first step forms an initial two-connected network, as shown in Figure 6.10(c).

Once all the nodes with connectivity of two are included in the two-connected subnetwork, the rest of the nodes with connectivity of one are linked by minimum spanning trees, as shown by dotted lines in Figure 6.10(d).

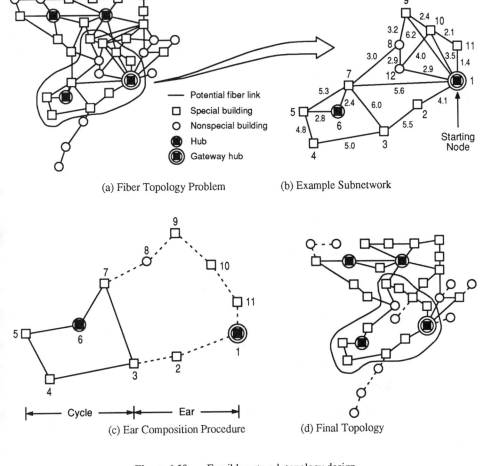

(a) Fiber Topology Problem

(b) Example Subnetwork

— Potential fiber link
☐ Special building
○ Nonspecial building
⬤ Hub
◉ Gateway hub

(c) Ear Composition Procedure

(d) Final Topology

Figure 6.10. Feasible network topology design.

6.3.2.2 Topology Perturbation

The topology perturbation heuristics commonly used in computer network design are also applied in the model discussed here. These heuristics apply local transformations to any feasible network to reduce the total network cost of placement, fiber material and splicing, and regenerators while maintaining feasibility. These transformations are applied until a locally optimal network is obtained, i.e., until no further reductions in

275

network cost are possible. The following discussion describes two perturbation heuristics: *add chords* and *one-optimal*. References [9,10,11] describe other perturbation heuristics.

To take into account the cost of fiber material and regenerator equipment, as well as placement cost, it is necessary to generate information on the first shortest paths for working spans and the second shortest diverse paths for protection spans for all node pairs that should be connected in the final solution. This information should be recomputed every time the candidate solution is modified. Simultaneously, the algorithm computes the fiber and regenerator costs. The regenerators are assumed to be available only at a node; thus, they are located at the last node along a certain path having a distance from the initial node (or the last regenerator) less than the maximum feasible span without a regenerator. Note that for dual-homing architectures, two shortest node-disjoint, rather than link-disjoint, paths are computed. Some algorithms for computing two shortest node-disjoint paths for each node pair can be found in Appendix C.

Add Chords Heuristic

Topologies with low placement cost tend to consist of a few interconnected rings, some of which can be quite long. This results in large distances between a building and its hub or between hubs, which consequently means that expensive regenerators may be needed, as shown in Figure 6.11(a). To overcome this problem, a heuristic is used to add a link (chord) to the current solution. The fiber paths are then recalculated as shown in Figure 6.11(b), and, if the total cost of fiber material, placement, and regenerator equipment is lowered, the chord is accepted into the solution. This heuristic repeatedly adds chords until no further cost reduction is possible.

One-Optimal Heuristic

The one-optimal heuristic removes a link (u,v) from the current feasible solution and replaces it with another link of the form (u,y) not in the current solution. Such an interchange is possible only if the resulting network is feasible and less costly than the original network. Figure 6.12 shows a one-optimal interchange. This approach considers each Node u. All links (u,v) incident to Node v in the solution are considered possible candidates for removal. To keep the computational effort manageable, a window size W is introduced to restrict the choices of the link (u,y) that may be added to the solution. The nodes are then sorted by distances from Node u, where the distance from a Node v to u is the minimum number of links that must be traversed to get from v to u. Node u is distance zero from itself, neighbors of u in the solution are distance 1, and so on. A window size of W implies that a link (u,y) can be considered only if y is one of the W closest nodes to u in terms of distance. The interchanges are continued until no further cost reduction is possible.

276

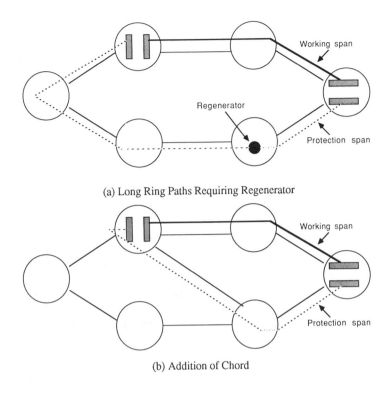

(a) Long Ring Paths Requiring Regenerator

(b) Addition of Chord

Figure 6.11. Adding chords.

6.3.3 Fiber Multiplexing Network Layout

The fiber multiplexing network layout phase designs a minimum-cost fiber multiplexing network to realize the topology obtained from the topology selection phase and to carry the DS3 demands obtained from the bundling phase. Cost factors considered in this phase include the fiber transport cost, which includes the fiber material and installation costs, the cost of any required regenerators, and fiber multiplex equipment costs. The total network cost is obtained by summing the transport cost from this phase and the multiplexing/DCS costs from the demand-bundling phase.

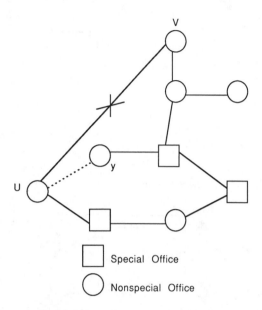

Figure 6.12. One-optimal interchange.

This design phase essentially uses two interrelated design procedures: multiplex span selection and DS3 demand routing. The core function of multiplex span selection is to determine which fiber system should bypass or terminate at which nodes (i.e., hubbing spans or point-to-point spans). The cost for realizing fiber systems on the selected multiplex span layout depends on the demand-routing scheme, which routes DS3 demands from one end to the other. The multiplex layout model described here uses a simpler heuristic method. This heuristic first builds a minimum-cost centralized span layout by routing required DS3s over spans and then performs a local perturbation procedure on the current span layout to reduce the network cost. The resulting span layout is a non-centralized span layout if at least one local perturbation is accepted in the final solution.

6.3.3.1 Centralized Multiplex Span Layout

As discussed in the Introduction (see Chapter 1), the economies of scale inherent in fiber capacity dictate that demand should be aggregated whenever possible. To accomplish this aggregation in the normal case, all demand from a CO is assigned to one fiber span going to its home hub, as shown in Figure 6.1, where a fiber span is composed of all the terminal electronics and fibers (working and protection). The fiber in a span may pass through intermediate COs on its way to the hub.

Other optimization options are available to try multiplexing at intermediate nodes to lower total network transport cost. (Section 6.3.3.2 describes one such option.) Dual homing is an exception to this rule because demand is split in the dual-homing case between a home hub and a foreign hub to ensure survivability for hub failures.

Fiber terminal cost is computed by assuming that it is a function of two factors: the number of DS3s carried by each terminal and the total number of working terminals. The first cost factor represents the cost of plug-ins, while the second represents the cost of frames and common control. Naturally, the protection system of the 1:N system is sized to carry the maximum number of DS3s carried by any of N working systems. The fiber material cost can be computed from the length of the working and protection paths as determined by the topology. Fiber placement cost is likewise determined by summing the length of fiber links being used.

All demand between hub buildings is sent via the gateway unless direct hub-hub spans are economically attractive. To determine the economics of hub-hub spans, the network is engineered first with all hub-hub demand sent via the gateway. Direct routing over hub-hub spans is then attempted if total network costs are lowered.

To engineer a fiber span for 1:N protection, the cost of using all possible line rates (for example, 565 Mbps, 1.2 Gbps, and so forth) and considering fiber and terminal costs is evaluated, and the least expensive alternative is chosen. For 1:1 protection, the situation is more complicated because multiple fiber spans of different rates between an origination and a destination can be used. For example, assume that there are 30 DS3s carried by a fiber span. This span could use 565-Mbps and 1.2-Gbps fiber systems to carry the required DS3 demand. In this case, one 565-Mbps and one 1.2-Gbps system would also be required for protection. To size fiber spans in this case, an integer programming approach can be used to optimize the fiber rates.

We now formulate the fiber span sizing problem. Let f be the total number of DS3 demands over the span. Also let f_{ij} be the DS3 demands carried on the j-th fiber system of the considered span using the i-th transmission rate, and let $Cost(f_{ij})$ be the cost including 1:1 protection for using the i-th transmission rate carrying demand f_{ij}. $Cost(f_{ij}) = c_i \times f_{ij}$, where c_i is per-DS3 unit cost for the i-th line rate. Let d_i be the DS3 capacity of the i-th transmission rate. The fiber capacity assignment problem for a 1:N/DP architecture can be formulated as the following constrained knapsack problem and can be solved with dynamic programming formulations [12]:

$$\min \sum_{i, j} (c_i \times f_{ij})$$

$$Subject\ to : \sum_{i, j} f_{ij} \geq f,$$

$$0 \leq f_{ij} \leq d_i \quad for\ all\ i\ and\ j.$$

If only the fully configured terminal cost is considered, the above sizing problem can be simplified as the following 0-1 knapsack problem and the fractional demand assigned to each line rate in the solution.

Let x_{ij} be a variable, which is equal to 1 if the j-th system with the i-th rate is deployed; otherwise, it is equal to 0. Also, let c_i be the terminal cost at the i-th rate (full configuration is assumed here). A simplified 0-1 knapsack formulation for fiber span sizing problem is as follows:

$$\min \sum_{i, j} (c_i \times x_{ij})$$

$$Subject\ to : \sum_{i, j} x_{ij}\ d_i \geq f,$$

$$x_{ij} = 1\ or\ 0 \quad for\ all\ i\ and\ j.$$

6.3.3.2 Span Layout Optimization

Although using dedicated fibers from each CO to its home hub in the centralized span layout results in a reasonable network cost, there are situations where it is possible to decrease network cost by demultiplexing at an intermediate CO and remultiplexing the total demand from that CO on another fiber span to the hub. Figure 6.13[2] shows this approach, which is most useful in eliminating the need for repeaters because both fiber spans are now shorter than the original fiber spans. This remultiplexing technique is applied at each CO in turn (i.e., CO visiting), and the nodes are ordered based on the cost of the regenerators used by the working and protection fiber spans passing through that CO. The process of CO visiting stops when no cost reduction is possible.

---

2. Node B in Figure 6.13(b) can be equipped by either two back-to-back OLTMs or one ADM device.

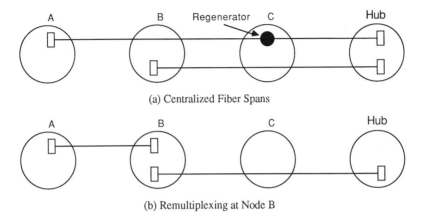

(a) Centralized Fiber Spans

(b) Remultiplexing at Node B

Figure 6.13. Remultiplexing at intermediate nodes.

6.3.3.3 Network Survivability Computation

Given the span layout, routing information for DS3 demand pairs, and DS1/circuit information within each DS3 (obtained from the demand-bundling process described in Section 6.3.1), network survivability can be calculated. The survivability calculation is performed on a link, CO, and hub basis. *Link (or CO) survivability* is defined as the portion of circuit demand for the entire network that is still intact when the fiber cable of that link is cut (or that CO fails) and the protection scheme is enforced. The CO and hub survivability computations are based on the same concept.

To calculate link survivability, the algorithm fails all fiber systems carried by that link, applies the protection strategy, and counts the number of circuits carried on DS1s that have not been counted yet for this link failure. It is necessary to count in this manner because the same DS1 can be carried in more than one DS3 in the same link, and double-counting of DS1s must be avoided. For example, a DS1 that terminates at CO-1 and CO-2 could route over a DS3 between CO-1 and the hub's DCS, and over a DS3 between the hub's DCS and CO-2. However, these two DS3s are routed over a common link between CO-3 and the hub building, as depicted in Figure 6.14.

281

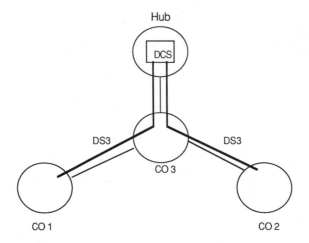

Figure 6.14. Example of DS1 double-count in survivability computation.

Calculating the ICSR [6] provides insight into the relative cost and survivability benefits of each survivable network architecture. The ICSR is simply the additional cost to provide a survivable network over that of a "base" architecture divided by the additional number of survivable circuits over the number of survivable circuits in the base architecture. Because the 1:N architecture is commonly implemented in today's networks and is the least expensive of available architectures, it is used as the base architecture to calculate the ICSR in Reference [6].

6.3.4 Network Design with Quantitative Survivability Constraint

The proposed model discussed in the previous sections can be easily adapted to a model that addresses the quantitative worst case survivability constraint. This is accomplished by adding several iterations to continually improve the worst link or node survivability performance using the available survivable network architectures described in Chapter 3. Designing the network to satisfy the the average network survivability quantitative constraint is far more complicated than that of the worst case constraint. For the average survivability case, a concept of *dominant node-pair* [13] was proposed to improve average network reliability for large data networks until the requirement has been fulfilled. A dominant node-pair in an average reliability enhancement algorithm [13] is a node pair for which any average reliability improvement for that node pair will improve the average network reliability. The

282

similar concept of dominant node-pair reliability enhancement may be used in fiber network design with the quantitative average survivability constraint.

6.4 MULTI-PERIOD CAPACITY ALLOCATION PROBLEM

6.4.1 Design Problem and Assumptions

The multiplexing network layout model discussed in Section 6.3.3 is a single-period model that addresses only the worst case scenario where all demands for a multi-period problem are summed over time into single-period demands. Worst case planning is usually used in long-term strategic planning that primarily looks for the target capacity or architecture needed given such a single-period solution. More detailed capacity transition planning is required to plan the network in a cost-effective manner over the multi-period horizon. Also, in most practical situations, the company budget for purchasing equipment is always deferred several years to relax the budget pressure and to take advantage of the equipment discount during the planning period. This section discusses a model for planning fiber span capacity for a multi-period demand requirement [14].

Given a set of available line rates, the problem is to determine which line rate should be used at which year in the network. It is easy to see that even for a short planning horizon (say five years), the problem becomes complicated for a reasonably large network. The key to reducing the problem size is to represent the fiber-hubbed network by a set of independent fiber spans. Thus, a much smaller capacity expansion problem is solved on a span-by-span basis, rather than on a network basis. Figure 6.15 shows this multi-period capacity assignment problem for each fiber span.

To make the model more manageable, the following assumptions are made:

1. Routing of DS3 systems over fiber spans is fixed during all periods.
2. Incremental demand on any fiber span in any single period is not greater than the maximum line rate.
3. Fibers and equipment that are already installed cannot be disconnected.
4. Multiple placements in the same period are not considered.

Note that the second assumption simplifies the model's complexity and can be removed by preprocessing the multi-period demand if this assumption does not apply to the given data. (See the discussion in Section 6.4.2.) The third assumption represents a least expensive service requirement (since equipment has already been paid for) specifying that the service disruption should be avoided when upgrading the system.

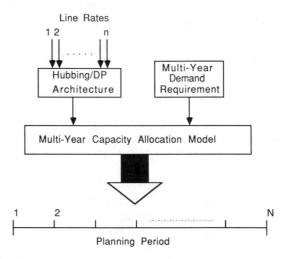

Goal: Find an optimum set of {i,j} (the i-th year implements the j-th line rate) such that the total cost in the entire planning period is minimized.

Figure 6.15.   Multi-period span capacity assignment problem.

The model for solving this multi-period capacity assignment problem is stated as follows:

1. Find all spans in the model network and compute the yearly demand (in terms of DS3s or STS-1s) for each span.
2. Apply the span capacity expansion algorithm (discussed below) to determine the number of fibers, amount of equipment, and their capacities over a planning period such that the span cost is minimized.
3. Compute the cost for each span separately, and compute the minimum cost for the model network by summing all span costs.

6.4.2 Multi-Period Span Capacity Assignment Algorithms

The problem associated with Step 2 in the model described above is sometimes referred to as the *capacity expansion problem*. An algorithm proposed in Reference [14] can be used to solve this capacity expansion problem for practical SONET networks.

This multi-stage decision problem can be formulated as a forward dynamic programming model and can be implemented efficiently using the tree and queue data structures. The tree data structure monitors the (partial) solutions developed during the process, while the queue data structure determines the next node in the tree to be processed and when the optimum solution is reached. In the following paragraphs, we use an example to explain this span capacity expansion algorithm. The detailed algorithm can be found in Reference [14].

Figure 6.16 depicts a decision tree for an example showing how the algorithm works. In Figure 6.16, a four-year planning period is assumed, and the cumulative demands in each of the planning periods are 9 DS3s (year-1), 23 DS3s (year-2), 38 DS3s (year-3), and 56 DS3s (year-4). Three candidate OC-N line rates are also assumed: OC-12, OC-24, and OC-48.

First, the algorithm creates a root and then tries the three line rate options. Capacity is assumed to be placed at the beginning of the period. For each option, the system carries the demand to the period following the period in which the capacity is exhausted. For example, the OC-12 system's capacity will be exhausted by the end of the second period, while the OC-48 system's capacity will be exhausted by the end of the fourth period. An interval period is created for each line rate that starts from the current period and continues to the year before the end of the period in which capacity is exhausted. (For example, the initial interval period for the OC-48 system is between the first and fourth periods.) The algorithm then computes the span cost for each line rate option over the interval period just obtained and inserts the costs into queue Q. The costs in Q are then sorted in increasing order, and the lowest cost is placed in the beginning of Q. The node with the lowest cost in Q is the next candidate for processing, and the algorithm repeats the same process with this node. The program stops when the minimum-cost node is at the end of the planning period. For example, the program stops when a node with a cost of 140 is found because 140 is the minimum cost in the queue and the node is in the fifth year. The optimal solution for the example depicted in Figure 6.16 is as follows: for the candidate fiber span, an OC-12 fiber system is used to carry demand to the end of the first year; an OC-48 fiber system is then added in the beginning of the second year to carry demand to the end of the planning period (i.e., end of the fifth year).

Numerical examples have shown that the algorithm can generate the optimum solution quickly (in less than 1 second), using a VAX 6420 computer, for planning period of 10 years and four transmission rates, which represents the most practical SONET planning scenario.

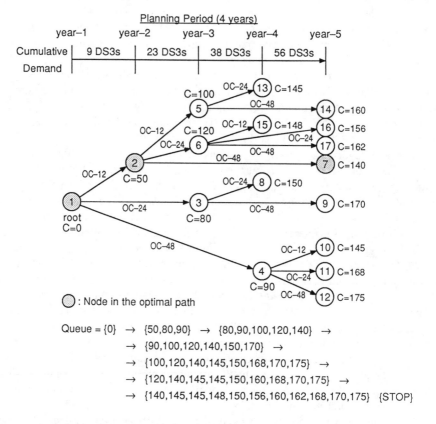

Figure 6.16. An example decision tree of a span capacity expansion algorithm.

As explained in Section 6.4.1, the first assumption for the proposed model, stating that the incremental demand for any single period cannot exceed the available maximum line rate, can be removed by preprocessing the given multi-period demand requirement. The demand preprocessing algorithm works as follows: when the incremental demand for any single period, d, exceeds the maximum line rate, r, the algorithm builds left ceiling $\dfrac{(d-r)}{r}$ right ceiling fiber systems with the maximum line rate to carry excessive demand where left ceiling $\times$ right ceiling is the smallest integer

286

that is greater than or equal to x. After building these excessive fiber systems, the incremental demand for that single period is updated as follows: $d \leftarrow d - r \times \left\lceil \frac{(d-r)}{r} \right\rceil$.

After multi-period demand has been preprocessed, the incremental demand for any single period should not be greater than the maximum line rate so the model discussed in this section can be applied. For example, given a five-year planning period with multi-period incremental STS-1 demands (45, 90, 38, 93, 29) and a maximum line rate of OC-48, the algorithm builds one OC-48 fiber system in the period between the second and third years, and the fourth and fifth years, respectively. Then, the multi-period incremental STS-1 demand is updated to become (45, 42, 38, 45, 29), which certainly satisfies the first assumption for the proposed model.

Reference [1] discusses another algorithm for solving the similar multi-period capacity expansion problem. This algorithm uses the branch-and-bound method to expand the capacity over a planning period for multimedia networks including radio, metallic cables, and fibers.

6.5 SUMMARY AND REMARKS

The network design model discussed in this chapter allows network engineers and planners to design a survivable network that includes DP and dual-homing architectures, as discussed in Chapter 3. The discussed design model is simple enough to be implemented in a PC and accurate enough to find an appropriate survivable fiber-hubbed network architecture and sizing from a strategic planning perspective. Many improvements can be added to make the model more efficient and optimal than the present methods (discussed in this chapter) in terms of computing times and network costs. For example, in the discussed methods, routing and capacity-sizing decisions are not made simultaneously for simplicity. This constraint can be relaxed by using an algorithm (proposed in Reference [15]) that couples routing decisions, including DP routing, with the capacity-sizing decision to achieve a more cost-effective and survivable network.

This chapter considered only a network topology design with a two-connectivity requirement because the probability of two links (cut) or two nodes failing simultaneously is very small in fiber interoffice communications networks. However, some non-interoffice networking applications may require k-connectivity protection (e.g., fiber networks for Navy aircraft carriers where multiple failures can occur during battle). For such applications, the methodology in this chapter is still applicable except that the topology design is extended from the two-connected network topology design to the k-connected network topology design where $k \geq 2$. Some efficient algorithms for designing a minimum-cost, k-connected topology and finding minimum-cost k-disjoint paths between a pair of nodes have been reported in the public literature, such as References [16,17].

The model discussed in this chapter is primarily taken from References [2,5]; other design models for different network requirements can be found in a special issue of *IEEE Journal on Selected Areas in Communications* [18] and in References [1,7,15]. The model described in this chapter primarily serves as a basic component for SONET network designs, because a possible SONET network architecture is likely to be composed of several types of survivable network architectures, including hubbing with DP, as discussed in this chapter. Chapter 7 will discuss design methodologies for SONET networks.

REFERENCES

[1] Tsai, Y., "Interoffice Transmission Network Planning with the Network Planning System (NPS)," *Proceedings of IEEE GLOBECOM*, December 1986, pp. 25.6.1-25.6.6.

[2] SP-ARH-000066, *FIBER OPTIONS*, Issue 1, Bellcore, December 1988.

[3] "FiberWord Planning Tools," Northern Telecom Ltd. Marketing Bulletin, April 16, 1990.

[4] Wu, T-H., and Cardwell, R. H., "Optimum Routing for Fiber Network Design: Model and Applications," *Proceedings of IEEE International Communications Conference (ICC)*, Philadelphia, PA, June 1988, pp. 2.5.1-2.5.7.

[5] Cardwell, R. H., Monma, C. L., and Wu, T-H., "Computer-Aided Design Procedures for Survivable Fiber Optic Telephone Networks," *IEEE Journal on Selected Areas in Communications*, Vol. 7, No. 8, October 1989, pp. 1188-1197.

[6] Wu, T-H., Kolar, D. J., and Cardwell, R. H., "Survivable Network Architectures for Broadband Fiber Optic Networks: Model and Performance Comparisons," *IEEE Journal of Lightwave Technology*, Vol. 6., No. 11, November 1988, pp. 1698-1709.

[7] Doverspike, R. D., "Algorithms for Multiplex Bundling in a Telecommunications Network," *Operations Research*, Vol. 39, No. 6, November-December 1991.

[8] Truitt, C. J., "Traffic Engineering Techniques for Determining Trunk Requirements in Alternate Routing Trunk Networks," *Bell System Technical Journal*, Vol. 33, No. 2, March 1954.

[9] Tanenbaum, A. S., *Computer Networks*, Prentice-Hall, 1981.

[10] Papadimitriou, C. H., and Steiglitz, K., *Combinatorial Optimization: Algorithms and Complexity*, Prenctice-Hall, 1982.

[11] Monma, C. L., and Shallcross, D. F., "Methods for Designing Communications Networks with Certain Two-Connected Survivability Constraints," *Operations Research*, Vol. 37, No. 4, July-August, 1989, pp. 531-541.

[12] Syslo, Deo, and Kowalik., *Discrete Optimization Algorithms*, Prentice Hall, 1983.

[13] Kundu, A. K., Stach, J., and Wu, T-H., "An Efficient Method for the Reinforcement of Topological Reliability in a Large Data Network," *Proceedings of IEEE INFOCOM'86,* Miami, FL, 1986, pp. 309-318.

[14] Wu, T-H., Cardwell, R. H., and Boyden, M., "A Multi-Period Architectural Selection and Optimum Capacity Allocation Model for Future SONET Interoffice Networks," *IEEE Transaction on Reliability,* Vol. 40, No. 4, October 1991, pp. 417-427.

[15] Sen, S., Doverspike, R. D., and Vohnout, S. I., "Diverse Routing in the Unified Facilities Optimizer," *Proceedings of GLOBECOM,* Dallas, TX, November, 1989, pp. 9.5.1-9.5.4.

[16] Suurballe, J. W., "Disjoint Paths in a Network," *Networks,* Vol. 4, 1974, pp. 125-145.

[17] Grotschel, M., and Monma, C. L., "Integer Polyhedra Arising from Certain Network Design Problems with Connectivity Constraints," SIAM, *Discrete Mathematics,* Vol. 3, No. 4, November 1990, pp. 502-523.

[18] Bonatti, M., Katz, S. S., Newman, C. F., and Varvaloucas, G. C. (Guest Editors), "Special Issue on Telecommunications Network Design and Planning," *IEEE Journal on Selected Areas in Communications,* Vol. SAC-7, No. 8, October 1989.

CHAPTER 7

Survivable SONET Network Design

7.1 SURVIVABLE SONET NETWORK ARCHITECTURES

SONET equipment's simpler control system and high-speed processing capability make the implementation of several survivable network architectures easier and more economical. Network survivability has become part of SONET network requirements because service interruptions due to high-speed SONET system failure are not affordable from a business perspective. Survivable SONET architectures include point-to-point/hubbing with APS, SHRs, and reconfigurable DCS mesh network architectures. These network architectures can be further divided into two network layers: the physical layer and the logical layer. At the physical layer, protection is provided by dedicated fiber facilities and/or equipment; at the logical layer, it is provided using the spare capacities within the working systems. Examples of architectures providing protection at the physical and logical layers are SHRs and reconfigurable mesh networks that use SONET DCSs, respectively.

Network reconfiguration time is another difference between physical layer and logical layer protection methods. As discussed in Chapters 3, 4, and 5, physical layer protection can typically restore services within 50 ms, whereas logical layer protection can restore services within seconds or minutes. For most practical environments, network providers may find that their network requirements do not match the best characteristics of any single restoration architecture. Thus, a mixture of restoration alternatives, or a hybrid approach, is usually required. A hybrid approach enables network providers to match tariff levels with restoration performance in a cost-effective manner. For example, crucial and priority services (e.g., 911 services) can be routed on rings or DP switching systems; other services can use the DCS's alternate routing capability for restoration.

Tradeoffs among different restoration architectures (see Section 5.4) have naturally resulted in a hierarchical restoration model for service protection, which is depicted in Figure 7.1(a). The hierarchy is structured (top to bottom) as follows: ring/route DP, DCS-based mesh restoration, and operator control. The manual operator control is initiated if the APS schemes cannot restore services completely or if sufficient bandwidth is unavailable for DCS restoration.

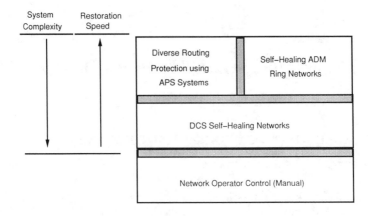

(a) An Integrated Network Restoration Model

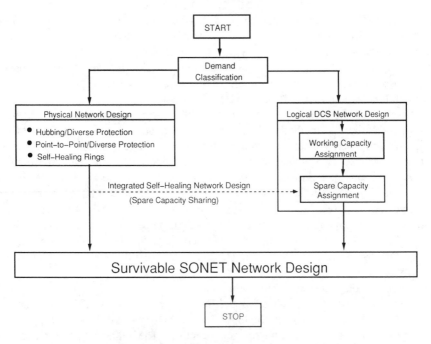

(b) A Survivable SONET Network Design Model

Figure 7.1. An integrated network restoration model and associated survivable network design model.

A survivable SONET network design model, as depicted in Figure 7.1(b), can be derived from the hierarchy of the integrated network restoration model [Figure 7.1(a)]. In this design model, the entire demand can be partitioned into two parts: one needing to be restored with high priority and the other with low priority. The high-priority demand uses physical layer protection, such as DP and SHRs, whereas the low-priority demand uses logical layer protection, such as DCS reconfiguration. Table 7-1 shows the network characteristics of each survivable SONET architecture. It is intended to illustrate the differences among survivable network architecture alternatives and, thus, to help develop reasonable assumptions for the design model.

Table 7-1. Comparison of Survivable SONET Architectures

Attributes	DP	SHR	DCS
Spare capacity needed	most	most/ moderate*	least
Restoration time	fastest	fastest	slowest
Mixed line rates	yes	no**	yes
Network size	two points	regional	global
Per-node cost	moderate (OLTM/APS)	lowest (ADM)	highest (DCS)
Fiber counts	highest	lowest	moderate
Connectivity needed	lowest	moderate	most (mesh)

* Depends on the demand requirement in the considered area
** Assume ADMs are needed for ring implementations.

According to the design model discussed herein, the physical and logical network layers can be designed separately, and the spare capacities within the physical network layer may be shared by the logical layer. This spare capacity sharing concept, which is called an *integrated self-healing system design,* aims to reduce the spare capacity required for the logical DCS network.

This chapter first discusses the physical and logical network designs separately, and then discusses an integrated self-healing system design based on spare capacity sharing. This chapter does not address network survivability enhancements for achieving a required quantitative level of survivability. Interested readers may refer to Reference [1] for a procedure to enhance network survivability.

7.2 SONET PHYSICAL LAYER NETWORK DESIGN

The major issue in designing survivable SONET physical networks is how best to utilize the unique characteristics of DP systems and SHRs to meet different demand requirements in a cost-effective manner. In general, when compared to DP systems, the ring architecture may be more economical because it shares facilities and equipment and has the potential to reduce the number of regenerators required. In particular, the SHR may offer a cost-effective solution to applications requiring dual-homing protection [2]. Such applications cannot be implemented economically using today's APS technology [3]. However, compared to the survivable hubbing or point-to-point network architecture, the SHR architecture may be difficult or expensive to upgrade once the SHR capacity is exhausted. This difficulty arises because system upgrade is conducted on a global basis for SHRs. Therefore, the choice of whether to use the SHR or the survivable fiber-hubbed (or point-to-point) architecture depends on economics, as well as on ease of network growth. A primary task in SONET physical network design is determining how best to utilize merits of these alternative survivable architectures (i.e., ring, hubbing/DP, point-to-point/DP) to minimize the cost of survivable network evolution.

Because SONET networks are a relatively new design area, virtually no complete design references related to this area exist in the public literature. (A few proprietary network design systems do exist.) Thus, the design philosophy and models described in this section are primarily taken from a survivable SONET physical layer network design system called STRATEGIC OPTIONS [4], which was developed by Bellcore's Applied Research area. This model may serve as a starting point for developing more powerful and optimal SONET network design systems. This section first describes the overall structure of STRATEGIC OPTIONS and then discusses each of the design modules that make up the design system. Note that the exact design methodology discussed in this section may not necessarily be implemented in STRATEGIC OPTIONS; however, the basic design philosophy is similar.

In the following subsections, we first discuss a multi-period architectural selection and capacity assignment model for a predetermined set of buildings or COs that can potentially be connected by a ring. Part of this model can be used as a multi-period multiplex cost module (see Section 7.2.2) in STRATEGIC OPTIONS, which will be discussed following the first multi-period architectural selection network design model. The purpose of STRATEGIC OPTIONS is to determine strategic locations and ring types for SONET ring placement.

7.2.1 Multi-Period Architectural Selection and Capacity Assignment Model

Given a set of buildings (or COs) that can potentially be connected by a ring, the survivable SONET network architecture selection problem for the multi-period planning scenario is to determine whether these buildings (or COs) should be connected using a ring architecture or 1:1/DP system. This architectural selection problem can be stated as follows:

Given: 1. A set of buildings (or COs) that can potentially be connected by
              a ring
              2. A list of link distances on the ring
              3. Candidate survivable network architectures
                 - rings, hubbing/point-to-point with 1:1/DP systems
              4. End-to-end multi-year demands for the considered network
              5. Cost associated with each network component
              6. A list of candidate SONET line rates.

Objective: Determine an appropriate set of survivable network architectures
              and associated capacities for a predetermined planning period such
              that the total network present worth cost for the considered area is
              minimized.

Subject to: End-to-end multi-year demand requirements.

The candidate survivable network architectures considered here include the ring
architecture, the fiber-hubbed architecture with DP, and the point-to-point architecture
with DP. The network cost in the model is the present worth cost,[1] which includes
facility and equipment costs, including the working and protection terminal cost, the
APS cost, the fiber material cost, the fiber placement cost, and the regenerator cost.
The demand requirement considered here is the STS-1 (51-Mbps) demand requirement
because it is the building block for transport in SONET networks.

The goal of the multiplex cost model is to evaluate the best combination of
available architectures and associated capacities that can be deployed in a cost-
effective manner over a planning period (for example, five years). As discussed earlier,
the major limitation of deploying the SHRs for the interoffice network application is
the capacity exhaust problem. Two types of planning methods are used to address this
problem: worst case planning and multi-period transition planning. Worst case
planning considers the sum of the given multi-year demand and plans it as a single-year
case. In this type of planning, ring exhaust is not a concern because the network can
always be engineered to meet the demand requirement. However, this method is
usually more expensive than the multi-period transition planning model because the
latter method can design the network based on a least-cost transition path. For
example, as reported in Reference [2], the network cost for using multiple SHRs could

---

1. The present worth cost takes into account the interest or discount rate in that planning period.

be higher than the cost of using the traditional 1:1/DP approach in a growth scenario. Thus, accounting for growth is crucial for SONET network planning when the SHR architecture is used. Therefore, a multi-year planning model is used to ensure that the right architecture is chosen at the right time in the candidate area to minimize the total network cost over a planning period. Also, in most practical situations, the company budget for purchasing equipment is always deferred several years to relax the budget pressure and to take advantage of the equipment discount during the planning period. Due to cost and budget-planning considerations, the multi-period transition planning method is usually chosen.

The design problem considered here is too complex to guarantee a tractable, optimum solution. Thus, the network growth assumptions shown below are made to simplify the design complexity [5]. These assumptions were derived from an application guideline provided by some local operating telephone companies. This guideline suggests that point-to-point systems can be used in conjunction with the ring and/or hubbing architectures to reduce survivable network costs for areas where only a few demand pairs have appreciably higher demand requirements than other demand pairs. A case study result, which suggests that this application guideline is a reasonably good engineering rule, will be discussed in Section 7.2.1.3. Figure 7.2 depicts the concept of these growth assumptions.

Network Growth Assumptions

1. Use point-to-point systems to carry point-to-point demands until the hubbing/DP and/or ring architectures become economical to carry the remaining demands.
2. The mixed use of hubbing/DP and ring architectures carrying the remaining demands is based on the following growth strategy:

    a. If the hubbing/DP structure is used as a startup architecture, it is used throughout the entire planning period.
    b. If the SHR architecture is used as the startup architecture, it is used until its capacity is exhausted. The remaining demand is carried by either another SHR or hubbing/DP architecture until further capacity is exhausted or the end of the planning period is reached.

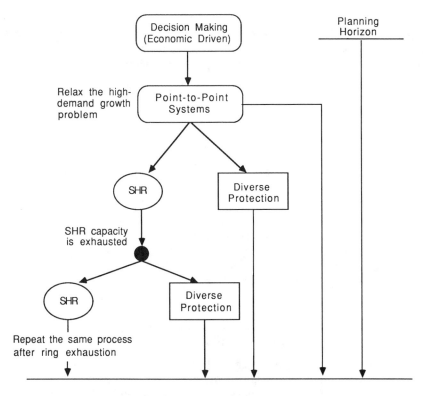

Figure 7.2. Multi-period network growth assumptions.

The first assumption promotes using point-to-point systems to delay the ring capacity exhaust if it is economical to do so. Assumption 2(a) is based on an observation that capacity exhaust does not significantly impact the hubbing architecture. As for Assumption 2(b), if the SHR capacity is enough to carry the growth demand, then it is most economical to let the SHR accommodate those growth demands. If the growth demand exhausts the SHR capacity, the following two strategies may be used to accommodate the growth demand:

a. Add one more overlay ring (but perhaps a different line rate).
b. Add fiber spans with DP to the hub.

Our planning model requires selecting a good starting architecture in year-1 and appropriating growth strategies in the n-th year to accommodate the growth demand in the n-th year. Thus, the total network incremental capital costs (where the capital costs

297

include costs for fiber and transmission equipment) over a planning period are minimized. Figure 7.3 depicts a structured view of the multiplex cost design model for selecting appropriate survivable network architectures and planning the capacity expansion of a fiber-optic network over a specified planning period.

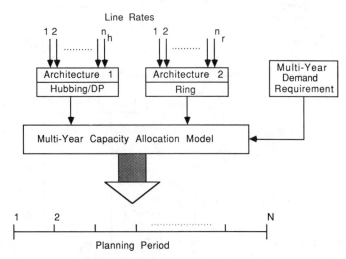

Goal: Find an optimum set of {i,j,k} (the i-th year implements the j-th architecture with the k-th line rate) such that the total cost in the entire planning period is minimized.

Figure 7.3. A structured view of the multi-period multiplex cost design model.

A set of offices and the demand requirements for office pairs over a planning period of T years are given as input to the model. The total demand between each pair of buildings increases each year; thus, the considered demand is the incremental demand. The model's objective is to minimize the present worth cost of network development over T years, while ensuring that sufficient fibers and equipment are installed in the network to accommodate the growth demand. Next, the high-speed SHR architecture is considered, and the resulting solution is compared with the one obtained using the hubbing/DP architecture. To maintain a 100 percent survivability capability for a single fiber cut, as for the SHRs, a 1:1/DP architecture is assumed for the hubbing/DP option. The 1:1/DP architecture is also applied to point-to-point systems considered in this model.

The model described here is based on the following assumptions:

1. The topology allows the ring to be built.
2. Incremental demand on any fiber span in any year is not greater than the maximum line rate considered in the hubbing/DP option.
3. Similarly, the incremental demand on any ring in any year is not greater than the maximum line rate considered in the ring option.
4. Fibers and equipment that are already installed are not rearranged.

Given a candidate ring area for analysis, the model is divided into three sub-modules, as depicted in Figure 7.2. The first part gives a method for determining how the point-to-point systems are to be built. The second part is used to solve the fiber-optic network capacity expansion problem only only a hubbing/DP architecture over a planning period. The third part describes an algorithm for solving the capacity expansion problem using a combination of SHR and hubbing/DP architectures. The following procedure describes how the above sub-modules interact [5]:

1. The architectural selection and capacity assignment algorithm determine the number of point-to-point systems (with DP) that should be built (see Section 7.2.1.1).
2. The model then subtracts the demand satisfied by the point-to-point spans from the original pair demands, starting from the outermost period and moving inward.
3. Based on the updated demand requirements from Step 2, the model runs a multi-period, multiplex cost algorithm for selecting a set of ring and/or DP network architectures and associated capacities so the total present worth cost is minimized over a planning period (see Section 7.2.1.2).
4. Last, the model adds the output cost of the SHR/DP model (Step 3) to the result of Step 1 (point-to-point systems) to obtain a total cost for the network.

7.2.1.1 A Point-to-Point System Design Algorithm

Determining the number of point-to-point systems that may be created requires the following three parameters: (1) the maximum number of point-to-point systems, (2) the STS-1 threshold, and (3) the percent fill. Users input these three parameters, which gives them flexibility to try as many combinations as they need to meet their planning requirements.

When the maximum number of point-to-point systems is set to zero, the model defaults to a SHR/DP solution (i.e., no point-to-point systems in the final solution). When it is set to the number of demand pairs, the result is all point-to-point systems with 1:1 DP if all point-to-point systems meet the STS-1 threshold and percent fill requirements. The STS-1 threshold is the minimum amount of total STS-1 demand necessary to trigger the building of a point-to-point system for a given demand pair. When the STS-1 threshold is set to zero, point-to-point systems can be generated in response to any demand; when it is set to the total demand for a given pair, no point-to-point system will be generated for that pair. Here the total demand is the sum of STS-1 demands throughout a planning period for a demand pair. The percent fill determines the minimum demand necessary to select the next higher rate for a point-to-point system. If the total demand does not meet the threshold for the lowest available line rate, no point-to-point system will be built. The decision is made in the following manner: if the [(line rate being considered - next lower line rate)×percent fill+next lower line rate] is greater than the total demand, then select the next lower line rate; otherwise, select the line rate under consideration. For example, suppose there is a total demand of 22 STS-1s, a 75 percent fill, and line rates of OC-12 and OC-24. The decision for the line rate would be OC-24 because [(24-12)×0.75]+12 = 21 (< 22).

After building point-to-point systems, the demand requirement matrix is updated by subtracting demands carried on point-to-point systems starting from the outermost time period and moving inward from the original demand requirement matrix. This simulates a planning strategy of identifying high-demand cross-sections and implementing direct point-to-point systems in these areas prior to designing a ring or hub/DP network. It directs growth demand into point-to-point systems, which allows rings that are less prone to near-term exhaust to be built at a lower (and more economical) rate.

Given the number of point-to-point systems desired, the STS-1 threshold, and the percent fill, the example described below shows how to build point-to-point systems according to an algorithm described in Reference [5]. This example uses a four-node network, and the bundled STS-1 demand for the area is shown in Table 7-2.

Table 7-2. Sample Area Demand

STS-1 demand requirement							
Demand pair #	Demand pair	Period (year)					Total
		1	2	3	4	5	
1	(1,4)	4	3	6	5	7	25
2	(1,2)	2	3	4	5	6	20
3	(2,4)	1	3	4	5	6	19
4	(1,3)	1	3	5	3	4	16
5	(2,3)	1	3	2	3	2	11

For this example, the available line rates are OC-3, OC-12, OC-24, and OC-48; the STS-1 threshold is 0; the percent fill is 50; and the number of point-to-point systems is 2. The algorithm then calculates the total demand for each demand pair and determines whether the value exceeds the threshold. (In this case, all demand pairs exceed the threshold of zero.) It then sorts the pairs by decreasing total demand and selects the number of point-to-point systems specified. In our example, demand pairs 1 and 2 have been selected.

Next, the algorithm builds working and protection span layouts for each of the selected demand pairs using an algorithm that minimizes link length and ensures disjoint paths for the working and protection spans (see Appendix C and Chapter 6). It then selects appropriate line rates and calculates the cost for each span. With a 50 percent fill, the demand of 25 STS-1s for span 1 does not qualify for an OC-48 rate. Therefore, an OC-24 is selected, and one STS-1 is returned to the demand matrix. Span 2, with demand 20, does qualify for the OC-24 rate over OC-12.

Last, the point-to-point span costs are summed together, and the demand requirement matrix is updated. A new demand matrix will show one STS-1 demand in year-1 for (1, 4), i.e., span 1. Span 2 (1, 2) will have no demand in any year, having been fully satisfied by a direct point-to-point system.

7.2.1.2 A Capacity Expansion Algorithm for a Combination of SHRs and DP Architectures

The algorithm for solving the capacity expansion problem for a combination of SHRs and hubbing/DP architectures, denoted by Algorithm CapExp_R&D [5], is similar to the capacity expansion algorithm for the hubbing/DP architecture (described in Section 6.4). As in the hubbing/DP case, a maximum number (n) of different line rates needs to be considered for a ring. At each period, a maximum of n+1 options needs to be considered: option 0 for hubbing and 1 through n for rings corresponding to the n line rates. Any of these options will be applied once the existing ring capacity is exhausted. Note that for option 0 (i.e., hubbing/DP), the capacity expansion layout can be obtained by the procedure described in Section 6.4. The capacity expansion model for networks using a combination of hubbing and SHR options is summarized below and corresponds to the growth model depicted in Figure 7.2. Again, a dynamic programming approach, which is similar to the capacity expansion algorithm discussed in Section 6.4, is used here to find an optimum solution for the architectural selection and capacity expansion problem.

Figure 7.4 depicts a decision tree created by using Algorithm CapExp_R&D; the detailed algorithm and analysis can be found in Reference [5]. In this example, a four-year planning period is assumed, and the cumulative demands in each of the planning periods are 40 STS-1s (year-1), 56 STS-1s (year-2), 70 STS-1s (year-3), and 98 STS-1s (year-4). Three candidate OC-N line rates are also assumed: OC-12, OC-24, and OC-48.

The algorithm first creates a root and then tries the hubbing/DP option and the three line rate options for rings. The capacity is assumed to be placed at the beginning of the period. If the algorithm starts from the DP option, it uses the same option to carry demands to the last planning year using the capacity expansion design algorithm discussed in Section 6.4. For each ring option, the system carries the demand to the period in which the capacity is exhausted. For example, the OC-48 ring capacity will be exhausted after the second year. Thus, a child node is created from the root for the OC-48 ring at the second year. An interval period is created for each line rate that starts from the current period and continues to the year that begins the period in which the capacity is exhausted. For example, the initial interval period for the OC-48 system is between the first and second years. The algorithm computes the cost for each selected line rate option for rings or for the DP option over the interval period just obtained and inserts the costs into the queue Q. The node cost is computed based on the demand-routing algorithm (e.g., the one described in Appendix D for bidirectional rings) and the cost model used. The costs in Q are then sorted in increasing order, and the lowest cost is placed in the beginning of Q. The node with the lowest cost in Q is the next candidate for processing. The algorithm then repeats the same process with this node. The program stops when the minimum-cost node is at the end of the planning period. For example, the program stops when a node with a cost of 170 is found, because 170 is the minimum cost in the queue and the node is in the fifth year. The optimal solution for the example depicted in Figure 7.4 is as follows: an OC-48 ring is installed to carry demand to the end of the first year, and a set of hubbing/DP spans is added in the beginning of the second year to carry demand to the end of the planning period (i.e., the end of the fourth year). The capacity arrangements for the hubbing/DP spans installed in the beginning of the second year are obtained by using the capacity expansion algorithm described in Section 6.4.

The computing efficiency of Algorithm CapExp_R&D has been reported in Reference [5]. In most practical applications (i.e., a 10-year planning period with five options: DP option with OC-3, OC-12, OC-24, and OC-48; and four ring options: OC-3 ring, OC-12 ring, OC-24 ring, and OC-48 ring), Algorithm CapExp_R&D can generate the optimum solution in less than five seconds using a VAX 6420 computer.

7.2.1.3 A Case Study: The Impact of Point-to-Point Systems on Ring Planning

The purpose of this case study is to see how a SONET engineering rule works using the algorithms and models described previously. This engineering rule suggests using point-to-point systems to carry enough demands to decrease the remaining demand to a level that can be economically carried by the ring and/or hubbing/DP networks. This should be done in areas where only very few demand pairs have appreciably higher demand requirements than other demand pairs.

302

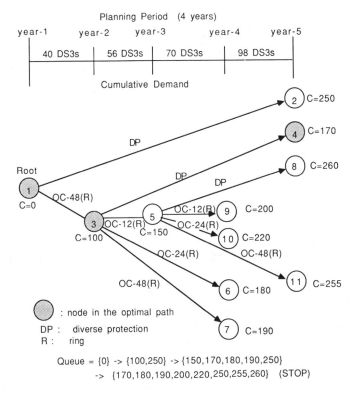

Figure 7.4.   An example of a SONET multi-period capacity expansion algorithm.

The model network of the case study is part of a metropolitan-area LATA network, which consists of nine COs (one of them is the hub). There are 16 demand pairs with a total of 82 STS-1s in a five-year planning period. Of these demand pairs, one has a significantly higher demand requirement than other demand pairs. Line rates of OC-3, OC-12, OC-24, and OC-48 were made available to the model. Figure 7.5 shows network costs as a function of the number of point-to-point systems assuming the percent fill is 75 percent [5]. In this particular example, a network using two point-to-point systems and an OC-48 ring generates a minimum-cost network design.

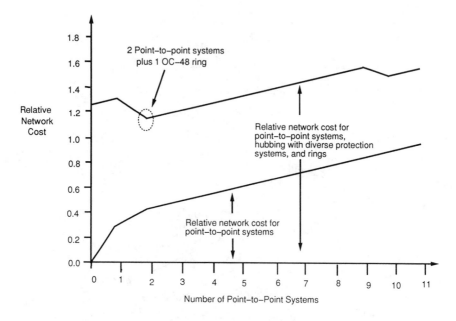

Figure 7.5. An engineering rule utilizing point-to-point systems with rings.

7.2.2 Design Architecture for STRATEGIC OPTIONS

The network design model discussed previously deals only with an area where buildings or COs can potentially be connected by a ring. In that model, buildings or COs on the ring are predetermined. In this section, we will discuss some methodologies that may help determine strategic locations for ring placement and types of ring architectures for these rings. These methodologies are primarily taken from STRATEGIC OPTIONS [4].

STRATEGIC OPTIONS is prototype software for the strategic planning of survivable interoffice networks. It is an extension of the FIBER OPTIONS software package [6], which is used to design survivable fiber-hubbed DP networks (also see Chapter 6). The software chooses a near-optimal set of SONET rings based on cost as compared to 1:1/DP. This section, which is a high-level description taken from Reference [4], discusses the basic functionality of STRATEGIC OPTIONS.

Figure 7.6 shows the structure of STRATEGIC OPTIONS and the modules it contains. The system is menu-driven and interacts with users via a keyboard and a graphical representation of the network. STRATEGIC OPTIONS considers two types

304

of survivable network architectures: 1:1/DP and SONET SHRs. The 1:1/DP architecture provides each CO with both a working fiber span and a diversely routed protection fiber span to its hub. STRATEGIC OPTIONS considers three types of SONET SHRs: USHRs, BSHR/2s, and BSHR/4s (see Chapter 4 or Reference [7]). It uses a multi-year cost model and considers the time value of money to decide when to use particular fiber spans and rings.

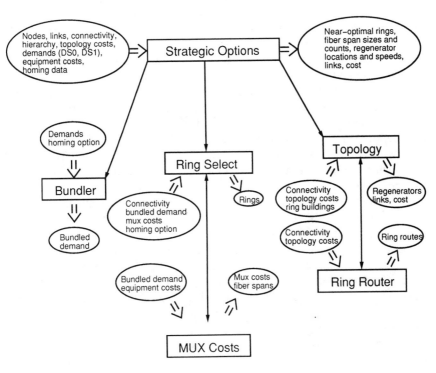

Figure 7.6. Functional diagram of STRATEGIC OPTIONS.

The top of Figure 7.6 shows the input and output of the STRATEGIC OPTIONS software system; the bottom shows the input and output of individual modules. The user must input the nodes, links, connectivity, facility hierarchy (e.g., hubs and their clusters), topology costs, and multi-year, point-to-point demands of the network. The topology costs include fiber (material and splicing), route mileage (installation), regenerator, and regenerator threshold. The demands should be in DS0s (circuits) or DS1s (24 DS0s). The user may optionally input dual-homing data, equipment costs, interest rate of money, and type(s) of ring to be considered. The dual-homing data indicates a foreign hub for each CO to be dual homed. When no dual-homing data is input, single homing is assumed. The equipment costs indicate the costs of multiplexers and regenerators of different rates, as well as costs corresponding to the different types of rings. The user may also specify different regenerator thresholds for different signal rates.

The STRATEGIC OPTIONS software assumes that two-connected COs that are not on rings will have 1:1/DP to their hubs. Output includes a set of near-optimal rings, fiber span sizes and counts, regenerator locations and speeds, the topology (set of links to be used), and the network cost. The type (USHR, BSHR/4, and BSHR/2) and size of each ring are also output. In addition, the time in the planning period that each ring and fiber span should be installed is output.

STRATEGIC OPTIONS has three main software modules: the bundler, the ring selector, and the topology optimizer. Every module uses the facility hierarchy information; therefore, hierarchy is not listed as input to the modules to reduce the complexity of Figure 7.6. The bundler takes the input multi-year, point-to-point demand, which is in DS0s or DS1s, and bundles it into multi-year STS-1 (28 DS1s) demand. Thus, the bundler takes into account any input dual-homing data to route the demand properly.

The ring selector uses the bundled demand, CO connectivity, and any dual-homing data to choose rings based on cost. It does this by comparing the costs of building rings with various groups of COs to the costs of providing 1:1/DP to those CO groups. Each ring contains the home hub of each building on the ring and also the foreign hub of any dual-homed building. The type(s) of ring considered depends on the constraint set by the user. The ring selector also proposes initial routes for the rings. To make the cost comparison, the ring selector calls the multiplex costs module, which uses the bundled demand and input or default equipment costs to determine the cost of any option that the ring selector is testing.

The topology optimizer considers connectivity, topology costs, and ring buildings to decide which links to use in the least costly topology that will preserve the rings and the 1:1/DP capability. As it tests different topologies, the topology optimizer calls the ring router to compute alternative ring routes. A ring route is a fiber route between offices multiplexed on a ring. The ring router considers connectivity and topology costs and outputs ring routes. The topology optimizer also includes regenerator cost in comparing alternative topologies. It outputs the links to be used, the regenerators required, and the total network cost. STRATEGIC OPTIONS also includes editorial features that allow the user to manually change the rings or topology to perform a sensitivity analysis.

7.2.3 Multi-Period Demand-Bundling Algorithm

The multi-period demand-bundling algorithm bundles multi-period, end-to-end DS1 (or VT1.5) demand requirements to multi-period STS-1 demand requirements in a cost-effective manner. The STS-1 channel is the building block for SONET transport. This algorithm uses an extension of a single-period demand-bundling algorithm discussed in Section 6.3.1 (or Reference [6]). The criterion used for determining a direct STS-1 path or indirect STS-1s using the SONET W-DCS may be either the minimum cost or the STS-1 threshold, as described in Chapter 6. A direct STS-1 path is a point-to-point STS-1 path that is not formed using a DCS (i.e., without grooming).

Figure 7.7 depicts a conceptual example that shows how this multi-period demand-bundling algorithm works. In Figure 7.7, nodes labeled "STS-1" represent direct STS-1 paths, and nodes labeled "Hub" represent indirect STS-1s using DCSs. The cost model is used in this example as the criterion for determining direct or indirect STS-1s. First, the algorithm creates two nodes with two possible options: using and not using a DCS. If the algorithm starts from a node using the DCS, it uses the same process to groom VT1.5 demands over the entire planning period. If the algorithm starts from a node with a direct STS-1 path, it continues to fill this STS-1 path until the STS-1 capacity is full. It then tries two options (i.e., with and without using the DCS) and repeats the same process until the end of the planning period is reached. Note that if the algorithm starts from a child node labeled "Hub," it continues to use the DCS for grooming until the end of the planning period is reached. Each node has an associated cost, which can be calculated using an algorithm similar to the one described in Section 6.3.1. The decision tree creating process is similar to the one that has been discussed in Section 7.2.1.2. The key idea for building this decision tree is to store the cost of each node being created to a queue that is then sorted in an increasing order. The candidate node for the next move is the node that is in the head of the queue (i.e., minimum-cost node). The algorithm stops when the considered node for the next move is at the end of the planning period. For example, in Figure 7.7, the solution is as follows: for the considered demand pair, a direct STS-1 path is formed to carry demands from the beginning of the planning period to the end of the second year, and the DCS is used for grooming demands from the beginning of the third year to the last year.

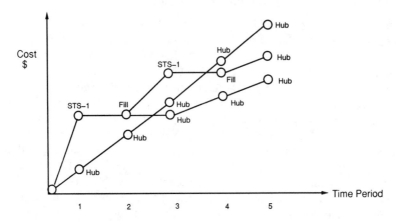

Figure 7.7. An example of a multi-period demand-bundling algorithm.

7.2.4 Ring Selection Algorithm

The design problem for ring selection is locating near minimum-cost rings in the network while considering network topology, network hierarchy, STS-1 demand requirements, the ring cost model, and the DP cost model. The network starts as a facility network with a fiber-hubbed architecture. A group of COs served by a single hub is called a *cluster*. Hubs may be fully interconnected or may be grouped to a higher hierarchy, as described in Chapter 3. We assume that economies-of-scale considerations dictate that each building will have only one piece of terminal electronics (ADMs in the case of rings or OLTMs in the case of 1:1/DP networks). Thus, all demand terminating in a building is placed on a single ring or single fiber span. With this approach, rings are utilized to share bandwidth among several buildings and decrease cost from the cost of the 1:1/DP network. The ring-selection algorithm uses the following simple heuristic to locate potential offices for ring deployment:

1. Choose any cluster as a starting point.
2. Find all cycles containing that hub. If dual homing is used, the cycle must contain the foreign hub of any dual-homing buildings in the cycle considered.
3. For each combination of buildings on each cycle (ring), calculate costs based on a multi-period multiplexing cost model (described in Section 7.2.1.2) and save if profitable.

308

4. Build the most profitable ring.
5. Remove all rings that are now illegal.
6. Repeat Steps 4 and 5 until there are no more rings for the cluster.
7. Pick another cluster if there is one available and go to Step 2; otherwise, stop.

The ring selector starts from the hub of the building with the least demand because this starting point may potentially allow more nodes on the ring. As reported in Reference [2], the ring may increase its economic benefit, as compared to the 1:1/DP systems, if the number of nodes on the ring can be increased without exhausting its capacity. Figure 7.8 depicts an example of the above ring-selection algorithm. This example includes six nodes, with Node 1 serving as the hub (or ring interconnection point). For this example, possible cycles are {12345, 1235}. In Cycle 12345, possible rings are 123, 124, 125, 134, 135, 145, 1234, 1235, 1245, 1345, and 12345. The algorithm does not consider two-node rings, under the assumption that it is less expensive to use OLTMs (see Chapter 6) for 1:1/DP than ADMs for a ring when only two nodes are involved. For each possible ring, network costs for the ring and the hubbing with a 1:1/DP system are calculated based on the multi-period multiplex cost model discussed in Section 7.2.1.2. If the ring cost is less than the 1:1/DP cost, the ring is called a "profitable" ring and is saved for further consideration; otherwise the ring is deleted from the candidate ring list. After selecting all profitable rings for a candidate cycle (e.g., Cycle 12345 in our example here), these candidate rings are compared, and the most profitable ring within that cycle is chosen. In this example, if ring 145 is chosen as the most profitable ring in this cycle, then all rings that include Node 4 or 5 must be removed from the candidate list. In this case, rings 124, 125, 134, 135, 145, 1234, 1235, 1245, and 12345 must be removed. Ring 123 remains and is retained for further consideration. Selecting final candidate rings is based on the comparison between the ring cost and DP costs, as described in Section 7.2.1.2.

Note that the described ring-selection algorithm is an exhaustive search with an intelligent starting point. Thus, this process may become a bottleneck, in terms of computing time, when the network becomes large (e.g., 200 nodes or more). Any computing improvement on this process will certainly speed up the entire process for obtaining the solution. For example, the ring-selection process can be shortened if the algorithm considers only economical rings.

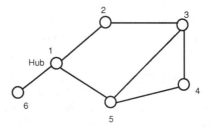

Possible Cycles – 12345, 1235

In Cycle 12345, possible rings are: 123, 124, 125, 134, 135, 145, 1234, 1235, 1245, 1345, and 12345.

If ring 145 is chosen, then rings 124, 125, 134, 135, 145, 1234, 1235, 1245, 1345, and 12345 must be removed. Ring 123 remains for further consideration.

Figure 7.8. Example of ring selection.

7.2.5 Ring Fiber Routing

The description, algorithm, and comparison in Sections 7.2.5.1 through 7.2.5.3 are taken from Reference [8].

7.2.5.1 Problem Description

After locations of ADM nodes on the ring are determined, the next step is to physically connect these ADM nodes by fiber systems. This problem is referred to as the *ring fiber routing problem*. Formally speaking, this problem involves routing fiber around a ring in a network when the network nodes, links, connectivity, and offices to be placed on that ring are known. In most cases, the ring routing problem cannot be solved by "traveling salesman" algorithms. However, under certain conditions, the problem degenerates to the traveling salesman problem, and the ring routing algorithm degenerates to the "nearest neighbor" method of solving that problem (see Section 7.2.5.3 or Reference [8]).

"A salesman, starting in his own city, has to visit each of $n-1$ other cities and return home by the shortest route [9]." This is a typical statement of the traveling salesman problem [10]. It is equivalent to finding a minimum-weight Hamiltonian cycle (a cycle containing all vertices) in a complete graph, where a complete graph is a

310

fully connected graph. Because this problem is NP-complete, exact solutions are computationally infeasible, whereas more efficient methods are merely approximations to the exact solution.

The problem considered here, ring routing, is distinct from the traveling salesman problem in several ways. For one thing, the classic traveling salesman problem is solved on a complete graph, whereas most interoffice networks are sparse networks, which are represented by incomplete graphs. In the traveling salesman problem, the cycle must contain all vertices of the graph; in the ring routing problem, only a subset of the offices is required to be in the cycle (the rest of the offices being optional). When routing rings in a network where conventional architectures are also used, the shortest ring may not be optimal, and the topology must be optimized as a whole. In contrast, traveling salesman solutions require the shortest cycle. The differences listed thus far are differences in problem definition, resulting from mathematical attributes in the graphs representing the problem and the solution cycle.

The conceptual, qualitative differences regarding the use of the solutions are even greater. The ring solution must be economical and reasonable. The cost is not a linear function of distance because there is a distance threshold between two ring offices beyond which a regenerator is required. For the ring to be reasonable, a qualitative attribute, a hop limit, is imposed. The hop count here is the number of links in a path. The solution is not practical unless it is both economical and reasonable, whereas the traveling salesman solution merely has to be economical.

The ring routing problem is a generalization of the traveling salesman problem, which is an NP-complete problem. One can therefore conjecture that the problem of finding the shortest ring route is also NP-complete [8]. Note that the algorithm described in the next section (Section 7.2.5.2) and in Reference [8] does not always find the shortest route.

7.2.5.2 A Ring Fiber Routing Algorithm

This section shows how the ring routing algorithm, called BUILD_RINGS, works by studying an example. Details of the algorithm can be found in Reference [11]. Figure 7.9 illustrates how BUILD_RINGS routes a ring in a simple example. Figure 7.9(a) shows a network with seven offices; Offices 1 and 5 are hubs. The example problem is to route a ring through Offices 1 through 4.

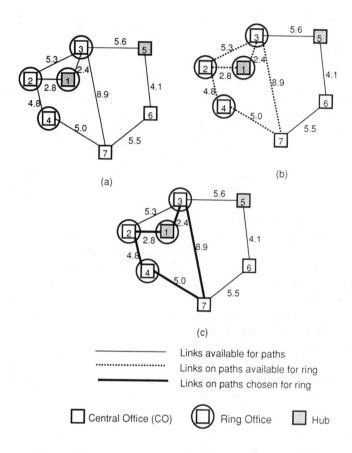

Figure 7.9. Example network for ring-routing algorithm.

————————— Links available for paths

.... Links on paths available for ring

━━━━━━━ Links on paths chosen for ring

☐ Central Office (CO)　◉ Ring Office　■ Hub

First, BUILD_RINGS uses a simple method to find a ring. If it fails, then a longer, more reliable algorithm is used. For the simple method, BUILD_RINGS finds a hub on the ring and the ring office furthest from it, and then uses two iterations of Dijkstra's shortest path algorithm to find the two shortest, link-disjoint paths (paths sharing no links) between the hub and the office. If the two paths are node-disjoint (share no nodes), and the cycle formed by them contains all the ring offices, then the ring is routed that way.

If the paths are not disjoint, or the cycle does not contain all the ring offices, then BUILD_RINGS begins a more intensive search for a ring route, beginning with a

312

depth-first search for paths between all pairs of ring offices and those containing only two ring offices. Some rules govern when to stop looking for these paths. A threshold upper-bounds the lengths of paths that will be considered in constructing the ring. In the beginning, this threshold is low so that BUILD_RINGS will look for short paths first; however, the threshold increases if necessary. There is also a limit on the number of hops (links) in a path. The default value of the hop limit is three, but the user can override this. Figure 7.10 shows four depth-first search trees that together contain all paths with fewer than three hops between ring offices in the network shown in Figure 7.9. The root and a subset of the leaves of each tree are shaded to indicate the paths found. In Figure 7.9(b), links corresponding to the paths in Figure 7.10 are shown as dotted lines.

After finding paths, BUILD_RINGS tries combinations of node-disjoint paths in an effort to construct a ring. The algorithm simply starts at some office and chooses the shortest path out of it, which terminates at some other office. It then chooses the shortest node-disjoint path out of the next office. The algorithm continues in this way until it either finishes the ring or has no more node-disjoint paths from which to choose for the next step. If it runs out of node-disjoint paths before completing the ring, it simply backs up and tries a different combination of paths. If, after trying every combination, there is no ring, the distance threshold increases and the algorithm finds more paths. Rings long enough to require two regenerators between a pair of ring offices are considered uneconomical. Therefore, if the threshold grows large enough to allow paths that require two regenerators, BUILD_RINGS stops without routing a ring, and the user sets the regenerator threshold. The algorithm stops after the first ring is found. Figure 7.9(c) shows the computed ring routing in bold lines.

The algorithm was programmed in C and run on a SPARCstation workstation. Computation times on 47 examples of feasible and infeasible rings were reasonable. Overall, the average, minimum, and maximum run-times were 0.41, 0.06, and 2.93 seconds, respectively [8]. Because the largest example network used in these results (167 offices and 240 links) is the size of a typical, large intraLATA network, it is judged that the algorithm runs fast enough for the intended application.

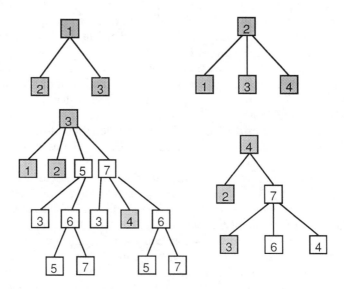

Figure 7.10. Depth-first search trees showing potential paths for constructing ring.

7.2.5.3 Relationship with Traveling Salesman Algorithms

Traveling salesman algorithms are designed not to stop until the least cost cycle (or an approximation to it in the case of approximate algorithms) is found. This cycle always exists in the classic version because the graph is complete. In contrast, BUILD_RINGS stops before a cycle is found if there is no useful cycle in the network, even if some cycle containing all of the ring offices exists. This behavior is desirable. It is better for the algorithm to quickly identify that no practical ring exists than to spend time finding one that will not be used. Naturally, if the algorithm stops due to insufficient room in the data arrays or insufficient computer memory, it may have missed a ring.

One approximation algorithm for the traveling salesman problem is the "nearest-neighbor" method. In this method, at each step, the shortest edge leading out of the current node that will keep the path acyclic is chosen as the next edge in the cycle. This continues until the n-th edge (if there are n nodes) leads back to Node 1. If we generalize this method by replacing links with paths between required nodes and requiring that the paths be disjoint (in addition to requiring that they combine acyclically until the last path is added), the resulting algorithm is BUILD_RINGS

314

without the distance and hop limits. In other words, when the distance and hop limits are infinite, the graph is complete, and all nodes are required, BUILD_RINGS degenerates to the nearest-neighbor method of approximating the traveling salesman solution.

7.2.6 Topology Optimization

The last module of the considered network design model is topology design, which requires choosing the most economical routing of the fibers for the rings and point-to-point systems. The following descriptions of the methodologies and algorithms to generate and optimize the topology of a survivable fiber network are taken from Reference [12]. The topology of such a network should be chosen to minimize the total topology cost, which includes route mileage (installation), fiber (material and splicing), and regenerator costs. Multiplexing costs are not considered in topology design because they do not change based on how fibers are routed; they depend on demand between COs and fiber-system rates.

Before beginning to design a topology, some decisions that will guide the topology-design process must be made. First, it must be determined which offices are to be protected (input of the model). Next, it must be determined which of these protected offices ought to be placed on rings together (determined by the ring-selection algorithm, see Section 7.2.4). Traffic considerations help to determine groups of offices within a cluster that could share a ring without exhausting it, whereas multiplexing cost considerations determine which of these rings is economical. (Section 7.2.4 addressed these aspects of network design.)

After these decisions are made, there are three steps in designing a topology. The first step is to determine whether the rings are topologically feasible, given the conduits available for running the fiber. Naturally, if the network is not two-connected, any ring will be infeasible. A two-connected topology is one in which each CO has two link-disjoint paths (i.e., two paths sharing no links) to each other CO. If the rings are feasible, the routings (paths around the rings) must be determined because it may be necessary to route fiber through offices that are not on the rings. (Section 7.2.5.2 discussed the ring fiber routing algorithm.)

The second step is to determine a two-connected topology that includes both the rings and all other protected offices. A two-connected topology provides for implementing rings and DP by ensuring that each protected CO has two disjoint paths to its hub. The final step is to optimize the topology based on the costs mentioned above.

The algorithms presented here determine the initial and optimized topologies, given the offices in the network, the offices on each ring, and the links available for the fiber. The topology optimization is based on route mileage, fiber, and regenerator costs. The

315

topology design module includes four algorithms [12]: BUILD_RINGS for routing rings [8], GREEDY_EARS for determining an initial two-connected network, and ADD_CHORDS and 1_OPT for optimizing a topology. Because Section 7.2.5.2 discussed BUILD_RINGS, the following two sections discuss only GREEDY_EARS, ADD_CHORDS, and 1_OPT.

7.2.6.1 Initial Topology

GREEDY_EARS, the algorithm used to compute a two-connected topology, is identical to the algorithm used for the same purpose in designing survivable fiber-hubbed networks [6] (or see Section 6.3.2.1). Its purpose is to compute an initial topology that has a two-connected subnetwork containing all protected offices. Recall that offices requiring protection were specified before beginning the topology design. Unprotected offices will be contained in the resultant topology, but not necessarily in the two-connected component. Containing all protected offices in a two-connected component ensures some level of survivability by providing for diverse working and protection spans for protected offices.

GREEDY_EARS begins by finding a cycle in the network that contains at least one protected office. It then chooses a protected office not on that cycle and builds an "ear" off of that cycle through the chosen office. That is, it finds a path from an office contained in the initial cycle, through the chosen protected office, to an office contained in the initial cycle. The cycle and the ear together comprise the current solution. The algorithm continues by choosing another protected office not contained in the current solution and building an ear off of the current solution containing the chosen office. When all protected offices are on the current solution, GREEDY_EARS connects any remaining unprotected offices to that solution via spanning trees.

To ensure that the initial topology contains the initial ring routings, links contained in the initial ring routings, but not in the GREEDY_EARS solution, are added to the GREEDY_EARS solution. Figure 7.11(a)-(c) shows an example network, initial ring routings, and GREEDY_EARS solution with ring routing links added.

7.2.6.2 Topology Optimization

Two algorithms are used to optimize the topology: ADD_CHORDS and 1_OPT. These algorithms are essentially the same as the design algorithms for survivable fiber-hubbed networks (see Section 6.3.2.2) except that they now account for rings, as described below.

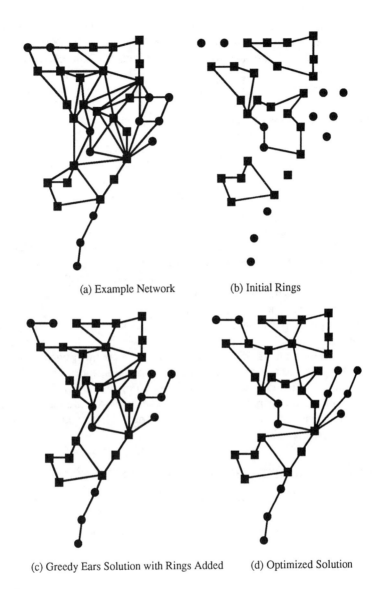

(a) Example Network   (b) Initial Rings

(c) Greedy Ears Solution with Rings Added   (d) Optimized Solution

Figure 7.11.   Topology construction.

317

ADD_CHORDS simply adds to the topology any links that will result in a lower cost. Candidate links to be added are those input as available for fiber. Although adding links will increase the route mileage cost, it can decrease the fiber and regenerator costs by providing shorter paths between offices and their hubs. Each time the algorithm adds a chord, it checks for lower cost ring routes by calling BUILD_RINGS.

1_OPT seeks to replace links in the solution with links not yet in the solution to lower the topology cost. It replaces a link (u,v) with a link (x,y), where x is one of the W offices closest to u, and W is the search window. The distance between two offices is defined here as the minimum number of links in a path between the two offices. Users input the search window W to limit the range in which 1_OPT searches for a link to replace a removed link. Again, only links input as available for fiber are considered. 1_OPT will not make changes that preclude routing the existing rings, although it will reroute rings by calling BUILD_RINGS. If the ring cannot be rerouted, or if the new ring route (in combination with the link exchange) increases the topology cost, then the links are not exchanged. Otherwise, both the link exchange and the ring reroute take place. After each exchange, 1_OPT calls BUILD_RINGS to check for lower-cost ring routes, even if no ring routes were disturbed by the exchange. Figure 7.11(d) shows an optimized topology for the example network.

The topology design programs were written in C and were run on a VAX model 785.[2] Numerical results for these algorithms (which run on a variety of example networks) show that the algorithms do reduce the costs mentioned above and run fast enough to be of use to a network planner. Computation times on 23 example rings in seven example networks were reasonable. For ADD_CHORDS, the average, minimum, and maximum run-times were 2, 0.04, and 14.5 minutes, respectively. The average, minimum, and maximum run-times for 1_OPT were 16.8, 0.2, and 101.5 minutes, respectively [12].

7.3 SONET LOGICAL LAYER NETWORK DESIGN USING DCS

The DCS network design problem involves minimizing the total number of working and spare channels required to meet (1) the end-to-end demand requirement carried on the DCS network and (2) a predetermined level of network survivability assurance. To make the design problem computationally feasible, the problem is generally divided

---

2. The computing speed of the VAX 785 is slower than that of the SPARCstation used to run BUILD_RINGS.

into two subproblems: working channel assignment and spare capacity assignment. Sections 7.3.1 and 7.3.2 discuss these problems.

7.3.1 Working Capacity Assignment for DCS Self-Healing Networks

The working capacity assignment problem involves minimizing the total number of working channels needed to carry the end-to-end DCS network demands, given the network topology and the number and locations of DCSs placed in the network. The number of DCSs and their strategic locations can be determined by network planners and engineers, based on their own requirements, or they can be determined by clustering algorithms that group several nodes together to form a cluster (which is served by a central node equipped with a DCS). The optimization criterion in the clustering algorithms may be the network costs or some other criterion. Reference [13] contains a general discussion of some clustering techniques.

After the topology, the number of DCSs, and their locations are determined, the end-to-end demand requirement on the DCS network may be distributed (routed) based on some routing criterion. The most common and simplest criterion is to find the shortest path, where the shortest path could mean the shortest-distance path, the minimum-cost path, or the minimum-hop path, depending on the definition of the link function. Let us assume the shortest-path criterion is used for routing. The demand routing can then be performed by creating a shortest-path tree for each node to all other DCS nodes (using the algorithm described in Appendix C or some other shortest-path routing algorithm). The end-to-end demands originating or terminating at the root node are then routed according to this shortest-path tree. After all end-to-end demands are routed, the working capacity required on each DCS link can be obtained.

7.3.2 Spare Capacity Assignment for DCS Self-Healing Networks

In the DCS self-healing network, the restoration performance of the self-healing protocol depends highly on the spare capacity engineered in the network. Thus, in most cases, spare capacity assignments have been proposed as part of DCS self-healing network systems [14,15]. The spare capacity assignment problem is to place the minimum amount of spare capacity needed to restore a predetermined percentage of demands (i.e., restoration ratio) from single or multiple network component failures, provided the network topology and the working channel assignment that satisfies all of the end-to-end demand requirement are given. The restoration ratio, which is the ratio of restored demands to affected demands, depends on the self-healing protocol and types of failures (e.g., single- link or multi- link). Note that the spare capacity assignment designed for line restoration[3] is also sufficient for path restoration because restoration paths for line restoration may also be restoration paths for path restoration, if other paths have insufficient spare capacity. Thus, this section discusses only the spare capacity assignment for line restoration.

As discussed in Section 5.3, the line restoration method identifies k alternate paths to restore all or some of the affected demands (depending on the preset restoration ratio) passing through the failed link as soon as possible. This design concept leads to some design heuristics for assigning the near-minimum spare capacity needed. The simplest and most efficient heuristic for spare capacity assignment is the algorithm reported in Reference [14]. This algorithm partitions the problem into an initial assignment problem for obtaining an initial feasible solution and an optimized assignment for minimizing the spare capacity obtained from the first feasible solution. Figure 7.12 depicts an example of how the algorithm works. In this example, there are five nodes and seven links with a total of 27 working STS-1 channels. The notation Ln(w,s) means that link n has w working channels and s spare channels.

In the initial assignment phase, the algorithm assigns the shortest alternate path for each failed link. The link's spare capacity in the alternate path is updated to the required restored demand if its present spare capacity is less than the required restored channel (i.e., affected working channels on the affected link). It remains unchanged if the present spare capacity is sufficient to restore the required demands. Figures 7.12(a) and 7.12(b) show the example network and a spare capacity assignment for link L6, respectively. In Figure 7.12(b), the alternate path for link L6 is chosen as L1-L5; thus, a spare capacity of three STS-1s is added to links L1 and L5. Figure 7.12(c) shows the initial spare capacity assignment (repeating the procedure described above). In Figure 7.12(c), the ratio of the spare capacity to the working capacity is 29/27. This spare-to-working ratio represents the maximum spare capacity required for restoring 100 percent demand and the fastest restoration scheme for each single-link failure scenario. However, its spare capacity is the highest among other spare capacity assignments.

---

3. Refer to Chapter 5 for more details about line and path restoration.

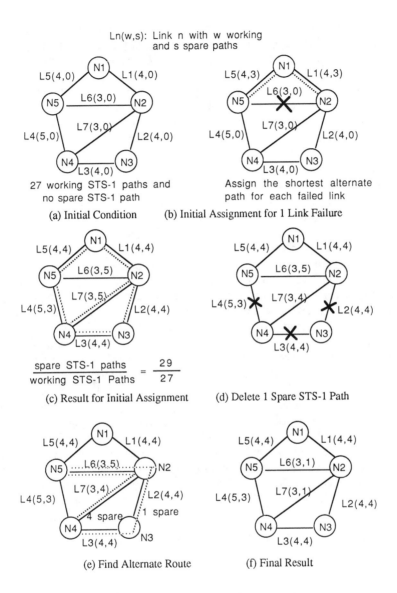

Ln(w,s): Link n with w working and s spare paths

(a) Initial Condition

27 working STS-1 paths and no spare STS-1 path

(b) Initial Assignment for 1 Link Failure

Assign the shortest alternate path for each failed link

(c) Result for Initial Assignment

$$\frac{\text{spare STS-1 paths}}{\text{working STS-1 Paths}} = \frac{29}{27}$$

(d) Delete 1 Spare STS-1 Path

(e) Find Alternate Route

(f) Final Result

Figure 7.12. An example of a spare capacity assignment algorithm (courtesy of Takafumi Chujo at Fujitsu Laboratories Ltd.).

The spare capacity assigned to one link for restoration may also be shared by other links for restoration according to the initial spare capacity assignment algorithm. Thus, the second phase (optimized phase) is to identify such spare capacity redundancies and delete as many as possible. The algorithm proposed in Reference [14] first chooses one spare channel from the highest spare capacity assignment and deletes it. It then identifies corresponding working channels that need the deleted spare channel. Figure 7.12(d) shows an example of the deletion of a spare channel from link L7 with five original spare channels. We find that L2, L3, and L4 require a spare channel on L7.

Because the algorithm deletes one spare channel, the alternate path for corresponding working channels does not have enough capacity. If the present spare capacity in L7 cannot restore required affected working demand, e.g., five STS-1s for L4, the algorithm finds another alternate path in addition to the first alternate path. In this example [see Figures 7.12(d) and (e)], the algorithm finds two alternate paths for link L4, as shown by the dotted lines in Figure 7.12(e). Thus, working demands for L4 can be restored via a path of L7-L6 with four spare demands and another path of L6-L2-L3 with one spare demand. The same process is repeated until some stopping criterion (e.g., the restoration ratio) is achieved.

Other heuristics for spare capacity assignment have also been proposed [15,16]. These spare capacity assignment algorithms essentially implement a design procedure similar to the one described above, but with more sophisticated heuristics for the initial spare capacity assignment problem. The algorithm reported in Reference [15] formulates the initial spare capacity assignment problem as a linear programming problem and uses the dual-simplex method to solve the problem. Compared to the heuristic discussed in Reference [15], the heuristic described in Reference [16] may achieve the same optimal solution for spare capacity assignment and can distribute the spare capacity of the network more uniformly over working DCS links [16].

Reference [17] discusses an even more sophisticated spare capacity assignment system for the DCS self-healing network. This system incorporates the path-level restoration method and the circuit-level restoration method, and uses a dynamic routing scheme to improve network survivability and fairness for telecommunications networks. Refer to Reference [17] for more details regarding its model and design.

7.4 INTEGRATED SELF-HEALING NETWORK DESIGN

To reduce the spare capacity needed for the integrated self-healing network, the concept of sharing spare capacity among different restoration schemes can be considered. In general, the 1:1/DP system and the ring do not share spare capacity because of a need to protect high-priority traffic. Because the DCS network is essentially a logical network, which may build on the physical network (a network composed of rings and DP systems), sharing spare capacity between the physical network and logical DCS network is possible. The DCS network may share spare capacity on protection rings and/or DP systems.

Sharing spare capacity reduces the transport cost for protection at the expense of a more complex control system and reduced architectural flexibility. The use of different equipment for different restoration methods (e.g., 1:1 APS systems use OLTMs, rings use ADMs, and DCS self-healing networks use DCSs) may complicate the integrated self-healing control system. To overcome the potential incompatibility problems among different equipment in the integrated self-healing system, it has been suggested that the DCS be used to support both the ring and the mesh self-healing functions [18], assuming the DCS network shares the spare capacity of rings.

Two spare capacity sharing schemes are possible in an integrated self-healing system: spare facility sharing and spare channel sharing. The first approach is to share the protection facility of the point-to-point system and/or the SHRs with the DCS self-healing mesh network. In this scheme, the demands carried by the DCS mesh network can be restored via the protection facility of the point-to-point system or the ring that is in the normal operations mode. If the point-to-point system or the ring is in the restoration mode, then its protection facility cannot be used by the DCS mesh network for restoration. In this scheme, 1:1 APS, the ring, and the DCS network must operate in the revertive mode to leave the protection facility available for use whenever needed. This may increase complexity for the DCS self-healing design, since current proposed DCS self-healing schemes operate only in the non-revertive mode.[4] It should be noted that the facility sharing between a USHR and the DCS network does not work due to the USHR's special self-healing system design (see Chapter 4 for more details).

---

4. Because the present proposed DCS self-healing control schemes operate only in the non-revertive mode, they will occupy the protection facility of the ring or DP systems until the next DCS network reconfiguration (i.e., next failure). Thus, during this period, if the network component of the ring or the DP system fails, the ring or the DP system cannot restore services because spare capacity has been occupied by the demand carried on the DCS network. Therefore, the DCS self-healing scheme must operate in the revertive mode to release spare capacity of the ring/DP system that it occupies during restoration. This revertive operation mode presents a technical challenge for the DCS self-healing design, especially for the distributed DCS self-healing network because it operates in an unpredictable environment.

The second spare capacity sharing approach is to share the "extra" spare channels in the protection facility of the DP system and/or the ring. Let x and y be the working channels and the line rate of the ring, respectively. The "extra" spare channels in the protection facility, which are available for use by the DCS self-healing network, are $y-x$, where x spare channels in the protection facility must be reserved for 1:1 protection for the ring or DP system. Thus, the DCS self-healing system uses only a portion of the ring protection facility, which is never used by the working ring for its 1:1 protection. Reference [18] describes an example of this second spare capacity sharing approach in which the DCS network uses the extra spare channels of the ring's protection facility as part of its spare capacity for restoration.

Compared to the first approach, the second approach does not interfere with the ring operations in both the normal and the failure situations. However, a more complex control system is needed because the spare capacity sharing is performed on the path level (e.g., end-to-end STS-1 paths) rather than on the line level (e.g., an entire protection facility), as in the first approach. The first approach may require less spare capacity than the second approach for the DCS self-healing network. However, it may lose some services carried on the DCS network if the shared protection facilities of the DP systems and/or rings are not available for use when the DCS network initiates restoration.

We next discuss a design model for the integrated self-healing network (see Reference [18]). First, 1:1 APS systems and rings are designed based on a design process such as the one discussed in Section 7.2. Then, the working capacity for each DCS link is designed using the algorithm described in Reference [18] or the algorithm described in Section 7.3.1. The spare capacity assignment for each DCS link is designed as follows. The working channels for each DCS link are split into two portions: those that can be restored via the extra spare channels of protection rings and those that cannot. Given the working channels, denoted by W, for the considered DCS link, $W = W0 + W1$, where W0 and W1 are the numbers of working channels that can and cannot be restored via the extra protection channels of rings, respectively. W0 can be computed as follows:

$$W0 = R_1 + 2 \times R_2 - R_3$$

where R_1 = total number of protection channels of rings, including the considered DCS link

R_2 = total number of protection channels of rings, including the two terminating nodes under consideration, but not the DCS link

R_3 = total number of working channels of rings that use the DCS link.

In the above equation, R_2 includes a factor of two because two link-disjoint paths are always available for the considered DCS link restoration (see the example in

Figure 7.13). The factor of R_3 preserves protection capacity for rings to restore all affected working channels, if the considered DCS link is used by those rings and that link is failed. Note that the procedure just discussed applies only to a single-link failure scenario.

Figure 7.13 depicts an example of how to compute the required spare capacity needed for the DCS network in the integrated self-healing system. This example shows a network of six nodes and its ring placement. In Figure 7.13, Nodes 1 and 4 use ADMs, while DCSs are placed in Nodes 2, 3, 5, and 6. Three rings have been deployed in this area: ring 1 is an OC-3 USHR carrying one working STS-1, ring 2 is an OC-12 BSHR/2 with three working STS-1s over link (2,6), and ring 3 is an OC-12 BSHR/2 that does not use link (2,6). For rings 2 and 3, only six STS-1s are available for protection for each ring due to its architectural constraint (see Chapter 4). The available spare capacity of ring 3 for DCS network restoration for link (2,6) can be doubled because two disjoint paths (path 1 = 2-1-6, and path 2 = 2-3-5-6) are available as alternate restoration paths for link (2,6). Thus, $R_1 = 9$ (3+6), $R_2 = 6$, $R_3 = 4$ (1+3), and $W0$ for link (2,6) is 17 [9 + (2×6) - 4]. If the number of working channels for the DCS link (2,6) (i.e., W) is nine STS-1s, the extra spare capacity of the rings (i.e., 17 STS-1s) is sufficient to restore these working channels (i.e., $W1=0$). If $W=20$, then $W1=3$. By repeating the same process for each DCS link, we obtain a set of $W1$ values for the DCS network. The spare capacity can be assigned for the DCS network with working channels $\{W1\}$ using the design process discussed in Section 7.3.2.

7.5 SUMMARY AND REMARKS

Future survivable SONET interoffice networks are expected to use a combination of automatic DP architectures, SHR architectures, and DCS self-healing mesh network architectures to provide an integrated and affordable network restoration system. Due to the difficulty of planning such a complex SONET integrated network restoration system, CAD planning tools are essential to ensure a reasonable and cost-effective network design that best utilizes the merits of each restoration architecture and meets required survivability and cost constraints. Because survivable SONET network design is a relatively new area and the need for planning tools has been recognized only in the last few years, virtually no complete system design reference exists in public literature. Thus, the philosophy and methods described in this chapter primarily serve as a first step toward a more complete and optimized network design tool. The model discussed in this chapter can also be used to study survivable architectural tradeoffs and associated network growth strategies. Several enhancement issues for the considered physical SONET network design have been suggested in Reference [6] for further study. More network design models can be found in Reference [19].

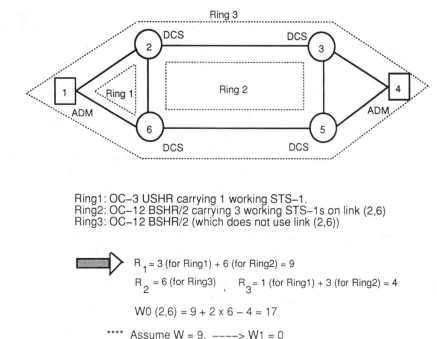

Ring1: OC–3 USHR carrying 1 working STS–1.
Ring2: OC–12 BSHR/2 carrying 3 working STS–1s on link (2,6)
Ring3: OC–12 BSHR/2 (which does not use link (2,6))

$R_1 = 3$ (for Ring1) + 6 (for Ring2) = 9

$R_2 = 6$ (for Ring3) , $R_3 = 1$ (for Ring1) + 3 (for Ring2) = 4

W0 (2,6) = 9 + 2 x 6 – 4 = 17

**** Assume W = 9, ----> W1 = 0
W = 20, ---> W1 = 3.

Figure 7.13. Computing shared spare capacity in an integrated self-healing system.

Many interesting engineering guidelines have been observed during the trials of methodologies described in this chapter. One such guideline is to use direct point-to-point systems in conjunction with SHRs to improve network solutions in a SONET environment where only a few demand pairs have appreciably higher demand requirements than other demand pairs.

The concept of the SONET integrated self-healing network, which reduces the required spare capacity of the DCS self-healing networks, is still in an early development stage. The concerns and issues associated with this concept include (1) the effects on network operations systems, (2) the effects on survivable architecture and equipment designs, (3) the effects on self-healing control design, and (4) the actual cost savings. The impact of the above issues, as determined by network planners and engineers, may determine the feasibility of implementing an integrated self-healing network architecture that shares spare capacity.

REFERENCES

[1] Roohy-Laleh, E., Abdou, E., Hopkins, J., and Wagner, M. A., "A Procedure for Designing a Low Connected Survivable Fiber Network," *IEEE Journal on Selected Areas in Communications*, Vol. SAC-4, No. 7, October 1986, pp. 1112-1117.

[2] Wu, T-H., and Burrowes, M., "Feasibility Study of a High-Speed SONET Self-Healing Ring Architecture in Future Interoffice Fiber Networks," *IEEE Communications Magazine*, Vol. 28, No. 11, November 1990, pp. 33-42.

[3] Wu, T-H., Kolar, D. J., and Cardwell, R. H., "Survivable Network Architectures for Broadband Fiber Optic Networks: Model and Performance Comparisons," *IEEE Journal of Lightwave Technology*, Vol. 6, No. 10, November 1988, pp. 1698-1709.

[4] Wasem, O. J., Cardwell, R. H., and Wu, T-H., "Software for Designing Survivable SONET Networks Using Self-Healing Rings" *Conference Record of the Second ORSA Telecommunications Conference*, Boca Raton, FL, March 1992.

[5] Wu, T-H., Cardwell, R. H., and Boyden, M., "A Multi-Period Architectural Selection and Optimum Capacity Allocation Model for Future SONET Interoffice Networks," *IEEE Transaction on Reliability*, Vol. 40, No. 4, October 1991, pp. 417-427.

[6] Cardwell, R. H., Monma, C. L., and Wu, T-H., "Computer-Aided Design Procedures for Survivable Fiber Optic Telephone Networks," *IEEE Journal on Selected Areas in Communications*, Vol. 7, No. 8, October 1989, pp. 1188-1197.

[7] Wu, T-H., and Lau, R. C., "A Class of Self-Healing Ring Architectures for SONET Network Applications," *IEEE GLOBECOM'90*, December 1990, San Diego, CA, pp. 403.2.1-403.2.8.

[8] Wasem, O. J., "An Algorithm for Designing Rings in Survivable Fiber Networks," *IEEE Transactions on Reliability*, Vol. 40, No. 4, October 1991, pp. 428-432.

[9] Gibbons, A., *Algorithmic Graph Theory*, Cambridge University Press, Great Britain, 1985.

[10] Lawler, E. L., Lenstra, J. K., Rinnooy-Kan, A. H., and Shmoys, D. B. (Editors), *The Traveling Salesman Problem: A Guided Tour of Combinatorial Optimization*, John Wiley, New York, 1985.

[11] Wasem, O. J., "An Algorithm for Designing Rings in Communication Networks," *NFOEC'91 Proceedings*, Nashville, TN, April 1991.

[12] Wasem, O. J., "Optimal Topologies for Survivable Fiber Optic Networks Using SONET Self-Healing Rings," *Proceedings of IEEE GLOBECOM'91*, December 1991.

[13] Nemhauser, G. L., Rinnooy Kan, A. H. G., and Todd, M. J., *Handbooks in Operations Research and Management Science: Volume 1- Optimization*, North-Holland Publishing Company, 1989.

[14] Chujo, T., Komine, H., Miyazaki, K., Ogura, T., and Soelima, T., "The Design and Simulation of an Intelligent Transport Network with Distributed Control," *Network Operations Management Symposium*, San Diego, CA, February 1990.

[15] Sakauchi, H., et al, "A Self-Healing Network with an Economical Spare-Channel Assignment," *Proceedings of IEEE GLOBECOM'90*, San Diego, CA, December 1990, pp. 403.1.1-403.1.6.

[16] Grover, W. D., Bllodeau, T. D., and Venables, B. D., "Near Optimal Spare Link Placement in a Mesh Restorable Network," *Proceedings of IEEE GLOBECOM'91*, Phoenix, AZ, December 1991, pp. 57.1.1-57.1.6.

[17] Yamada, J., and Inoue, A., "Intelligent Path Assignment Control for Network Survivability and Fairness," *Proceedings of IEEE ICC'91*, Denver, CO, June 1991, pp. 22.3.1-22.3.5.

[18] Okanoue, Y., Sakauchi, H., and Haseqawa, S., "Design and Control Issues of Integrated Self-Healing Networks in SONET," *Proceedings of IEEE GLOBECOM'91*, Phoenix, AZ, December 199, pp. 22.1.1-22.1.61.

[19] Reibman, A., and Raghavendra, C. S. (Guest Editors), "Special Issue on Design for Reliability of Telecommunication System and Services," *IEEE Transactions on Reliability*, Vol. 40, No. 4, October 1991.

CHAPTER 8

Network Survivability in Fiber Loop Networks

8.1 FIBER LOOP NETWORK ARCHITECTURES

Fiber in the local loop has progressed from exploratory and trial stages to planning and development of implementation strategies. Fiber facilities have been actively deployed in the feeder segment of local loop networks to reduce operating costs of present copper-based networks and to provide a fiber-optic infrastructure that will support new high-bandwidth telecommunications services, such as broadband integrated switching services. Figure 8.1 depicts a CO-serving area that forms a loop network having single-star and double-star subscriber loops. In Figure 8.1, subscribers relatively close to the serving CO are served by dedicated facilities (referred to as the *single-star topology*), and subscribers located beyond a critical distance from the serving CO are served by *Remote Nodes* (RNs) that connect subscribers to the serving CO via shared feeder facilities (referred to as the *double-star topology*). The general architecture shown in Figure 8.1 is applied to both the conventional copper-based loop network and the fiber-based loop network.

The double-star fiber-based loop network depicted in Figure 8.2 is composed of two segments: feeder and distribution. The fiber distribution network includes one *Remote Digital Terminal* (RDT), which is located at the RN,[1] and several *Optical Network Units* (ONUs), which are connected by a *Passive Distribution Network* (PDN). The RDT terminates optical signals on the network side of the PDN, and its functions include multiplexing and/or concentrating upstream traffic, demultiplexing and/or routing downstream traffic, service grooming, conveying signaling for services, remote-service provisioning and testing, and facility alarming and testing. ONUs are positioned at or near *Customer Premises* (CP). The ONU terminates the optical signal from the PDN and develops tariffed service interfaces for delivery to the customer's inside wiring. The ONUs and corresponding terminations on the host RDT may be viewed as a form of a small *Digital Loop Carrier* (DLC) system similar to the DLC

---

1. The RDT may be located at the CO for the single-star loop network.

system used in present copper-based loop networks. A *Fiber in the Loop* (FITL) system [1,2] is composed of an RDT[2] and all subtending ONUs connected by optical paths through the PDN.

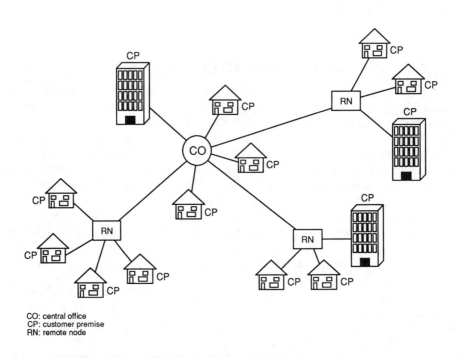

CO: central office
CP: customer premise
RN: remote node

Figure 8.1. A general architecture for a local loop serving area.

---

2. Note that Reference [1] (Bellcore TA-NWT-000909) uses the *Host Digital Terminal* (HDT) to distinguish from the conventional RDT. Several functions (e.g., signaling) originally performed in the RDT are moved to the ONU; thus, the HDT is simpler than the RDT in terms of functionalities.

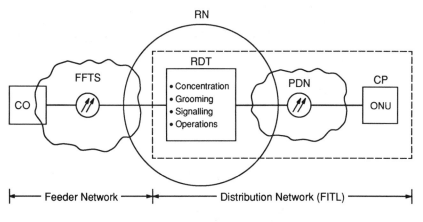

CP: customer premise
FFTS: fiber feeder transport system
FITL: fiber in the loop
ONU: optical network unit
PDN: passive distribution network
RN: remote node

Figure 8.2.   A conventional view of the double-star loop architecture.

The *Fiber Feeder Transport System* (FFTS) shown in Figure 8.2 is a conventional view[3] of the double-star architecture, which provides access and cross-connect functions between the RDT and its serving CO. The FFTS includes asynchronous-based or SONET-based multiplexer transport and cross-connection at various digital rates. The topology used for the FFTS depends on multiplexing technology, economics, and survivability considerations. Figure 8.3 depicts two examples of the FFTS that include a SONET W-DCS, ADM devices, and diverse routing of transport facilities. The topology used in Figure 8.3(a) is usually a tree where the W-DCS in the CO aggregates and grooms traffic from all OLTMs in RNs. Figure 8.3(b) depicts a survivable fiber feeder transport system that uses ADMs to implement an SHR to provide protection for fiber cable cuts.

---

3. Note that the *Passive Optical Network* (PON) for the subscriber loop (as will be discussed in Section 8.1.2) does not functionally look like Figure 8.2 because a PON has a totally passive RN. Hence, in a PON, the RDT would be in the CO, and the FFTS and PDN would be merged.

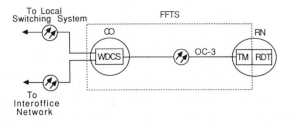

(a) SONET Terminal Multiplexers

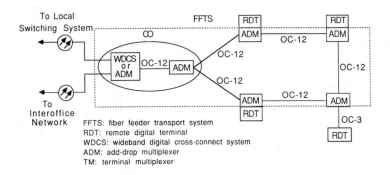

FFTS: fiber feeder transport system
RDT: remote digital terminal
WDCS: wideband digital cross-connect system
ADM: add-drop multiplexer
TM: terminal multiplexer

(b) Add-Drop Multiplexers and Route Diversity

Figure 8.3. Examples of fiber feeder transport systems.

The choice of which fiber feeder multiplexing technique to use in the RN of the double-star architecture is a fundamental design issue because it determines the type of equipment required at the RNs of the loop network. Critical features of the RN include its cost, size, power dissipation, maintenance requirements, and system upgradability. Network survivability is also very important.

TDM and WDM are two major multiplexing techniques used in the fiber feeder transport network. TDM is a signal multiplexing scheme that shares a transmission link among multiple channels by assigning time intervals to individual channels during which they have the entire bandwidth of the system for use. In contrast, WDM is a signal multiplexing scheme that combines (or separates) two or more channels with different wavelengths to (or from) a common optical waveguide; each channel may access the entire bandwidth of the system. For the TDM scheme, traffic concentration on the loop feeder relies on high-speed electronic processing at the RN to achieve higher fiber utilization and uses optical technologies only for point-to-point transmission. In contrast, the WDM scheme processes signals directly in the optical domain using different wavelengths for different channels rather than via traditional high-speed electronics. Chapter 4 discussed a relative comparison between the TDM and WDM schemes (see Section 4.6).

Two major architectures may be used for the double-star fiber loop network depending on the multiplexing technique used in the RN: the double-star architecture with active RNs and the double-star architecture with passive RNs. The following two sections describe these techniques.

8.1.1 Double-Star Active RN Loop Architecture

The double-star active RN network uses high-speed TDM electronics in the outside plant environment to concentrate traffic from CP and possibly to provide service selection. These active RNs are fed via feeder fibers from the CO with further fiber, copper pairs, or coaxial cable connected to the CP. Figure 8.4 depicts a simplified view of the active RN implementation. One example of the active RN loop network is the SONET-based, double-star loop network (see Figure 8.3).

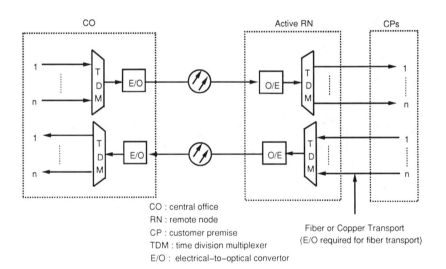

CO : central office
RN : remote node
CP : customer premise
TDM : time division multiplexer
E/O : electrical-to-optical convertor

Fiber or Copper Transport
(E/O required for fiber transport)

Figure 8.4.   A simplified view of active RN implementation.

8.1.2 Double-Star Passive RN Loop Architecture

As an alternative design to the TDM RN implementation, the multiplexing and routing function may be performed in the *optical* domain. The network that uses this design is commonly referred to as a PON. The PON architecture represents a broad class of new network architectures that process signals directly in the optical domain. One example of the PON is the *Passive Photonic Loop* (PPL) architecture [3] (see Section 8.1.2.2). The PON can be implemented using optical couplers or passive WDM components. Figure 8.5 depicts these two common, passive fiber-optic components that are used in PON implementations. The optical power combiner and splitter are passive fiber-optic components that combine and separate optical powers among fibers, respectively. The WDM and demultiplexer are passive fiber-optic components that combine and separate optical channels with different wavelengths, respectively. For WDMs, power is distributed based on wavelength, permitting multiple wavelength transmission over a single fiber. Bellcore Technical Advisory TA-NWT-000442 [4] (optical couplers) and Bellcore Technical Reference TR-TSY-000901 [5] (WDM) contain generic criteria for these passive fiber-optic components.

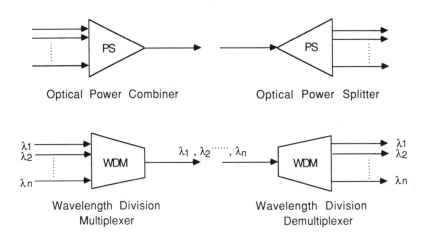

Figure 8.5. Passive components for passive RN implementation.

8.1.2.1 PON/OC Architecture

Figure 8.6 depicts a PON architecture using optical couplers [6], which is denoted by PON/OC. In this architecture, PSs and combiners are used to perform optical multiplexing and routing functions at the RN. The PON/OC consists of a single fiber from the serving CO that is fanned out via passive optical splitters at the RN location to feed a number of individual customers. The TDM signal is broadcast on a single optical wavelength to all terminals connected to the PON via the PS. After an optical receiver at the ONU (located at the curbside or the CP) detects the signal, the *Bit Transport System* (BTS) in the RN (or CP) accesses the time-multiplexed channels intended for that destination and delivers the data, plus some necessary signals, to a further service access unit in the CP. In the return direction, data from the customer's terminal is inserted at a predetermined time to arrive at the CO within the assigned time slot.

A *Time Division Multiple Access* (TDMA) protocol implements the time management of the system. It periodically determines the path delay between the RN's terminal and the serving CO and updates a programmable path delay in the RN's equipment. Thus, the TDMA protocol ensures that each channel is correctly timed at the CO end of the system. Reference [6] details this TDMA protocol and the PON/OC frame structure. The whole system is operated synchronously in a master-slave manner, and data in the downstream direction is scrambled to ensure good clock recovery at the RN's terminal. This system can also add an optical filter in the RN's equipment that passes only the PON system wavelength. This enables other wavelengths to be added at a later date to provide new services without disturbing the existing voice and data services.

Each RN needs a 1×N or 2×N PS to broadcast the signals to end customers. These PSs may be assembled from arrays of 2×2 couplers or other arrays of devices depending, upon the manufacturing technique. For example, a 1×7 splitter reported in Reference [6], which has excess loss less than 0.3 dB, is made by bundling seven fibers and melting their cores together.

In 1990, as part of Ameritech's fiber-to-the-curb field trial, the first PON/OC network was installed in the U.S. at the Jefferson Meadows residential development near Columbus, Ohio. This network has a physical star, logical bus structure that uses a passive splitter between the network and residential development. At the splitter, the signal is passively divided among a set of fiber groups, which are then run to curbside pedestals in a star configuration. All channels are broadcast to all pedestal units.

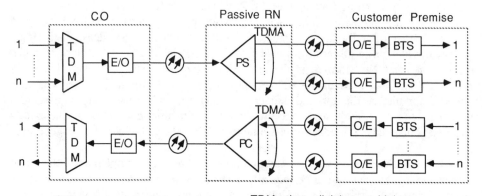

BTS: bit transport system   TDM: time division multiplexer
PC: power combiner   TDMA: time division multiple access
PS: power splitter

Figure 8.6.   A passive optical network architecture using optical couplers (PON/OC).

8.1.2.2 PON/WDM Architecture

Figure 8.7 depicts an alternative PON implementation that uses WDM technology and is denoted by PON/WDM. This architecture is also referred to as the PPL architecture [3]. In this architecture, each of N subscribers served by a particular feeder is assigned to two different wavelengths: one for upstream and the other for downstream transmission. Thus, the PON/WDM architecture is essentially a high-capacity, dedicated point-to-point system. At the CO, signals destined for different subscribers modulate their respective downstream wavelengths, which are then multiplexed onto the feeder fiber. At the RN, the wavelengths are then demultiplexed onto the appropriate distribution fibers. Upstream transmission is performed in a similar way: different subscribers transmit in unique WDM bands that are multiplexed at the RN, transmitted over the feeder, and demultiplexed at the CO. A major advantage of WDM is that upgrading can occur with minimal disruption to existing services and can provide high-capacity services planned for in a B-ISDN system. Each added wavelength provides a new transparent optical path that is independent of the other wavelengths carried on the system. Thus, any combination of services can be provided at any time with arbitrary bandwidth and modulation format.

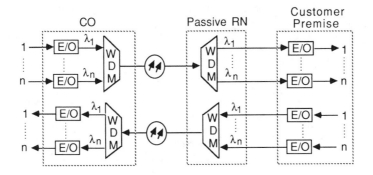

WDM: Wavelength Division Multiplexer/Demultiplexer

Figure 8.7. A passive optical network architecture using WDM devices (PON/WDM).

The PPL architecture employing 32-channel WDM (i.e., 16 wavelengths) for a line rate of 600 Mbps has been demonstrated, and the result has been reported in Reference [3]. This experimental demonstration is viewed as a practical, cost-effective application for broadband services because it uses commercially available components and reduces insurmountable barriers.[4]

Through its allocation of dedicated wavelengths to each subscriber, the PON/WDM architecture shown in Figure 8.7 avoids the power-budget and per-subscriber bandwidth limitations of single-wavelength, splitter-based PON architectures like the one depicted in Figure 8.6. However, the PON/WDM relies on aggressive optical technologies that may not be commercially available or economical for several years. Thus, a near-term PPL architecture has been proposed [3] that combines architectural aspects relying solely on splitters and those relying solely on WDM components. Figure 8.8 depicts one such architecture. In this near-term architecture, a group of subscribers uses a common wavelength, and different groups use different wavelengths. In the downstream direction, each wavelength is distributed to a particular group of subscribers via a PS. Upstream transmission is similar, with that group of subscribers

4. The arguments of cost-effectiveness and reducing insurmountable barriers for the PON/WDM architecture are based on two assumptions: the costs of the WDM technologies will drop significantly, and broadband services that the PON/WDM was designed to deliver will emerge in the near future.

337

using the same wavelength based on TDMA protocols (e.g., the protocol described in Reference [6]).

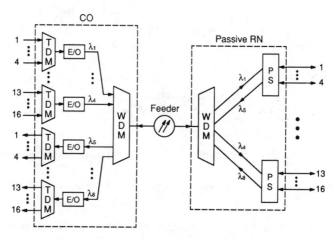

Figure 8.8.   A hybrid PON using WDM and optical couplers.

8.1.3 Relative Comparisons between Active and Passive RN Implementations

Table 8-1 summarizes a relative comparison between the active and passive RN implementations. The active TDM network relies on high-speed electronic processing to achieve the higher fiber utilization and uses optical technologies only for point-to-point transmission. These high-speed electronics at the RN may require power and backup systems, which may necessitate the use of a *Controlled-Environment Vault* (CEV) to house the RN. Craftsperson dispatch is needed for maintenance at the remote location, and back-to-back *optical-to-electrical* (O/E) and E/O conversions are also needed at the RN.

The PON architecture relies solely on passive WDM components or optical couplers to perform routing and multiplexing at the RN. The all-optical nature of WDM devices or optical couplers eliminates the need for back-to-back O/E and E/O conversion, air-conditioned CEVs, and remote power backup at the RN.[5] High-speed

---

5. In either the TDM or PON approach, there are significant electronics and optics at the curbside or CP's location. The maintenance action to the controlled temperature environment of a CEV may not be smaller than the maintenance required to the curbside units, where temperature is not controlled.

feeder multiplexing electronics are repositioned back in the CO where the floor space and maintenance are easier and less expensive to provide. Remote maintenance requirements for the passive WDM components or optical couplers should be minimal. Furthermore, in contrast to electronic TDM, the WDM channels are independent of one another and are transparent to data format. The channel transparency allows the architecture to be completely compatible with future network standards (e.g., B-ISDN). Thus, physical fiber loop infrastructure can be installed before detailed higher-level standards without risk of future incompatibility. Finally, eliminating CEVs and the relative compactness of WDMs permits much greater flexibility in RN deployment.

Table 8-1. Relative Comparisons between Active and Passive RN Implementations

Network attributes	Double-star with active RNs	Double-star with passive RNs
Equipment cost at RN	higher	lower
Space requirement	higher	lower
Maintenance cost	higher	lower
Power supply needed at RN	yes	no
Flexibility for upgrading to add new services	yes/no*	yes
Network control flexibility	higher	lower
SHR implementation	easy	difficult
Interoffice and loop network integration	yes+	yes++
Number of users by each RN	higher	lower
Optical technology	simple	complex

+ Assume the SONET technology is used.
++ A passive interoffice and loop-integrated network was recently proposed in Reference [7].
* Only if adding new service requires significant extra capacity that may exhaust the RN electronic equipment.

On the other hand, the TDM network offers flexibility of processing tributary signals, which results in more flexible network control features and more services that can be supported, and may support cost-effective transport architectures (e.g., SHRs) more easily than its passive counterparts. From an operations perspective, integrating interoffice and loop networks may reduce operations costs and difficulties associated with growth in loop areas. This interoffice and loop network integration can be implemented by using both active (e.g., SONET [8,9,10]) and passive technologies [7]. The optical technology needed for the TDM network is simpler than that needed for the PON. Because the TDM network uses high-speed electronic equipment, the number of subscribers that each RN can support is much greater than that supported by the PON. For example, the current capacity of grating-based WDM technology or wavelength-flattened optical splitters may limit the number of supported subscribers to a few tens; in comparison, each active RN can support 1000 or more subscribers. Hence, the PON approach requires considerably more feeder fibers than its active counterparts. However, the much shorter distribution fibers[6] in the PON architecture may compensate for higher fiber count in the feeder, thus allowing the PON to achieve excellent fiber savings for the number of subscribers [3]. Some studies on the relative economics between the PONs and their active counterparts have been reported in References [11,12,13]. In References [11,12], the PPL is compared to the active double-star architecture. Both are similar in installed first cost, but the PPL may have life cycle cost advantages due to the passive nature of the RN. In Reference [13], the PPL is compared to the PON/OC. In long-term time frames, the PPL should be 10 to 20 percent more costly than the PON/OC, with the extra cost providing substantially greater per-subscriber transmission bandwidths, particularly in the upstream direction.

8.1.4 Survivability Considerations for Fiber Loop Networks

Like interoffice networks, fiber loop survivability has become an increasing concern to both the telephone companies and business customers. This is due to the star architecture's vulnerability and the very high volume of data being carried by a very few fiber strands in the loop networks. However, the progress of enhancing loop network survivability is much slower than its interoffice counterparts partly because of economics. The cost of loop survivability enhancement may be shared by both the service provider and its customers or paid by the customers. Also, it is not expected that loop survivability enhancement will necessarily be used within the distribution network (between the ONU and the RDT) (see Figure 8.2) because (1) a single ONU can serve only a small number of customers, and (2) providing this capability within the distribution network is costly and complex. Thus, the loop survivability concerns

---

6. The ONU in the PON is placed at the curbside or the CP (see Figure 8.6), rather than at the RN, as in the active double-star architecture.

presented in this chapter focus on the feeder transport network. Note that large business customers served by private lines or a single-star topology may require highly survivable loop networks. In this case, survivable architectures designed for the fiber feeder network are also applicable for these large business customer-based loop networks. Because this chapter focuses on SONET technology, unless specified elsewhere, the active feeder network discussed in the subsequent sections is referred to as a *SONET feeder network*.

Several approaches that have been proposed for enhancing interoffice network survivability can also be applied to the feeder transport network. These approaches include route diverse protection and dual-homing protection, where route diverse protection is used only for cable cuts and dual homing, which connects each RN to two serving COs and provides partial protection for a serving CO failure. According to a recent reliability study for loop networks [14], network reliability, measured in terms of *Mean Time To Failure* (MTTF) and years, for a 1:1 protection system increases on the average by factors of 3 and 150 by deploying fiber with DP paths and dual homing, respectively.

The route diverse protection capability can be implemented using simple electronically or optically controlled APS systems [15,16]. The dual-homing protection capability can be implemented by deploying two diverse fiber spans from each RN (or from the CP if the single-star topology is used) to two serving COs using the dedicated [15] or broadcast fiber spanning approach [17] (see Section 3.5 of Chapter 3). The SHR is another cost-effective survivable network that can provide 100 percent protection for both a single cable cut and a serving CO failure[7] [8,9,10]. As will be discussed in Section 8.2, the SHR architecture is more difficult to implement for the PON feeder networks than for its SONET counterparts because (1) signal adding-dropping for the PON feeder network has to be processed at the optical domain, and (2) the self-healing capability requires a more complex protection control scheme than its electronic counterpart.

The following sections discuss loop network survivability enhancement in more detail. Section 8.2 discusses active fiber feeder transport networks and focuses on SONET SHR implementations; this is because the architecture design and operations for route diverse protection and dual homing are the same as those systems discussed in Chapter 3. Section 8.3, which addresses passive feeder transport networks, first discusses passive point-to-point route diversity protection and passive dual-homing

---

7. However, demands terminating at that failed CO are lost.

protection, which can be implemented using commercially available components. It then discusses a passive SHR architecture for the feeder network.

8.2 SONET SHR ARCHITECTURE FOR ACTIVE FEEDER TRANSPORT

Figure 8.3(b) depicted an example of a SONET OC-12 SHR implementation to enhance survivability for the SONET feeder transport network. The SONET ring shown in Figure 8.3(b) is composed of SONET ADMs on a ring topology, which provides a highly survivable transparent channel transport between RDTs and their serving COs. Channel management for customers is performed at RDTs of RNs before entering the SONET feeder ring transport system.

USHRs and BSHRs[8] may be used to implement feeder ring transport, depending upon demand routing in normal conditions. As discussed in Section 4.3.1, the selection of the SONET ring architecture depends on the demand pattern and demand requirements. The demand pattern in the loop network is likely to be the centralized demand pattern in which all demands go to a single central site (e.g., serving CO), thus perhaps favoring the USHR architecture (see Section 4.3.2).

The protection control scheme for USHRs can be implemented using the line protection switching scheme or the path protection switching scheme.[9] Also as discussed in Section 4.3.2, the path protection switching scheme is simpler than the line protection switching scheme in terms of control complexity for protection switching. Thus, this section discusses only SONET USHRs with path protection switching for the survivable active feeder network application. TR-TSY-000496 [9] documents the ADM equipment and operations criteria associated with the implementation of the SONET USHR architecture using path protection switching. For convenience, this architecture is denoted by USHR/P.

This section presents single-CO and dual-CO SONET USHR/P architectures for active fiber feeder networks based on References [8,9]. Single-CO access architectures provide cable cut protection but do not provide protection against serving CO failures. Dual-CO access architectures provide cable cut and CO failure protection. Figure 8.9 depicts a generic architecture of the OC-N USHR/P, which consists of one ADM at each office and a pair of fibers with signals going in opposite directions. In Figure 8.9,

---

8. Refer to Section 4.2 of Chapter 4 for more details regarding SONET ring alternatives.

9. The line and path protection switching systems are systems that are triggered by the SONET line overhead (e.g., K1 and K2 bytes) and the path-layer signal (e.g., Path AIS), respectively (see Section 4.2.1 of Chapter 4).

the duplex working channel is carried around the working ring (e.g., outer ring) in one direction only (e.g., counterclockwise [CCW]), and the duplex protection channel is carried around the protection ring (e.g., inner ring) in an opposite direction. In the normal ring state, one or more STS-M ($M \leq N$) signals are transmitted onto both the clockwise (CW) and the CCW directions of the ring. It should be mentioned that VT rings are also possible and are discussed later. When duplicate signals are transmitted in both directions on the ring, it is considered a "dual-fed" or "1+1 protection" approach. These two duplicated signals propagate along the working ring and the protection ring in opposite directions and are finally dropped at one of the offices on the ring. Thus, at the receiving office, two path signals are available for signal selection.

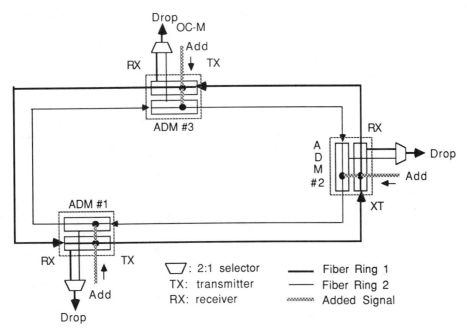

Figure 8.9. A generic network architecture for a SONET USHR/P.

Suppose these two signals are designated fiber path 1 and fiber path 2. During normal operation, only one of the path signals is used, although both signals are monitored for alarms and maintenance signals. If a fiber cable on the ring is cut, the service will not be interrupted as path protection switching will be used to select the alternate path signal.

8.2.1 Single-CO Access Ring Architecture

Figure 8.10 shows a physical USHR/P network in a feeder application that is compatible with the present *Integrated Digital Loop Carrier* (IDLC) system [18]. In this application, three RDTs communicate with the local switch in the CO. These RDTs connect to the ring through a chain of SONET ADMs that collect traffic headed to the CO; connection to the RDT can be made via DS1s or an OC-M ($M \leq N$). The ring topology is completed by connecting the last SONET ADM in the chain back to the CO. This provides increased assurance of uninterrupted service by providing two facility paths back to the CO. If a cable cut occurs between any two nodes (the OC-N link), redundancy exists to reroute circuits to the CO in the opposite direction, as explained above.

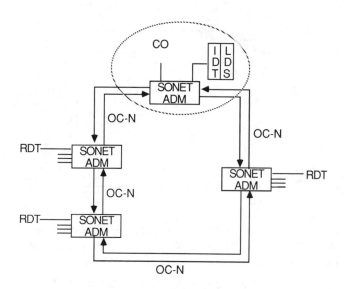

IDT: integrated digital terminal
LDS: local digital switch
RDT: remote digital terminal

Figure 8.10.   An example of a USHR/P architecture in an FFTS.

344

A facility interconnection between an access ring and an interoffice network can be established via a multiple-ring, single-CO configuration (discussed in Reference [9]) if the connecting interoffice network is a USHR/P. In this configuration, each ring has an ADM at the ring interconnection point, and these ADMs are interconnected at the low-speed interface level (i.e., back-to-back ADMs). This multiple-ring configuration provides the transport of DS1 and DS3 services between the access ring and the interoffice ring, but does not provide protection against serving CO failures. The SONET multi-ring interworking system reflects a SONET-integrated planning concept that eliminates the boundary between the access network and the interoffice network.

Another multi-ring interworking system, which includes at least one bidirectional ring, is also possible and has been discussed in Reference [10]. Refer to Reference [10] for more details on this system.

8.2.2 Dual-CO Access Ring Architectures

The dual-CO access architecture, which network providers are increasingly considering, is designed for access networks requiring service protection from a serving CO failure. Figure 8.11 depicts a ring network architecture composed of multiple, interconnected USHR/Ps [8,9]. It is assumed that the interoffice ring operates at a higher speed than the access ring in a given hierarchical network. Each ring comprises two unidirectional communication fiber paths transmitting in opposing directions. This multi-ring interworking system assumes that channels between customers and a *Point of Presence* (POP)[10] need to be protected to provide enhanced service survivability. This discussion assumes the network shown in Figure 8.11 supports both local and IXC services. The main issue involved in Figure 8.11 is how to provide a protected connection between USHR/Ps. As shown in Figure 8.11, the rings are interconnected with a pair of serving nodes (serving node 1 and serving node 2), where a serving node provides paths between rings. Any demand that passes between rings is fully and automatically protected against the loss of a serving node. For service survivability, protection against the loss of a serving node is essential in addition to having protection against a fiber cable cut in the ring.

---

10. A POP is a physical location within a LATA at which an *Interexchange Carrier* (IXC) establishes itself (for the purpose of obtaining LATA access) and to which the local telephone company provides access services.

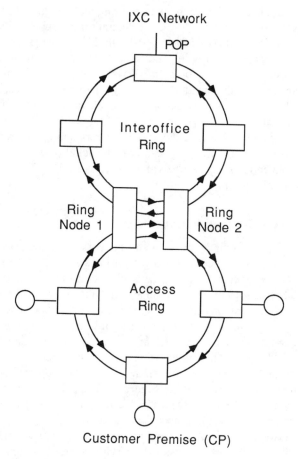

IXC Network

POP

Interoffice Ring

Ring Node 1

Ring Node 2

Access Ring

Customer Premise (CP)

Figure 8.11. Hierarchical ring interworking architecture.

Figure 8.12 shows, in detail, how two USHR/P rings can be interconnected to provide protection against serving node failures. Starting at the POP location, channels (e.g., DS1 services) from an IXC are carried via the interoffice ring to the designated serving nodes (i.e., serving node 1 and serving node 2), which distribute channels on the access ring to various customer terminations on the access ring. Channel assignments onto the access ring are made at the serving nodes. Note that the ADMs in both serving nodes are connected with an electrical STS or optical OC-M intraoffice interconnection to preserve alarms and maintenance signals carried in path overhead. At the POP location, the DS1s are mapped into 28 floating VT1.5s on the interoffice

346

ring. For a channelized (i.e., VT-organized) STS-1 from the IXC, the STS-1 is carried on the interoffice ring to the designated serving nodes and terminated. This STS-1 is then demultiplexed to VT signals for channel assignment onto the access ring. Higher-speed channels (e.g., DS3 rate services) may also be delivered directly from an IXC to a customer termination on the access ring (carried via an STS-1).

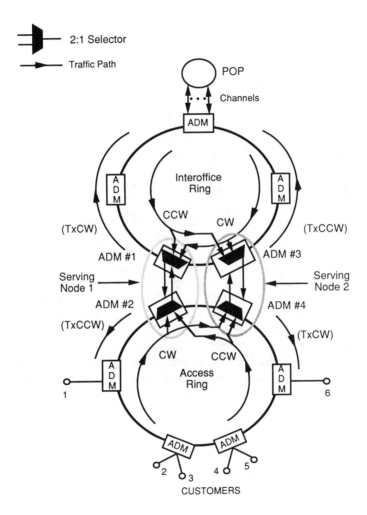

Figure 8.12. Protected USHR/P ring interconnection architecture.

The SONET ADM at the POP location transmits its VTs in both the CW and CCW directions on the USHR/P interoffice ring. This VT signal from the POP is routed to both serving nodes via a drop-continue feature (see Reference [19]), which permits the same signal to be delivered to the two serving nodes from the same direction. Note that the data moving in the return direction is the signal from the opposite direction. For instance, CCW traffic is dropped off at SONET ADM #1 and passed to SONET ADM #3. Likewise, CW traffic is dropped off at SONET ADM # 3 and passed to SONET ADM #1. Thus, with the drop-continue feature, the continued signal is forwarded like a pass-through signal; in this way, two signals (one from each direction) are available for signal selection at either serving node. The ADM drop-continue feature is expected to be provisionable at the path level. During normal operating conditions (i.e., no ring failures), 2:1 selectors in the serving nodes can be set up such that the preferred (or default) signals for traffic selection are opposite each other, ensuring that a cable cut in the interoffice ring will not cause channel switching in both serving nodes. It is assumed that the USHR/P rings are revertive. Hence, at serving node 1, the 2:1 selector in SONET ADM #1 selects either the CW or the CCW channel received from the POP as the dropped signal, and this signal serves as the input to SONET ADM #2. SONET ADM #2 transmits this signal only in the CCW direction on the access ring. This implies that a 2:1 selector is needed only for each channel received from the POP.[11] The same procedure occurs at serving node 2, except that SONET ADM #4 transmits the preselected signal from SONET ADM #3 only in the CW direction on the access ring. This implies that the ADMs in a USHR/P ring configuration are provisionable to disable the bridging function to either fiber ring. Signals entering the ring would then be routed to only one fiber. This procedure therefore satisfies the USHR/P ring requirement of having duplicate signals in the CW and CCW directions.

In the same manner, VT signals from customers on the access ring (heading back to the POP) are routed to both serving nodes 1 and 2 via the drop-continue feature. Channel assignments onto the interoffice ring are made at the serving nodes. Signals from the serving nodes 1 and 2 are transmitted in only one direction on the interoffice ring, opposite each other (ADM #1 transmits CW and ADM #3 transmits CCW), thus satisfying the USHR/P ring requirement of having duplicate signals in the CW and CCW directions.

---

11. Because all VTs are transported directly between the POP and the serving node in a single STS-1 signal, VT-level protection may be accomplished by providing protection at the higher STS-1 path level.

The demand-routing diagrams in Figures 8.13(a), (b), (c), and (d) illustrate channel signal routing between the interoffice ring and the access ring under different scenarios. In particular, Figure 8.13(a) shows normal channel signal routing to customers from the interoffice ring and from the access ring to the POP. The diagrams in Figure 8.13 assume revertive switching (default service is routed through the CW direction). Figure 8.13(b) shows the channel signal routing affected by a link failure in the interoffice ring, and Figure 8.13(c) shows the channel signal routing affected by a link failure in the access ring. Figure 8.13(d) shows the channel signal routing affected by a serving node failure.

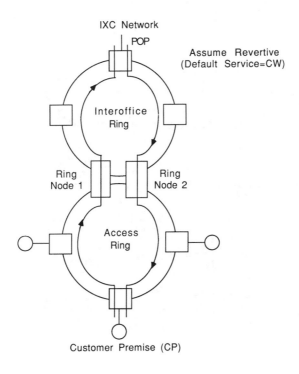

(a) Normal Channel Routing

Figure 8.13.   Channel routing on the interconnection ring.

349

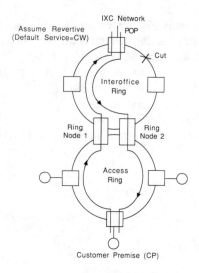

(b) Example of Interoffice Failure

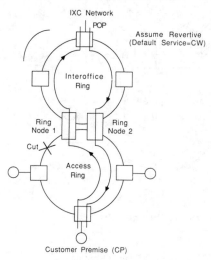

(c) Example of Access Failure

Figure 8.13. (Continued)

350

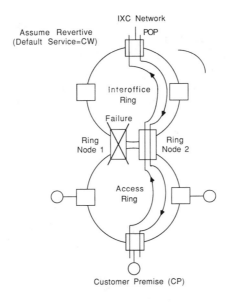

IXC Network

POP

Interoffice
Ring

Failure

Ring
Node 1

Ring
Node 2

Access
Ring

Customer Premise (CP)

(d) Example of Node Failure

Figure 8.13. (Continued)

It is important to emphasize that the operation of serving nodes 1 and 2 in Figure 8.12 is symmetrical. ADM #1 and ADM #2 in serving node 1 serve one of the unidirectional fibers in each USHR/P ring. Similarly, ADM #3 and ADM #4 in serving node 2 serve the other unidirectional fiber in each USHR/P. With this arrangement, service continues between the interoffice ring and access ring if serving node 1 or serving node 2 is lost, including interconnecting links within each serving node, and a path between interoffice and access rings always exists with the loss of either node. Restoration is automatic in the event of a node failure — no rerouting of traffic between serving nodes is necessary. Also, both the interoffice and the access rings are treated as autonomous, independent rings, which provides a robust network. Any fault recovery is confined to the ring with the fault. This means the interoffice ring can operate at the STS-1 level, and the access ring can operate at the VT level. More notably, this configuration can survive a simultaneous cable cut in both the interoffice and access ring networks without loss of communication between rings.

Simpler serving node arrangements are possible [see Figures 8.14(a) and 8.14(b)] using basic SONET ADMs without the need for the drop-continue feature and 2:1

selectors. Also, DCSs may be used at the serving nodes in place of ADMs, depending upon economics and the DCS deployment strategy. In the arrangement shown in Figure 8.14(a), channels between rings do not use bandwidth between the two serving nodes, as does the arrangement shown in Figure 8.12. Thus, spare bandwidth is available for point-to-point transmission between the two serving nodes. In essence, one larger logical ring (encircling both the interoffice and access rings) is created to connect the POP and the customer. Normal channel routing is the same as shown in Figure 8.13(a). The arrangement in Figure 8.14(a) is survivable against a single point failure in the interoffice ring or access ring, or against a failure in one of the two serving nodes; a diverse signal through the alternate serving node is available to restore service. The main disadvantage of this simpler arrangement is that it may not survive all simultaneous failures (i.e., cable cuts) that may occur in both the interoffice and access rings for which restoration is possible. Unlike the arrangement shown in Figure 8.12, the interoffice and access rings shown in Figure 8.14(a) are no longer independent. This means that the POP must also switch at the VT-path level, which is the same as the access ring.

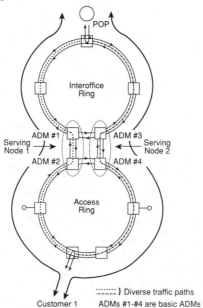

(a) Service Node Arrangements Using Basic ADMs

Figure 8.14.   Serving node arrangements.

352

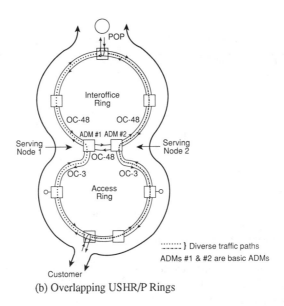

(b) Overlapping USHR/P Rings

Figure 8.14. (Continued)

Similarly, the arrangement shown in Figure 8.14(b) creates one larger logical ring connecting the POP and the customer, but only a single SONET ADM is used at the two serving nodes. In this example, a VT-based ring optimized for distributing small numbers of DS1s from the interoffice ring to various customer locations in the loop is formed using OC-3 drops from the interoffice OC-48 capacity ring. The low-speed OC-3 interfaces need STS-1 and/or VT add-drop capability for selectively adding-dropping STS-1/VTs between the OC-48 and OC-3 optical signals. This arrangement can be viewed as two overlapping rings, whereby both rings share the spare bandwidth between the two serving nodes. Sharing bandwidth may necessitate the need for careful planning because any traffic common to each ring will use portions of this bandwidth, and, hence, it could potentially become a bottleneck. This arrangement has the same disadvantage as the arrangement shown in Figure 8.14(a) — it does not have independent interoffice and access rings. However, the simplicity of the arrangements shown in Figures 8.14(a) and (b) may be desirable in some applications, and their suitability for network providers needs to be addressed. Table 8-2 summarizes a relative comparison among different SONET ring architectures discussed in this section.

Table 8-2. Comparing Different SONET Ring Architectures

Ring architecture	Single cable cut	Single CO failure	Two cuts (one per ring)	ADM feature	Spare bandwidth between serving COs	Costs
Multiple Single CO (Figure 8.10)	yes	no	conditional*	basic ADM	--	low
Dual CO (Figure 8.12)	yes	yes	yes	drop-continue	no	high
Dual CO (Figure 8.14)	yes	yes	conditional*	basic ADM	yes+/ limited++	high+/ medium++

* Depends on the locations of two cable cuts
\+ Dual-CO interconnection of rings depicted in Figure 8.14(a)
\++ Dual-CO interconnection of rings depicted in Figure 8.14(b)

Another multi-ring interworking system, which includes at least one bidirectional ring, is also possible and has been discussed in Reference [10]. Refer to Reference [10] for more details on this system.

8.2.3 Economic Merit and Survivability Enhancement for Feeder Networks Using Rings

Like the SONET interoffice ring networks discussed in Chapter 4, the economical merit of the feeder ring over the conventional 1:1/DP system depends upon many economic and engineering factors. These factors include network size, traffic data, relative ADM and OLTM costs, and the cost model. The cost model for the SONET feeder network depends upon the way the network adds-drops the DS1 signals. Note that the DS1 demand is the major type of demand in today's fiber facility networks. Figure 8.15 depicts two methods for adding-dropping DS1 signals from STS ring transport. The first method [see Figure 8.15(a)] uses the VT ring to add-drop VT1.5 signals directly from STS-N signals using VT pointer processors. Note that the VT ring is usually more expensive than the STS ring because a significant number of DS1 signals (compared to STS-1 signals) are needed to be processed and monitored. For example, the cost of an OC-12 fully configured VT ADM is at least twice the cost of an OC-12 fully configured STS-1 ADM. The second method uses an STS-1 ring to add-drop STS-1 signals and M13-like components to demultiplex the DS3 to DS1s and vice versa.

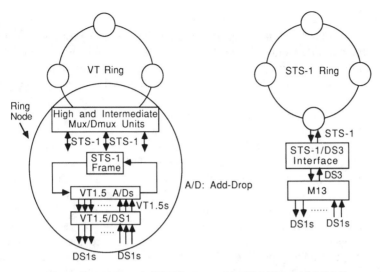

(a) Direct VP Add-Drop for VT Ring   (b) STS-1 Ring with M13 Components

Figure 8.15.   Two engineering models for DS1 add-drop.

We next discuss a case study based on the second add-drop method [Figure 8.15(b)] that was reported in Reference [8]. This case study investigated the economic feasibility of using ring architectures, described in Sections 8.2.1 and 8.2.2, in a LATA feeder network. The economic study for LATA networks is usually divided into two independent portions: an interoffice network and a loop feeder network. Although the architectures depicted in Figures 8.12 and 8.14(a) are integrated architectures for interoffice and loop networks, it may be preferable that traffic planning for the interoffice network and the loop network be separate because these networks have some very different considerations. For example, it is difficult to predict or project traffic growth in the loop network because nodes are added or removed much more frequently than in the interoffice network. For this reason, this section examines only demand in the loop area.

The ADM demand (i.e., add-drop) requirements for these two architectures are essentially the same, and the ADM with the drop-continue feature used in the first architecture (Figure 8.12) is not much different from the basic ADM used in the second architecture [Figure 8.14(a)]. The difference in hardware configurations[12] between the basic ADM and the ADM with the drop-continue feature is that the former uses a full-duplex circuit for the added and dropped STS-1s, whereas the latter uses one circuit to convey the dropped STS-1 to the 2:1 selector and the other circuit to convey the continued STS-1 (which is split from the dropped STS-1) back to the STS-3 multiplexer. Thus, from a generic equipment point of view, the cost of an ADM with the drop-continue feature and a basic ADM may not be much different.

Reference [8] reports a case study for a feeder network based on these assumptions. This case study compared single-CO and dual-CO USHR/Ps with corresponding single-CO and dual-CO 1:1/DP networks because 1:1/DP has the same 100 percent survivability for a single cable cut as the SHR architecture. The model network is a 10-node feeder network, including two serving COs and eight RNs, as depicted in Figure 8.16. The total demand requirement within this area is 165 DS1s, and the candidate line rates considered here are OC-3, OC-9, OC-12, OC-24, and OC-48. The end-to-end DS1 demand requirement was directly converted to the STS-1 demand requirement without demand grooming. For example, if there were five DS1s between Node A and Node B, then one STS-1 path was directly assigned to this node pair. Also, in this study, the 1:1/DP network used a back-to-back configuration for working and protection terminal multiplexers at each node.

The case study results [8] showed that the ring architecture is more economical than the 1:1/DP network (cost savings up to 60 percent for the dual-CO application) with improved hub survivability (that increased about 49 percent for the dual-CO application). Here, the hub survivability is defined as the percentage of the total DS1 demand on the ring that is still intact when that hub fails. The cost and survivability merits of the ring architecture shown in this particular feeder network are primarily because all RNs can be connected by a single high-speed ring.

---

12. The equipment models of the SONET basic ADM and the SONET ADM with the drop-continue feature have been shown in Figures 4.25 and 4.31 of Chapter 4, respectively.

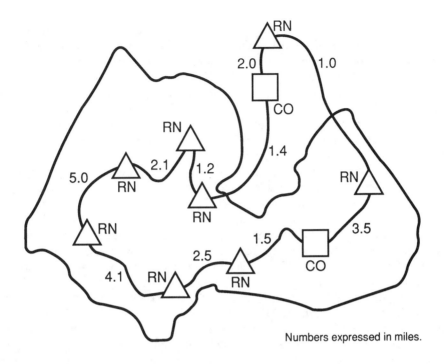

Figure 8.16.   Loop model network.

8.3 SURVIVABLE PASSIVE OPTICAL LOOP NETWORK ARCHITECTURES

Service protection in the PON system can be provided at the passive RN or at the individual ONU level. For protection at the ONU level, optoelectronic devices must be duplicated and a protection switching unit, which may be remotely controlled by its serving CO or manually controlled by the customer, must be added in the ONU. The ONU protection level may be too expensive for the non-business or smaller business customer.

Service protection at the ONU level also requires 1:1 equipment protection at the serving CO. To reduce the installed first cost and increase availability (as compared to the 1:1 protection systems just described) at the serving CO, the r:N system, in which N working components share r protection components, has been proposed for the PON/WDM architecture [20]. The r:N system has been shown to be particularly attractive in providing protection for multi-wavelength video transport systems. Reference [20] details this r:N protected video transport system.

For service protection at the RN level, the primary concern is to protect the network from the cut or failure of the fiber cable between the passive RN and its serving CO. Survivability implementations for PON systems at the passive RN level are different from their SONET counterparts. In the SONET feeder network, terminal multiplexers at the CO and the RN have intelligence to detect the failure message and trigger protection switching. In contrast, the passive RN is designed to transport optical signals transparently; thus, no signal processing and monitoring functions exist in the passive RN. This implies that an additional control system is needed at each passive RN to perform optical signal protection switching whenever necessary.

Like the protection provided in its SONET counterparts, service protection for passive RNs (i.e., protection at the RN level) in the PON systems may be provided via DP routing, dual-access protection, or SHRs. However, the APS system used to provide the route DP capability for the SONET feeder network cannot be used to provide protection in the PON system at the RN level due to lack of signal processing and monitoring capabilities in the passive RN. But a power loss detection system designed for point-to-point optical protection switching systems [16] can be used for service protection at the RN level in the PON system.

As described in Section 3.5 (Chapter 3), the optical dual-homing implementation without the rerouting capability [15,17] can also be applied to the PON dual-homing application. The passive ring implementation for the PON feeder is more complex than the SONET ring because the node of the PON ring requires more complicated optical technologies to add-drop optical channels and perform protection switching when a network component fails.

8.3.1 PON with DP at RN Level

Figure 8.17 depicts a 1+1/DP switching[13] PON/WDM architecture. In this architecture, a PS is equipped at the transmitter side (at the CO or RN) to split optical signals to both the primary and alternate routes. A 2:1 optical switch is needed at the receiver side (at the RN or CO) to select the best optical signal from both routes based on a power loss criterion implemented in an *Optical Power Loss Detection System* (OPLDS), such as the one described in Chapter 3 (see Figure 3.8). The OPLDS is a protection switching system that can detect the fiber facility failure based on the level

---

13. The bridging of the head-end office (i.e., the serving CO) of the 1+1 protection switching system is permanent. The tail end (RN) always receives two identical signals and selects a best signal based on criteria programmed in the micro-controller.

of optical power on that facility; it has a micro-controller that changes the configurations of attached 1×2 optical switches if the detected power level is below some threshold (see Section 3.4.1).

For downstream communications in the normal situation, an optical PS splits optical signals emitted from the WDM device in the serving CO. These signals are then sent to the RN via the working and DP fiber routes. In the passive RN, the 2:1 selector implemented by a 1×2 optical switch always terminates two identical signals with different delays and selects the signal from the primary route in normal conditions. If the fiber facility in the primary route fails, the power loss detection system detects this failure and sends a control signal to force all connected 2:1 selectors to select signals from the DP route. The same operations process is applied to upstream communications.

A similar optical protection switching system can be applied to the PON/OC system at the RN protection level, as depicted in Figure 8.18.

8.3.2 Economic Comparison between Point-to-Point Passive and Active Feeder Systems

When comparing the PON with its active counterpart using route DP and dual homing, the point-to-point PON system is expected to show an economical advantage. Costs of the passive components duplicated at the RN are insignificant compared to the cost of duplicating the OLTM at the active RN. Reference [14] describes a case study that compares the PON/WDM architecture with its active counterpart using DP and dual homing. The model network used in this case study is an urban loop network having two COs and a ring topology with a total length of 7850 feet. The PON/WDM implementation uses two separate 16-channel WDM devices for upstream and downstream fibers with the RN. The results[14] of this case study indicate that the PON/WDM implementation for DP and dual homing may have capital cost savings of 22 percent and 11 percent, respectively, compared to its active counterparts.

---

14. This case study considers only the component and fiber costs; it does not include the protection control system cost.

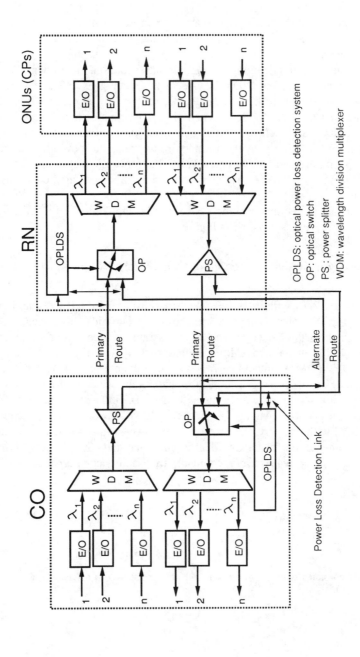

Figure 8.17. A 1+1/DP PON/WDM (PPL) architecture.

360

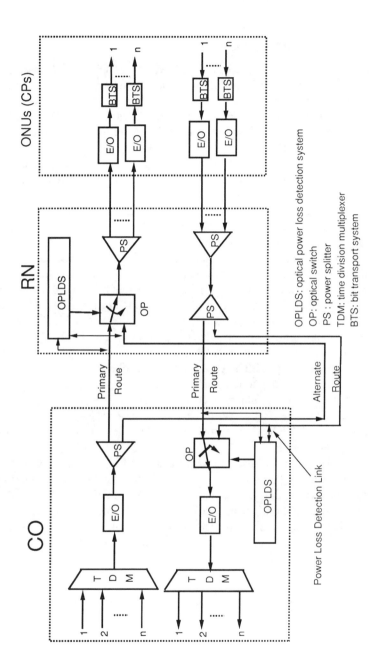

Figure 8.18. A 1+1/DP PON/OC architecture.

361

8.3.3 A Passive SHR Architecture for FFTSs

This section discusses a passive SHR architecture for FFTSs. This passive ring architecture uses WDM and optical filter technologies to add-drop optical channels in the all-optical domain. The passive SHR architecture for the feeder network can be realized by combining technologies of the PON/OC and PON/WDM (i.e., PPL) architectures. It also uses a special characteristic of the feeder network — a centralized demand pattern (i.e., each RN has to go to the serving CO before reaching its destination). The described passive ring architecture is essentially a simplified version of a passive SHR architecture that uses WDM devices for interoffice networking applications (see Section 4.6.1 and Figure 4.29). This passive ring architecture does not require as many wavelengths as its interoffice counterpart. The passive ring architecture adds-drops channels using WDM technology and delivers them to customers using PSs, just like the PON/OC architecture.

Figure 8.19(a) depicts an example of channel add-drop in the ring configuration. This architecture assigns each RN a wavelength, and different RNs are assigned different wavelengths. Thus, only N-1 wavelengths are enough for the feeder network of N-1 RNs served by a CO. This differs greatly from the interoffice passive ring using WDM devices, which needs (N)(N-1)/2 wavelengths (see Section 4.6.3) because two COs may have direct communications between them. As shown in Figure 8.19(a), each RN equips a wavelength filter pair, a WDM device pair, and a PS. The tunable wavelength filter[15] drops the optical channels with the wavelength destined to this RN. The WDM device multiplexes the added channels with the wavelength assigned to that RN to transmit channels, and the multiplexed optical signal is sent to the next node. When the optical channels are dropped at the RN, they are broadcast to users through the PS (as is done in the PON/OC). Optical amplifiers may be needed if the ring size is too large.

Several SHR concepts can be used to provide the self-healing capability for the passive feeder rings. Among them, the USHR using path protection switching may be appropriate because it has the simplest protection switching scheme (see Section 4.3.1). This self-healing implementation, as depicted in Figure 8.19(b), is similar to that used for interoffice passive, multi-wavelength SHRs (see Section 4.5). For the downstream traffic, optical signals added to the ring are duplicated using PSs and sent to the destinations node via two opposite optical paths. At the receiving end, the node always

---

15. A tunable wavelength filter transmits the WDM channels continuously while allowing random selection of one wavelength.

receives two identical signals with the same wavelength; one signal is selected to drop to the PS via a 2:1 optical selector, which may be implemented using a 1×2 optical switch. Again, each RN must equip a power loss detection system that detects the fiber facility failure and sends a voltage to the associated optical switches to select the alternative signal if a single cable cut affects the primary working signals.

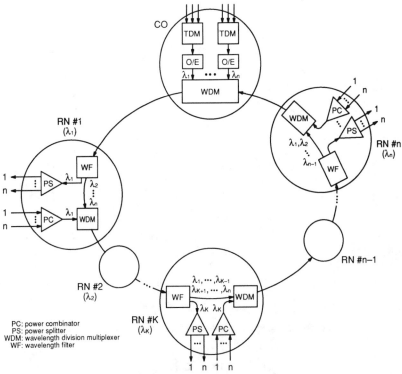

(a) Optical Channel Add-Drop in a Passive Feeder Ring

Figure 8.19. A multi-wavelength passive SHR architecture for feeder networks.

363

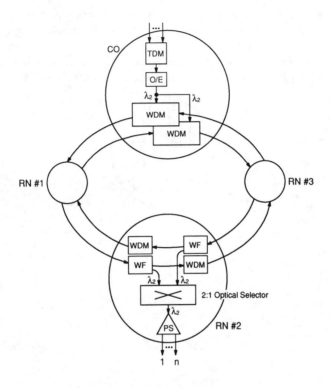

(b) Self-Healing Arrangement for a Passive Feeder Ring

Figure 8.19. (Continued)

8.4 SUMMARY AND REMARKS

Many network providers have actively deployed fiber systems in local loop networks to reduce the operating costs of today's cooper-based networks and to provide a fiber-optic infrastructure that will support new high-bandwidth telecommunications services, such as B-ISDN services [21,22]. As with interoffice networks, fiber loop network survivability has become an increasing concern because today's business environment relies heavily on telecommunications services and cannot afford any potential service disruption. The survivability concern for loop networks is usually greater than for interoffice networks from a customer point of view, because customers usually have very limited connectivity to access the transport network. This survivability concern may grow when the loop network that uses an economical star architecture is implemented; with this architecture, a single failure in the fiber facility or the node (serving CO or the remote node) would disrupt customer services. As a result, route diversity and dual access to two serving COs have become part of service assurance requirements for many large business customers.

Loop survivability is likely to be provided in the shared fiber feeder network because the cost of survivability enhancement in the distribution network may not justify its capital investment, except for private lines or large business. The choice of which survivable fiber feeder architecture should be used depends on the signal multiplexing method used in the RN. For the feeder network with active RNs, SONET technology is desirable because of its simpler control scheme, flexibility, and standard format.

Survivable SONET network architectures used for interoffice networks can also be used for feeder networks. Particularly, the loop network can be integrated with the interoffice network as a single operation unit; eventually, this will provide fast provisioning and reduce the operations cost. This chapter discussed a SONET unidirectional ring architecture using path protection switching because the loop demand pattern usually favors the unidirectional ring architecture, and the considered ring architecture has a simpler system complexity than other ring alternatives.

For feeder networks with passive RNs, survivable network architectures are somewhat different from their active counterparts — the passive RN cannot generate the control signal to trigger protection switching (when it is necessary). For the point-to-point configuration, the 1+1 optical protection switching system can be used to protect the network from a fiber cable cut and to minimize requirements for providing loop survivability. The self-healing architecture for passive feeder networks may be more difficult to implement than its active counterpart because it requires centralized control and more complex optical technology. The passive ring architecture discussed in this chapter processes and adds-drops wavelength division multiplexed channels. Other types of passive ring architectures have also been discussed in the public literature. For example, Reference [23] describes a passive ring architecture that processes and adds-drops time division multiplexed signals in the optical domain, and Reference [24] discusses three types of multi-wavelength ring architectures for consolidation of switching resources at a single node on the ring. These passive ring architectures are usually not as flexible as the SONET rings in terms of service arrangement; however, they may provide high-speed transport for future B-ISDN services.

REFERENCES

[1]   TA-NWT-000909, *Generic Requirements & Objectives for Fiber in the Loop (FITL) Systems*, Issue 1, Bellcore, December 1990.

[2]   Boyer, G. R., "A Perspective on Fiber in the Loop Systems," *IEEE Magazine of Lightwave Communications Systems (LCS)*, Vol. 1, No. 3, August 1990, pp. 6-11.

[3]   Wagner, S. S., and Lemberg, H. L., "Technology and System Issues for a WDM-Based Fiber Loop Architecture," *IEEE Journal of Lightwave Technology*, Vol. 7, No. 11, November 1989, pp. 1759-1768.

[4]   TA-NWT-000442, *Generic Criteria for Fiber Optic Couplers*, Issue 2, Bellcore, November 1990.

[5]   TR-TSY-000901, *Generic Requirements for WDM (Wavelength-Division-Multiplexing) Components*, Issue 1, Bellcore, August 1989.

[6]   Faulkner, D. W., et al., "Optical Networks for Local Loop Applications," *IEEE Journal of Lightwave Technology*, Vol. 7, No. 11, November 1989, pp. 1741-1751.

[7]   Wagner, S. S., et al., "Demonstration of an All-Optical Loop/Interoffice Architecture Passive Optical Networks (PONs) to a Broadband Switching Hub," *Technical Digests of Optical Fiber Communication Conference (OFC)*, San Diego, CA, February 1990, Paper ThN2.

[8]   Sosnosky, J., and Wu, T.-H., "SONET Ring Applications for Survivable Fiber Loop Networks," *IEEE Communications Magazine*, Vol. 29, No. 6, June 1991, pp. 51-58.

[9]   TR-TSY-000496, *SONET Add-Drop Multiplex Equipment (SONET ADM) Generic Criteria for a Unidirectional, Dual-Fed, Path Protection Switched, Self-Healing Ring Implementation*, Issue 2, September 1989, Supplement 1, September 1991.

[10] Kremer, W., "Ring Interworking with a Bidirectional Ring," AT&T Network Systems, T1X1.5/91-043, April 15, 1991.

[11] Lu, K. W., Eiger, M. I., and Lemberg, H. L., "System and Cost Analyses of Broadband Fiber Loop Architectures," *IEEE Journal of Selected Areas in Communications*, August 1990, pp. 1058-1067.

[12] Eiger, M. I., Lemberg, H. L., Lu, K. W., and Wagner, S. S., "Cost Analyses of Emerging Broadband Fiber Loop Architectures," *Proceedings of IEEE ICC*, Boston, MA, June 1989, pp. 5.5.1-5.5.6.

[13] Wagner, S. S., and Lu, K. W., "Comparative Video Distribution Techniques for Broadband Fiber-Optic Subscriber Loops," *Proceedings of IEEE ICC*, Atlanta, GA, April 1990, pp. 313.4.1-313.4.6.

[14] Liew, S. C., and Lu, K. W., "New Architectures for Diversity in Fiber Loop Networks," *IEEE Communications Magazine*, December 1989, pp. 31-37.

[15] Wu, T-H., Kolar, D. J., and Cardwell, R. H., "Survivable Network Architectures for Broadband Fiber Optic Networks: Model and Performance Comparisons," *IEEE Journal of Lightwave Technology*, Vol. 6., No. 11, November 1988, pp. 1698-1709.

[16] Wu, T-H., and Habiby, S., "Strategies and Technologies for Planning a Cost-Effective Survivable Fiber Network Architecture Using Optical Switches," *IEEE Journal of Lightwave Technology*, Vol. 8, No. 2, February 1990, pp. 152-159.

[17] Wu, T-H., "A Novel Architecture for Optical Dual Homing Survivable Fiber Networks," *Proceedings of IEEE ICC'90*, April 1990, pp. 309.3.1-309.3.6.

[18] TR-TSY-000303, *Integrated Digital Loop Carrier System Generic Requirements, Objectives, and Interface*, Issue 1, Bellcore, September 1986, plus Revisions and Supplements.

[19] TR-TSY-000496, *SONET Add-Drop Multiplex Equipment (SONET ADM) Generic Criteria*, Issue 2, Bellcore, September 1989.

[20] Lu, K. W., and Liew, S. C., "Analysis and Applications of r-for-N Protection Systems," *Proceedings of IEEE GLOBECOM*, San Diego, CA, December 1990.

[21] Shumate, P.W., Krumpholz, O., and Yamaguchi, K. (Guest Editors), "Special Issue on Subscriber Loop Technology," *IEEE Journal of Lightwave Technology*, Vol. 7, No. 11, November 1989.

[22] Burpee, D. (Guest Editor), "Special Issue on Fiber in the Local Loop," *IEEE Magazine of Lightwave Communications Systems (LCS)*, Vol. 1, No. 3, August 1990.

[23] Fujimoto, N., Rokugawa, H., Yamaguchi, K., Masuda, S., and Tamakoshi, S., "Photonic Highway: Broadband Ring Subscriber Loops Using Optical Signal Processing," *IEEE Journal of Lightwave Technology*, Vol. 7, No. 11, November 1989, pp. 1798-1805.

[24] Wagner, S., and Chapuran, T. E., "Multiwavelength-Ring Networks for Switch Consolidation and Interconnection," submitted to ICC'92.

CHAPTER 9

ATM Transport Network Architectures and Survivability

9.1 ASYNCHRONOUS TRANSFER MODE TECHNOLOGY AND LAYERED MODEL

ATM has been recognized as a future technology that would support a wide variety of broadband services requiring a wide range of bandwidths in a uniform multiplexing and switching network. As used herein, *transfer mode* refers to the switching and multiplexing process. In the past few years, many experiments that use ATM technology to implement B-ISDN have been conducted. These efforts include developing ATM standards, designing and implementing ATM switches and cross-connect systems, and studying aspects of ATM network performance. However, the debate about competitive applications of full-scale ATM technology continues.

Because of the uncertainty of application feasibility and traffic behaviors, most network control schemes, including self-healing control, are still in the early development stage, especially the ATM *Virtual Path* (VP) concept for network restoration. Thus, like Chapter 2 of this book, this chapter provides a background review of ATM transport architectures and the current progress on ATM/SONET network survivability for readers who may be interested in research and development in this area.

9.1.1 ATM vs. STM Technology

SONET transmission combined with synchronous time division switching (i.e., *Synchronous Transfer Mode* [STM[1]]) provides a high-capacity transport system with powerful operations and maintenance capabilities. However, traditional synchronous time division techniques used in SONET networks, which divide bandwidth into a number of fixed capacity channels, require all data to be transported at a standard transmission rate (e.g., OC-3, OC-12, and so on). These characteristics make SONET

---

1. STM is basically an underlying circuit-switched transport structure based on TDM. Examples of SONET equipment that use STM technology include SONET ADMs and SONET DCSs.

efficient only for *Constant Bit Rate* (CBR) (or isochronous) services, which map well into the available fixed-capacity channels, such as the voice service. Thus, the present SONET/STM system alone may not be an adequate choice of technology for B-ISDN[2] because it does not provide sufficient flexibility to manage a variety of bandwidth requirements within the same network.

To support this future B-ISDN service requirement, ATM has been proposed as the target technology to carry B-ISDN services [1,2]. ATM is a packet-oriented transfer mode in which information is organized into a fixed-size entity known as a *cell*. It is asynchronous in the sense that the recurrence of cells containing information from an individual user is not necessarily periodic. ATM technology combines the flexibility of traditional packet-switching technology, which is efficient for supporting services with various bandwidth requirements, with the determinism of TDM, which is suitable for supporting services requiring stringent delay requirements.

Conventional packet-switching techniques allow variable length packets to support various services by flexibly managing bandwidth. However, these techniques may not meet the stringent delay requirements needed for some services (e.g., voice) due to a complex hardware design for various-length packet processing and a complex end-to-end reliable communication protocol. Adoption of TDM techniques in ATM systems fixes the cell size, thus simplifying the hardware-processing system and producing predictable system behavior. A simpler end-to-end communication protocol coupled with tight error specifications (i.e., no retransmission at the link level) also makes ATM practical for supporting services that require stringent delay requirements.

The multiplexed information flow in ATM systems is organized into cells of a fixed size that contain a user information field and a header. The header contains an explicit label that identifies the channel. If the cells are assigned in a regular and periodic manner, a fixed bandwidth channel is derived because the cells have a fixed length. Thus, ATM can support both CBR services and *Variable Bit Rate* (VBR) (i.e., bursty) services. ATM is termed a connection-oriented technique because each end and the switching point of the ATM connection must recognize the cell labels. However, ATM also offers a flexible transfer capability common to connectionless services.

Table 9-1 shows a relative comparison between ATM and STM systems. ATM is a label-oriented switching and multiplexing principle that uses queueing or admission blocking to resolve short-term contention. On the other hand, STM is a position-

---

2. B-ISDN will allow for the efficient provisioning of telecommunications services to business and residences by providing a common, high-speed interface that can integrate voice, data, graphics, and video information.

oriented (i.e., based on time slot positions) switching and multiplexing principle that uses admission blocking to resolve contention. STM uses dedicated time slots to identify connections, which makes its path structure dependent on the physical multiplexing hierarchy. On the other hand, ATM uses labels to identify channels and is thus independent of the physical multiplexing hierarchy. Therefore, ATM path capacity is not hierarchical when compared to the hierarchical path capacity structure of STM (e.g., VT1.5, VT6, STS-1, STS-3c, STS-12c, and so on). A hierarchical path capacity structure also implies that a hierarchical multiplexing stage is needed to form a high-speed channel (e.g., STS-12) from low-speed channels (e.g., VT1.5s). The path allocation in STM networks is deterministic because each path has a fixed capacity (e.g., the STS-1 path has the capacity of 51 Mbps). In contrast, the path capacity of ATM networks can accommodate any capacity up to the transmission link capacity. Thus, it can be allocated either deterministically or statistically depending upon tradeoffs between network utilization and system complexity.

Table 9-1. Comparisons between STM-Based and ATM-Based Networks

Attribute	STM	ATM
Path identification	positioned	labeled
Frame structure	time slots	cell
Path capacity	hierarchical	non-hierarchical
Multiplexing stages for a STS-Nc+	multiple	single/two*
Path capacity allocation	deterministic (fixed)	deterministic or statistical
Contention resolution	blocking	queueing or blocking

* Single multiplexing stage for ATM VP-based networks and two
  multiplexing stages (VC and VP) for ATM VC-based networks
+ N = 3, 12, 48, and so forth

9.1.2 ATM Cell Structure

Information carried by ATM networks is organized as a stream of fixed-length cells. Figure 9.1 depicts the ATM cell format defined in CCITT Recommendation I.361 [1].

371

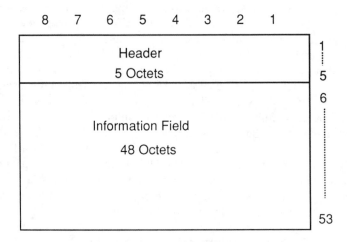

Figure 9.1. ATM cell structure.

The ATM cell has two parts: the header and the payload. The header has five bytes and the payload has 48 bytes. Within the header, there are fields to perform routing, flow control, and other functions. The transmission sequence for the ATM cell is to (1) process bits within an octet in decreasing order, starting with bit 8, and then to (2) process octets in increasing order, staring with octet 1. For all fields, the first bit sent or processed is called the *Most Significant Bit* (MSB).

The header's primary role is to identify cells belonging to the same *Virtual Channel*(VC) on an asynchronous time division multiplexed stream. As depicted in Figure 9.2, two different header coding structures are defined: the *User-Network Interface* (UNI) [see Figure 9.2(a)] and the *Network-Node Interface* (NNI) [see Figure 9.2(b)].

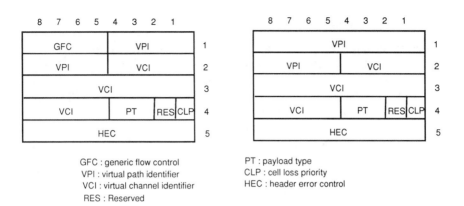

GFC : generic flow control
VPI : virtual path identifier
VCI : virtual channel identifier
RES : Reserved

PT : payload type
CLP : cell loss priority
HEC : header error control

(a) Header Structure at UNI (b) Header Structure at NNI

Figure 9.2. ATM cell header structures.

In the UNI header format, four bits are assigned to the *Generic Flow Control* (GFC) field, which is primarily responsible for shared-media local-access flow control. Twenty-four bits are assigned to routing fields, which are identified by a *Virtual Channel Identifier* (VCI) (16 bits) and a *Virtual Path Identifier* (VPI) (eight bits). The VCI and VPI are used for routing that is established by negotiation between the user and the network. Two bits are also assigned to the *Payload Type* (PT) field to identify whether information is user information or network information,[3] and one bit is assigned to the *Cell Loss Priority* (CLP) field to determine if the cell should be discarded based on network conditions. Finally, eight bits are assigned to the *Header Error Control* (HEC) field to monitor header correctness and perform single bit error correction. One bit is reserved for future use.[4]

The NNI header structure is almost the same as the UNI's except that it has (1) no GFC field and (2) 28 bits for routing (12 bits for VPI and 16 bits for VCI).

---

3. An ATM cell carrying network control information for a particular VC connection is called an *Operations and Maintenance* (OAM) cell.
4. There is a proposal to merge the PT field with the RES field. This proposal has been adopted in the U.S.

9.1.3 Switching System Evolution

As discussed in the previous section, ATM technology evolved from conventional packet-switching technology, which was invented in the early 1960s. This packet-switching technology has been used to implement public packet-switched data *Wide Area Networks* (WANs) that are based on the X.25 protocol [3],[5] such as TELENET[TM], and many LANs, such as the token ring defined in IEEE 802.5. It offers flexibility to manage different bandwidth requirements and provides efficient resource utilization. However, today's X.25 packet-switched networks suffer from high complexity due to a fundamental design assumption that transmission links have relatively low transmission quality (e.g., high BER). Under this design assumption, packet retransmission is needed to ensure end-to-end quality. This results in higher delay, which eliminates support of many delay-sensitive services.

On the other hand, the ATM network operates with high-quality fiber links (e.g., SONET systems have a BER of 10^{-10} [4]). Thus, many functions designed for low-quality transmission links in X.25 packet-switched networks become unnecessary. These functions include link-to-link error control,[6] link-to-link congestion control, and link-to-link packet retransmission. Additionally, the ATM network's fixed cell size makes the ATM cell switch simpler than the X.25 packet switch. Thus, ATM networks can operate at a much higher data rate than conventional X.25 networks. They can also support delay-sensitive services, as well as data services, due to a simpler switching nodal complexity and a simpler end-to-end protocol. Flow control on an end-to-end basis for ATM networks has not yet been defined by CCITT and ANSI T1S1.5 standards groups. For ATM networks, continuous bit rate services (e.g., voice) are offered via a method called *circuit emulation*. Circuit emulation means that the continuous bit rate information is transmitted in frames, and all induced network jitter is removed at the receiver node to restore the continuous bit rate information.

Packet-switching technology has also been applied successfully to LANs and MANs. Recently, some high-speed LANs that use fiber as a transmission medium have been proposed and implemented. These high-speed fiber LANs include the FDDI [5] and the *Distributed Queue Dual Bus* (DQDB) [6]. The FDDI was initially proposed as a high-speed LAN running at 100 Mbps, but with a span up to 100 km. The initial

---

5. X.25 is an *International Standards Organization* (ISO) and CCITT standardized network layer interface protocol that provides computers with a direct access to a packet-switching network. TELENET is a trademark of Telenet Communications Corporation.

6. In ATM networks, the ATM layer performs some error monitoring on a per-link basis or on a portion of a connection (i.e., multiple links).

requirements for FDDI were data services, and the initial protocol was based on a 4-Mbps token ring protocol defined in IEEE 802.5.

With large-scale deployment of LANs, the need for LAN interconnection is increasing. A MAN is a regional network that can provide high-speed, switched, end-to-end connectivity across distances typically ranging between 5 and 50 km (i.e., a metropolitan area). This allows the MAN to become a bridge for interconnecting LANs and allows the simultaneous transport of different types of traffic, such as data, voice, and video. Among MAN proposals, DQDB was chosen as the IEEE 802.6 standard to support both the non-isochronous and isochronous services. SMDS [7] is one high-speed data switching service that can be provided using this standard.

MANs based on IEEE 802.6 are viewed as the earliest version of B-ISDN because the DQDB's frame structure is consistent with the one currently defined in CCITT Recommendations for B-ISDN. Thus, ATM technology may find its first applications in MANs and in interconnecting MANs to support high-speed, switched data services.

Like the X.25 network, frame relay is a standard interface protocol (CCITT I.441) that specifies an interface between a LAN or data terminal and a public or private WAN. The term "relay" implies that the layer 2 (the link layer) data frame is not terminated or processed at the endpoints of each link in the network but is relayed to the destination, as is the case in a LAN. Frame relay is similar to the X.25 interface protocol but has the advantage of reducing transmission delay by no error correction (packet retransmission) and no congestion control at intermediate nodes. So far, equipment vendors have offered frame-relay equipment that operates at speeds ranging from 64 kbps to 2 Mbps. However, frame relay has the potential to operate at speeds up to 45 Mbps. The primary application of frame relay is interfacing high-speed LANs to WANs, where very large LAN packets must be quickly injected into WANs. More discussions on frame relay can be found in Reference [8].

Table 9-2 summarizes relative comparisons of X.25 data networks, frame relay, FDDI, DQDB, and ATM networks. Reference [9] reviews several network switching technologies, including X.25- and DQDB-supporting SMDS.

9.2 ATM LAYERED MODEL

Figure 9.3 shows the B-ISDN protocol reference model for ATM systems defined in CCITT I.321 [1]. This model reflects the principles of layered communication defined in CCITT Recommendation X.200 [the reference model of *Open Systems Interconnection* (OSI) for CCITT applications]. It consists of three planes: user, control, and management.

Table 9-2. Comparing ATM Networks with X.25, Frame Relay, FDDI, and DQDB

Attribute	X.25	Frame relay	FDDI	DQDB	ATM
Area	WAN	WAN	LAN/MAN	MAN	WAN
Services supported	data	data	data	non-iso. & iso-chronous	non-iso. & iso-chronous
Standards	ISO 8208	CCITT I.122	ASC X3T9.5#	IEEE 802.6	CCITT/ T1S1.5
Data rate	<Mbps	<100 Mbps	100 Mbps	150 Mbps	>150 Mbps++
Packet or cell size	variable	variable	variable	fixed (segment) (53 octets)	fixed (53 octets)
Topology	various	various	dual ring	dual bus	various
Connection type	connection-oriented	connection-oriented	connectionless-oriented	connectionless-oriented	connection-oriented
Circuit emulation* for isochronous traffic	N/A	N/A	N/A	no	yes
Link congestion control	yes	no	yes+	yes+	not yet defined
End-to-end flow control	yes	yes	no	no	not yet defined
Link-error control	yes**	limited##	no	no	limited***
Link-packet retransmission (Error recovery)	yes	no	no	no	no
Switching-node complexity	high	median	median	median	lower

+   Through *Media Access Control* (MAC) protocol

++   Note that there are some proposals to map ATM cells onto DS1 and DS3 signals.

\#   *Accredited Standards Committee* (ASC) X3T9, which is chartered to develop computer *input/output* (I/O) interface standards

*   Circuit emulation is a service offering over the ATM network, which is a real-time connection-oriented service with a constant bit rate (see Section 9.2.3).

**   X.25 networks use *High-Level Data Link Control* (HDLC) for each link. The core functions of HDLC include frame delimiting, multiplexing, error detection, and error recovery (retransmission).

***   There would be a capability to monitor link errors.

\#\#   Implement only the core function of the HDLC without link-error recovery (retransmission). The packet retransmission is performed on an end-to-end basis.

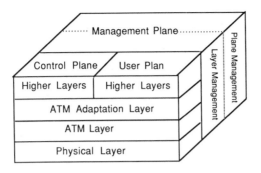

Figure 9.3. B-ISDN reference model.

The user plane primarily transports user information; the control plane processes signaling information; and the management plane performs network operations and maintenance functions. The management plane provides management and coordination functions between the user plane and the control plane.

The reference model depicted in Figure 9.3 can be further divided into several layers, as depicted in Figure 9.4, to understand signal transport. Three layers have been defined: physical, ATM, and *ATM Adaptation Layer* (AAL), where the physical and ATM layers form the ATM transport hierarchy.

9.2.1 Physical Layer

The physical layer transports data (bits or cells) and consists of two sublayers: *Physical Medium* (PM) and *Transmission Convergence* (TC). The PM sublayer, which depends greatly on the PM used (e.g., fiber or coaxial cable), performs only physical medium-dependent functions, such as bit timing. The TC sublayer performs all functions required to transform an ATM cell stream into a data stream (e.g., bits) that can be transmitted and received over a physical medium. The functions included in the TC sublayer are cell-rate decoupling, HEC header sequence generation/verification, cell delineation, transmission frame adaptation, and transmission frame generation and recovery.

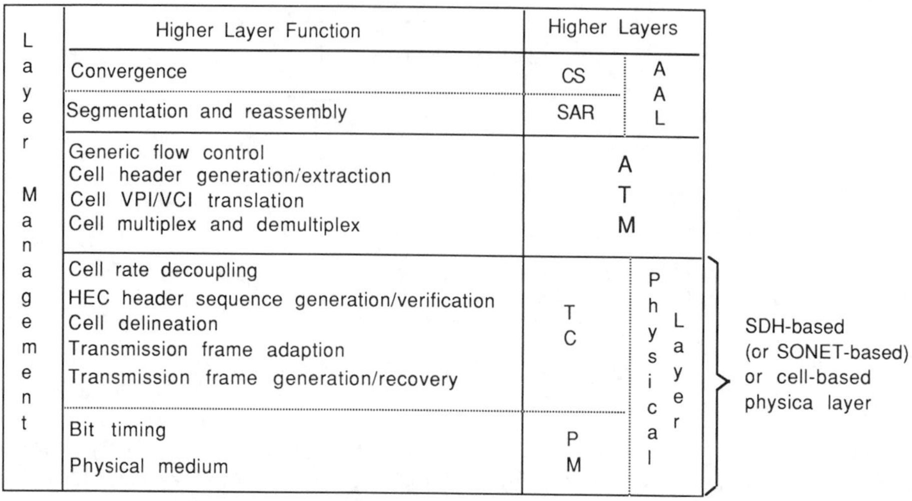

	Higher Layer Function	Higher Layers	
L	Convergence	CS	A A L
a			
y	Segmentation and reassembly	SAR	
e			
r	Generic flow control		
	Cell header generation/extraction		A
M	Cell VPI/VCI translation		T
a	Cell multiplex and demultiplex		M
n			
a	Cell rate decoupling		
g	HEC header sequence generation/verification		
e	Cell delineation	T C	Physical layer
m	Transmission frame adaption		
e	Transmission frame generation/recovery		
n			
t	Bit timing	P M	
	Physical medium		

SDH-based (or SONET-based) or cell-based physica layer

Figure 9.4. Functions of B-ISDN in relation to the CCITT protocol reference model.

Each ATM cell contains information in the header that indicates whether the cell is occupied with user information. The bit rate available for user information cells, signaling cells, and OAM cells is 149.760 Mbps in the STS-3c payload of 155.520 Mbps.[7] Idle cells[8] are also processed (inserted and/or extracted) in the physical layer to adapt the cell-flow rate at the boundary between the ATM and physical layers to the available payload capacity of the transmission system used.

CCITT recommends two kinds of transmission frame adaptation: cell-based interface and SONET-based interface. In the cell-based physical layer, cells are transported continuously without any framing (e.g., 125-μs framing used for SONET). In the SONET-based physical layer, the payload of the SONET 125-μs frame carries ATM cells.

The physical layer based on SONET technology provides functions such as scrambling, error monitoring, and protection switching.[9] All broadband transmission

---

7. CCITT has not yet specified the case for 622.080 Mbps (i.e., payload of STS-12c). However, the ATM cell mapping at the STS-12c termination has been specified in ANSI T1S1/91-634 "Draft for Broadband ISDN User-Network Interfaces: Rates and Formats Specifications."

8. In SR-NWT-001763 [10], idle cells are not used because the ATM layer has unassigned cells.

9. Protection switching is not used across the UNI.

streams are SONET signals operating at the STS-3n(c) rate, where n=1, 4, 16, and so forth. The payload of the SONET signal carries ATM cells, as shown in Figure 9.5(a). Figure 9.5(b) shows an example of cascaded SONET multiplexing of STS-3c and STS-12c frames carrying ATM cells. In this example, after ATM cells are mapped to both STS-3c and STS-12c frames, the STS-3c and STS-12c frames are multiplexed together to form a STS-48 frame according to the SONET multiplexing process.

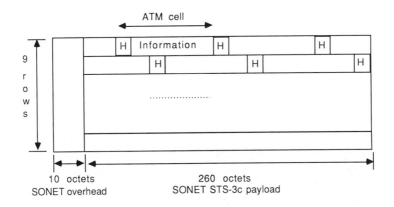

(a) ATM Cell Mapping on SONET STS-3c

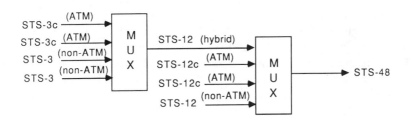

(b) Cascaded SONET Multiplexing

Figure 9.5.   ATM cell mapping on SONET STS-3c and multiplexing.

In the approach shown in Figure 9.5(a), all types of services, including CBR and VBR services, are transformed into ATM cells and then placed in the SONET frame payload for transport. Alteratively, a hybrid ATM/SONET frame structure may be used for network transition from SONET networks to ATM networks. This hybrid framing approach reserves a portion of the STS-N payload for isochronous services (e.g., DS1 services), using SONET STS-1 or VT frames, and another portion of capacity for non-isochronous services (e.g., data or switched DS1 services) using ATM cells. Once the STS-N signal is demultiplexed, STS frames carrying isochronous services are processed by the STM equipment (e.g., SONET DCS or ADM), and STS frames carrying ATM cells are processed by ATM switches or ATM DCSs.

9.2.2 ATM Layer

The ATM layer provides transparent transfer of fixed-size data units between a source and the corresponding destination(s) with an agreed *Grade of Service* (GOS). This transfer is performed in a "best effort" manner, i.e., lost or corrupted data units are not retransmitted. The ATM layer performs multiplexing and switching functions for cells and is built independently of the physical layer. Each ATM cell contains a label in its header to explicitly identify the VC to which the cell belongs. A VC is a generic term used to describe a unidirectional communication capability for the transport of ATM cells. This label consists of two parts: a VCI and a VPI. A VC provides transport of ATM cells between two or more endpoints for user-user, user-network, or network-network information transfer. The points at which the ATM cell information payload is passed to a higher level for processing are the endpoints of a VC. VC routing functions are performed at a VC switch. This routing involves translating the VCI values of the incoming VC links into the VCI values of the outgoing VC links. The ATM network preserves cell sequence integrity within a VC.

A VP is a group of VCs, and *Virtual Path Terminators* (VPTs) delimit the VP boundaries. At VPTs, both VPI and VCI values are processed. Between VPTs associated with the same VP at intermediate nodes, only the VPI values are processed (and translated) at ATM *Network Elements* (NEs). The VCI values are processed only at VPTs and are not translated at intermediate ATM NEs. Cell sequence integrity needs to be maintained within a VC, but the sequence of cells for different VCs within a VP is not yet specified in CCITT Recommendations [1].

Each ATM cell header contains an ATM label that uniquely identifies the VP and the VC to which the cell belongs. The ATM cell label is the concatenation of the VPI and VCI. VCs on different VPs may have the same VCI value though they have different VPI values. Cells associated with a particular VP are identified by a VPI value in the ATM cell label and by the physical link over which the cell is carried.

Between two VPTs associated with the same VP, the VPI value may be translated at ATM NEs (i.e., VP cross-connect systems).

Figure 9.6 shows the relationship among the VC, VP, physical link, and physical layer. The physical layer provides a bit stream to carry ATM cells, and its physical layer termination points are signified by ATM cell header processing. A physical layer may include multiple physical links, and each physical link may carry multiple VPs, each having multiple VCs.

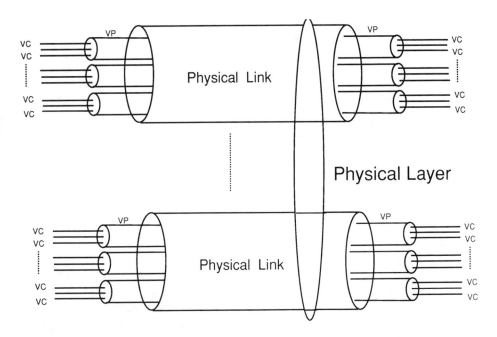

Figure 9.6. Relationship among the VC, VP, physical link, and physical layer.

9.2.3 ATM Adaptation Layer (AAL)

The AAL performs the functions necessary to match the services provided by the ATM layer to services required by its service-user and is divided into two layers: *Segmentation and Reassembly* (SAR) and *Convergence Sublayer* (CS). The primary function of the SAR is to segment the service information into ATM cells and reassemble ATM cells into service information. The CS sublayer represents the set of service-related functions.

The AAL enhances the services provided by the ATM layer to support functions required by the next higher layer. It performs functions required by the user and control, and supports the mapping between the ATM layer and the next higher layer. Thus, the AAL is primarily responsible for adapting service information to the ATM cell stream. Figure 9.7 depicts service classes for AAL, as defined in Reference [1], and service examples for these AAL classes. Three parameters forming the AAL service classes are delay, bit rates, and connection modes.

Service Classes for AAL

Parameters	Class A	Class B	Class C	Class D
Timing relation*	Required		Not required	
Bit rate	Constant	Variable		
Connection mode	Connection oriented			Connectionless
Examples	DS1/DS3**	Variable bit rate video	Data file transfer	SMDS

* Timing relation is between the source and the receiver.
** Provided through circuit emulation on the ATM layer.

Figure 9.7.   Service classes for AAL.

Class A deals with connection-oriented services with constant bit rates; their timing at the source and receiver are related. These services involve an uninterrupted flow of digital information (e.g., DS1/DS3) at a constant bit rate. They are time-sensitive, need a constant and predetermined amount of capacity, and will be primarily supported through circuit emulation on the ATM layer. Circuit emulation is a method to emulate the circuit-switching capability on ATM networks. A timing-recovery scheme is needed to recover timing at the receiver side in order not to lose received information. Some timing-recovery proposals that are under CCITT consideration can be found in References [11-12]. Because the network provider guarantees certain GOSs for these services, this class of service imposes stringent constraints on ATM networks. In addition to the guaranteed constant capacity, the key performance constraints for Class A services are small mean cell delay, small cell delay variation, and, in some cases, small cell loss.

Class B represents connection-oriented services with variable bit rates, and related source and receiver timing. These services are real-time. An example of Class B services is the VBR video. The required capacity of a VC with Class B services varies between a maximum and a minimum. Like Class A services, this class typically has stringent cell delay, cell delay variation, and cell loss requirements.

Class C deals with bursty connection-oriented services with variable bit rates that do not require a timing relation between the source and the receiver. The main performance metrics of this class of services are the mean cell delay and/or a percentile of cell delay (e.g., 95 percentile), throughput, and service data unit loss rate. An example of Class C services is connection-oriented data services.

Class D is essentially the same as Class C except the services are connectionless. When services are transported through the network, traffic belonging to Class A and Class B services is assigned a higher delay priority than that belonging to Class C and Class D services. More details on protocols and formats for these four types of AAL can be found in References [1,10].

9.3 ATM SWITCHED NETWORK ARCHITECTURES

9.3.1 ATM Network Transport Hierarchy

An ATM transport network is structured as two layers, ATM and physical, and the transport functions of the ATM layer are further subdivided into two levels: VC and VP. The transport functions of the physical layer are subdivided into three levels: transmission path, digital section, and regenerator section. The transport functions of the ATM layer are independent of the physical-layer implementation. Figure 9.8(a) and (b) depict a hierarchy of the ATM transport network and NEs associated with this functional hierarchy, and Figure 9.8(c) lists network equipment and associated layer terminations for the ATM transport network.

Higher Layers		
ATM Layer	Virtual Channel Level	
	Virtual Path Level	
Physical Layer (SONET)	SONET Path Level	
	SONET Line Level	
	SONET (Regenerator) Section Level	

(a) ATM Transport Functional Hierarchy

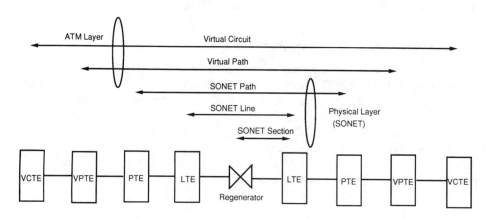

(b) Layered Concept and Associated NEs

Figure 9.8. ATM transport network architecture.

Network node	Highest-terminated layer
SONET Regenerator	SONET Section
SONET B-DCS, ADM	SONET Line
SONET W-DCS, OLTM	SONET Path
ATM VP DCS	Virtual Path
ATM switch	Virtual Circuit

(c) Layering and Network Termination

Figure 9.8. (Continued)

9.3.2 ATM Cell Switching Architectures

Figure 9.9 shows an ATM cell switching system model.[10] An ATM cell switching system includes a physical switch for cell routing, a queue management system for contention control, and a VCI/VPI table translation system for controlling the physical switch configuration. Incoming ATM cells are physically switched from input port I_i to output port O_j, and at the same time, their VCI/VPI values are translated from incoming values to outgoing values. On each incoming and outgoing ATM cell stream associated with a pair of input and output ports, the VCI values are unique, but identical VCIs may be found on different ATM cell streams terminating at different input ports (e.g., x on links I_1 and I_n).

In the Figure 9.9 translation tables, all cells having a VCI equal to x on input port I_1 are switched to output port O_1, and their VCI is translated to value k. All cells with a VCI x on link I_n are also switched to output port O_1, but their VCI is translated to value n.

---

10. Figure 9.9 shows the output queue model; another alternative may be the input queue model.

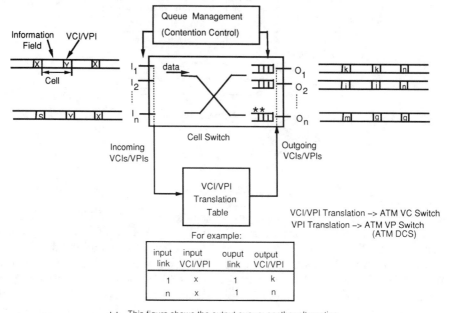

Figure 9.9. ATM cell switching architecture.

Figure 9.9 also shows two cells of different input ports (e.g., from I_1 and I_n) arriving simultaneously at the ATM switch and destined to the same output port (e.g., O_1). Thus, the cells cannot be placed on the output port at the same time, and a buffer has to be provided somewhere in the switch to store the cell(s) that cannot be served. This is typical for an ATM switch because it statistically multiplexes cells. Therefore, a queue management system must be provided to ensure that ATM cells destined to the same output port are stored rather than discarded, according to some contention control algorithms.

Reference [13] discusses tradeoffs between throughput and utilization in the input queue model versus the output queue model (see Figure 9.10) in the ATM cell switching system. In general, mean queue lengths are always greater for queueing on input than for queueing on output, and the output queues saturate only as the utilization approaches unity. Input queues, on the other hand, saturate at a utilization that depends on the switch size, but is approximately 0.586 when the switch size is large, assuming uncorrelated traffic is among input ports and is uniformly distributed to all output ports [13]. The maximum throughput of 58 percent for the input buffer model is due to a so-called *head-of-line* (HOL) effect caused by input buffer switches implementing the *First-In-First-Out* (FIFO) discipline. Reference [14] discusses three possible methods to increase throughput of input buffer switches (i.e., reduce the HOL effect): doubling the operating line speed, looking around not only the head but also the succeeding cells in the buffer, and scheduling the output timing of each cell from the input buffer. The last method (using scheduling control) has been used in an ATM DCS system [15,16], as will be discussed in Section 9.7.1. A class of ATM switch architectures and their comparisons can be found in References [14,17,18].

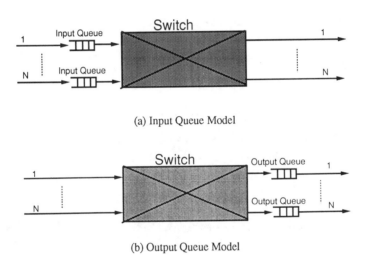

(a) Input Queue Model

(b) Output Queue Model

Figure 9.10. Input and output queue models for ATM cell switches.

If map translation is performed at the VC level, the ATM cell switch becomes the ATM VC switch, which is usually associated with call processing and path bandwidth management capabilities. The ATM VC switch is commonly referred to as an *ATM switch*. If the map translation is performed only at the VP level, the ATM cell switch becomes the ATM VP switch, which is sometimes called the *ATM Cross-connect System* (denoted by ATM/DCS). The ATM/DCS does not have call processing, VP capacity allocation, and VP route-selection functions; it simply transports signals transparently. The way in which these ATM switching functions are implemented and

387

where they are located in the switch determine their switching architecture and applications. References [17,18] provide detailed switching architectures and traffic theory for ATM cell switches.

The relationship between offered traffic characteristics (e.g., voice, video, and so forth) and resulting service quality (e.g., cell transmission delay and cell loss rate) is a complex issue. It may be desirable for service quality parameters to be controlled so the network can meet its service requirements while operating efficiently. An ATM network controller based on neural network technology was proposed [19] to study the relationship between offered traffic characteristics and service quality parameters. Refer to Reference [19] for more details on this neural network-based ATM network controller.

9.3.3 ATM Switched Network Architecture Alternatives

As shown in Figure 9.11, three possible switched network architectures can deliver ATM cells on an end-to-end basis.

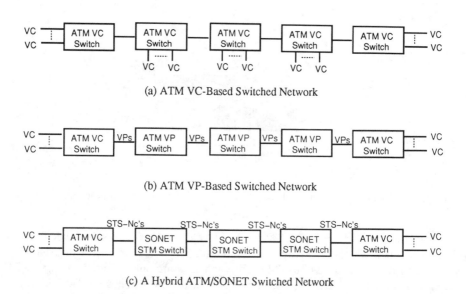

(a) ATM VC-Based Switched Network

(b) ATM VP-Based Switched Network

(c) A Hybrid ATM/SONET Switched Network

Figure 9.11. ATM switched network architecture alternatives.

388

Figure 9.11(a) illustrates the first alternative, the ATM VC-based switched network, which utilizes ATM VC switches as key NEs and processes the ATM connection on a VC-by-VC basis. This architecture is sometimes referred to as the *full ATM switched network*.

Figure 9.11(b) illustrates the second alternative, the ATM VP-based switched network, which uses two ATM VC switches at two ends of an ATM VC/VP connection to perform VC/VP call setup and other VC/VP bandwidth management functions. This architecture uses ATM VP switches at intermediate (transit) nodes to transport ATM signals on the VP basis. The ATM VP switch at the transit node does not perform VP or VC bandwidth management and only cross-connects VPs in a transparent manner. Compared to the VC-based switched network, the ATM VP-based switched network has less flexibility in efficient network bandwidth use but has a simpler, faster signal transport system and requires simpler NEs at transit nodes.

The third switched network alternative is similar to the VP-based network architecture except that it uses the SONET/STM transport system, rather than the VP-based transport system, to transparently transport ATM signals after they are processed at the end nodes of the ATM connection. Figure 9.11(c) illustrates this third architecture, which is also called the *hybrid ATM/SONET network architecture*. Among these three architectures, the hybrid ATM/SONET switched network architecture is the least flexible in terms of bandwidth utilization but is the simplest in terms of signal transport. It evolves easily from the SONET network infrastructure.

It may be expected that the VP-based and hybrid ATM/SONET switched networks may be easier, faster, and less expensive than the VC-based switched network in providing the real-time network self-healing capability. Thus, due to the focus of this book, we will not discuss the VC-based switched network further except in the next section (Section 9.3.4). Because the transport system at transit nodes in the hybrid ATM/SONET switched network is the same as those described in Chapters 3, 4, and 5, the rest of this chapter will focus only on ATM VP-based switched networks.

9.3.4 Call Setup and Bandwidth Management for ATM VC-Based Networks

Both switching (real-time) and cross-connect (non-real-time) functions may be provided by the ATM switch using VCs/VPs. For switching functions, the call control service module listens on a signaling channel VCI and assigns a VCI (or a VPI) value to a user connection for the duration of a call. Because there is real-time contention among call requests, some call requests may be delayed or rejected. This results in blocking or queueing of calls. On the other hand, for cross-connect functions, VCI or VPI values are assigned to the user through service orders, or they are updated via network management functions.

In ATM VC-based networks, a VC connection can be established in the following way [20]. When a new call (VC) arrives at a local ATM switch, the total capacity of the new call plus existing circuits carried by the same VP is calculated. If the resulting capacity requirement does not exceed the VP capacity, the call is accepted; otherwise, the call is rejected. This call setup scheme requires ways to determine the required capacity of the new call. Reference [21] discusses several parameters that can be used to allow the network to determine if the incoming call will exhaust the path capacity or not and, thus, determine whether to accept the call. If the network knows these parameters for new calls in advance, it can use them to decide whether to accept calls. If the network does not know these parameters in advance, the user must specify or declare them, along with other necessary information such as GOS, so the network can determine whether to accept or reject the call. See Reference [21] for more details on these performance parameters. In cases where statistical multiplexing characteristics are evaluated based on user-declared parameters, the network must ensure that cells from the signal source are transmitted according to the declared parameters. If the declared parameters are violated, the network must restrict cell transmission to ensure cell transfer quality for other sources. This cell-flow restriction process is called the policing function or *Usage Parametric Control* (UPC). In other words, policing ensures that cell arrival at the multiplexing node is within the specified value.

ATM bandwidth enforcement and management may be provided at the VC level. When a VC is set up, a specific amount of bandwidth is requested and supported by the network after negotiation. In this case, the VP provides only a logical connection between two users. As the users transmit VCs within the VP, bandwidth needs to be negotiated for each individual VC. Bandwidth enforcement is exercised over each individual VC, and real-time bandwidth switching is performed at an ATM VC switch.

9.4 ATM VP-BASED NETWORK ARCHITECTURE

References [22-24] propose a VP-based network architecture to provide efficient and cost-effective ATM signal transport. This VP-based network architecture essentially incorporates the concept of system simplicity from SONET networks and bandwidth management flexibility from ATM technology. The philosophy behind this ATM VP-based network architecture is similar to that of a SONET/STM network architecture for switched services [25]. Both the SONET/STM network architecture for switched services and the ATM VP-based network architecture are made up of two switches at the two ends of a path and a transport network conveying signals through intermediate nodes of a path. In both cases, all major call processing and management functions are performed at two end nodes of a path, and the transport network is designed to deliver

390

signals as quickly and as cost-effectively as possible with a minimum amount of processing at intermediate (or transit) nodes. Figure 9.12 depicts a relative comparison between these two architectures.

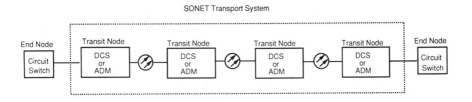

(a) SONET/STM Network Architecture for Switched Services

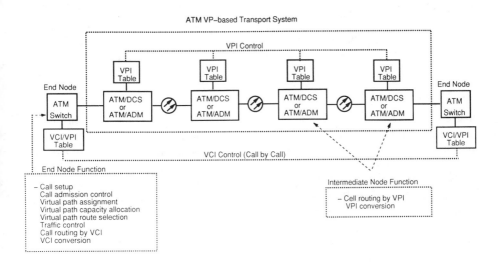

(b) ATM VP-Based Network Architecture and Nodal Functions

Figure 9.12. SONET/STM network architecture versus ATM VP-based network architecture.

391

In the STM/SONET network architecture for switched services [25] [see Figure 9.12(a)], digital switches (not SONET/STM equipment) at two end nodes of the path connection perform call processing and management functions. Intermediate transit nodes along the path perform only path cross-connection and/or add-drop functions using SONET DCSs and/or ADMs. In the ATM VP-based network architecture, VCs are processed and assigned to a VP connecting the VC switches at either end (analogous to assigning calls to STM paths) and are transported transparently via an ATM VP-based transport network. An ATM VP-based transport network is made up of ATM VP switches (i.e., ATM DCSs) and/or ATM ADMs, depending on the transport network configuration. The end node of a VP connection executes more processing than the transit node within the ATM VP-based transport network.

Figure 9.12(b) also depicts nodal functions of a VP-based network architecture. As depicted in Figure 9.12(b), the ATM switch, which is composed of a high-throughput packet switch and a high-performance processor, handles VP assignment setup processing, VP capacity allocation, and VP route selection. The ATM/DCS also has a packet or space switch (see Section 9.7.1). However, it is much simpler (and thus less expensive) equipment than the ATM switch because its control processor does not perform VP assignment setup, VP capacity allocation, and VP route selection. This low-cost processor is also used in the ATM/ADM, which is primarily used in a ring topology to add-drop or pass through ATM cells within the ring node. In practice, both the ATM/DCS and the ATM/ADM may be used if the transport network is large.

9.4.1 VP Assignment Setup

The use of VPs reduces the VC control mechanism's load because the functions needed to set up a path through the network are executed only once for all VCs subsequently using that path. By reserving capacity on VPs, VCs may be established quickly and simply. The VP assignment setup protocol is not yet specified in CCITT Recommendations [1]. The following description of this VP assignment setup protocol is taken from Reference [23]. Figure 9.13 depicts this VP assignment establishment procedure.

First, a setup cell is sent from the customer terminal to its local network node. Control intelligence in the node analyzes the source and destination addresses and determines if a core network VP exists between the source and destination nodes. If a VP exists, its available capacity is checked. If the capacity is sufficient to support the VCs within that VP and the new VC is at an appropriate GOS, the VC connection is established. Note that, in this case, no call processing is needed at transit network nodes. VCs supported on core network VPs with reserved capacity can be established with less effort than VCs carried on networks using only VC switching.

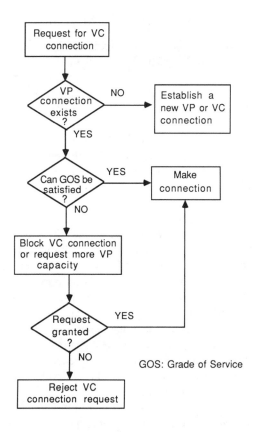

Figure 9.13.   A VP assignment establishment procedure.

Note that the process of setting up a VP is decoupled from the process of setting up an individual VC. This means that the setup time for individual VCs using VPs does not include the time required to set up the VPs. Thus, the network control mechanism can be seen as operating at two levels:

1. For groups of VCs corresponding to VPs, the VP control mechanisms are needed to calculate routes, allocate capacity, and store connection state information.
2. For individual VCs, processors at network nodes merely determine if there is a VP to the required destination node with sufficient available capacity to support the connection (with an appropriate GOS) and then store the required state information (VPI, VCI mapping).

393

9.4.2 Virtual Path Capacity Allocation for ATM VP-Based Networks

A VP carried within a physical channel is identified by the VPI value. It is possible for different VPs associated with different physical channels to have the same VPI value. Thus, a VPI translation function is needed in the ATM switch or cross-connect system to preserve VPI uniqueness at multiplexing points. For a VP that is not terminated at the switch, the VCI values of the VCs contained in the VP are not changed by the switch.

Bandwidth enforcement may be provided at the VP level. As a VP is set up, a specific amount of bandwidth is requested and supported by the network for the VP. Information may be transmitted on any VC within the VP, as long as the VP's bandwidth is not exceeded. In this case, bandwidth enforcement is exercised over the VP and not over the VC.

The capacity of VPs can be allocated deterministically or statistically. Deterministic VP capacity is allocated using the peak cell rate, which can be multiplexed over the VP. For the statistical VP capacity allocation, the VP capacity is specified by a set of parameters, such as peak rates and cell rates averaged over one or more time intervals. If the parameter values are within the specified path values, the specified cell-transmission quality (delay and/or cell-loss probability) is guaranteed. Figure 9.14 shows a statistical multiplexing scheme for cells. The statistical path allocation scheme needs less capacity than the deterministic path allocation scheme for supporting services but requires a more complex control and policing system.[11] More details on a VP capacity design can be found in Reference [21].

Where capacity is reserved on VPs, it may be desirable to dynamically adjust the allocation to track changes in network traffic flow. The VP capacity management functions can be controlled in a centralized or distributed manner. The centralized VP capacity allocation system suggested in Reference [23] uses a central database and a central processor of modest capacity to allocate VP capacity. Like SONET DCS networks (see Chapter 5), the centralized control scheme may make global optimum VP capacity allocation easier than the distributed control scheme but may suffer a fatal failure of the central VP controller. However, such a failure results only in the inability to alter capacity allocations to VPs; new VC connections can still be established using the existing VP capacity allocations.

---

11. The policing function is also needed for the deterministic path allocation scheme, but it may be simpler than that used in the statistical path allocation scheme.

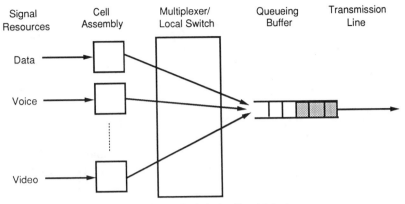

Figure 9.14. Statistical cell multiplexing.

9.4.3 VP Applications

Several applications of the VP concept exist. First of all, VPs provide a logical direct link between two nodes. This application simplifies signal transport (and thus reduces transport costs) between two nodes. For example, Reference [24] reports a case study that suggests the ATM/DCS networks can even support existing DS1 services more cost-effectively than SONET DCS networks. This case study[12] indicates 43 percent savings in node (switching) cost, 18 percent savings in link (transmission) cost, and 36 percent savings in the total network cost for an eight-node model. The logical direct link concept has also been applied to interconnect IEEE 802.6 MANs, as described in Reference [26]. Second, VPs provide protection switching or alternate routing in case of failure. Because a VP's route and capacity can be controlled independently, two or more routes can be preassigned or dynamically assigned for one VP without preallocating capacities; the loopback route can then be selected immediately by changing the VPI or its translation at the originating node. Third, VPs may provide a tool within CPs that segregates customers and services, which allows multiplexed use of a large-capacity subscriber line by many users and services.

---

12. The case study results were presented at IEEE GLOBECOM'91.

9.5 ATM NETWORK SURVIVABILITY

9.5.1 Service Protection Level

As discussed in previous chapters, services carried on STM/SONET networks can be protected at the path or physical-link level, where the path level includes circuit, VT (or VT groups), or STS-1/STS-3c's. Because fiber systems carry a high volume of data and need fast and cost-effective service restoration, the service restoration level for the STM/SONET network is usually at the STS path or physical-link level.

Similarly, ATM network restoration can be performed at the physical (SONET) or ATM layer, according to its transport network hierarchy. For physical-layer protection, service restoration can be performed at the STS-3nc level (where n is a positive integer) or at the physical-transmission link level. For ATM layer protection, restoration can be performed at the VC or VP (or group of VPs [27]) level. Service restoration at the VC level is slower and more expensive than at the VP level due to greater VC-based ATM network complexity. The restoration system complexity for ATM/SONET networks to restore services at the VP level is similar to that used by STM/SONET networks to restore services at the STS-1/STS-3c level. Determining which layer (SONET, VP, or VC) is an appropriate restoration layer depends on the ATM network architecture and other factors, and remains to be studied. Figure 9.15 depicts the SONET service and ATM service protection levels.

ATM Transport Network Hierarchy			Possible Service Protection Level	
			STM/SONET	ATM/SONET
A T M	L A Y E R	Virtual Circuit (VC)	—	VC
		Virtual Path (VP)	—	VP
S O N E T	L A Y E R	Path	STS–1 STS–3c VT1.5	STS–3c
		Transmission Link	Line	Line

Figure 9.15. STM/SONET and ATM/SONET service protection levels.

396

SONET physical-level protection may be used when a dedicated redundant facility is used, and VP-level restoration may be used when the redundant channels or capacities (rather than dedicated redundant facilities) are used. If the SONET physical layer is chosen as the service restoration level for ATM networks, restoration systems designed for STM/SONET networks (discussed in Chapters 3, 4, and 5) may also be applicable (with minor modifications). Service restoration schemes at the VP layer in ATM/SONET networks are still in the early stage of development.

9.5.2 ATM VP Technology Impact on Network Survivability Design

There is a growing consensus that the target STM/SONET transport network architecture is likely to be composed of various survivable SONET network architectures (see Section 5.5). SONET survivable network architectures include point-to-point systems with automatic DP routing, SHR architectures, and self-healing mesh network architectures. The DP routing system and the SHRs use dedicated redundant facilities for service protection.[13] The SONET self-healing mesh network architectures with SONET DCSs reserve a portion of fiber capacity as protection channels for service protection. Most survivable STM/SONET network architectures are expected to be implemented in 1993-1995.

Compared to SONET technology, features of ATM VP technology may simplify the survivable network architecture design. These features include non-hierarchical path multiplexing (see Table 9-1), the hidden path, the OAM cell, and the idle cell. A *hidden path* is a VP that can be preassigned without reserving capacity for it. Unlike the STM (i.e., SONET) DCS, which can support only path cross-connection with a single bit rate, the ATM/DCS can cross-connect VPs with any bandwidth. This feature may make control and management of the path layer easier and faster than in STM-based networks without penalizing bandwidth utilization for normal conditions.

For rapid network control in network restoration, more computational capabilities and high-speed control data channels are required. In the STM-based network, the control/management signal is transferred (via section or path overhead) with a periodic cycle of 125 μs. In the ATM network, OAM signals can be transferred via OAM cells, which are transferred within a cycle of 125 μs, because OAM cells can be placed in the STS-Nc payload. The OAM cells can be inserted into the main ATM cell stream at any time as long as there is an available capacity. OAM cells can be given higher priority

---

13. The BSHR/2 architecture is an exception; it uses dedicated redundant channels for service protection. Refer to Chapter 4 for details.

397

than user information cells; thus, they may pass through the network even if the network may be experiencing congestion.

9.5.3 Survivable ATM/SONET VP-Based Network Architectures

Like their STM/SONET counterparts, survivable ATM/SONET network architectures could include ATM/SONET SHR architectures and ATM/SONET self-healing mesh network architectures. ATM/SONET SHR architectures can be implemented using ATM/ADMs, and ATM self-healing mesh network architectures can be implemented using ATM/DCSs. Like its STM/SONET counterpart, the hubbed network is likely to implement point-to-point with automatic DP using ATM DCSs for service and facility grooming.

Like STM/SONET SHRs and mesh networks, the ATM VP-based SHRs and mesh networks may perform service restoration at the VP or SONET physical-line level and the VP (or group of VPs) level, respectively. Table 9-3 shows a comparison between survivable STM/SONET and ATM/SONET network architectures and associated equipment.

Table 9-3. Comparing Survivable STM/SONET Networks and Survivable ATM/SONET Networks

Survivable network architecture	STM/SONET		ATM/SONET	
	Protection level	Equipment	Protection level	Equipment
Point-to-point/ diverse routing	line	OLTM/APS*	line	ATM TM/ APS**
Self-healing ring	line or STS/VT path	SONET ADM	line or VP	ATM ADM
Self-healing mesh architecture	STS-1/STS-3c path	SONET DCS	VP (or group of VPs)	ATM DCS

* OLTM/APS: Optical Line Terminating Multiplexer with an APS system
** TM/APS: Terminal (statistical) multiplexer with APS

The following sections discuss only ATM VP-based SHR architectures and ATM VP-based self-healing mesh network architectures. Service protection for the ATM hubbed network is similar to that for the SONET/STM hubbed network if the SONET layer is chosen as the protection layer for the ATM network.

9.6 ATM VP-BASED RING ARCHITECTURE

This section first reviews VP applications for the ATM VP-based ring and then discusses how these VP concepts can be used to build an ATM/ADM VP-based ring that supports different types of services. Work in this area is still in the early stage of development. Thus, this discussion will focus only on conceptual architecture designs.

9.6.1 VP Applications for ATM VP-Based Rings

Compared to mesh network architectures, ring architectures have the following benefits: economical signal processing, simpler control schemes, and survivability at the expense of limited growth potential. STM/SONET SHRs have been well accepted as a good network architecture for existing voice and leased line data services from both economic and survivability points of view. However, the probability of the ring capacity being exhausted increases significantly (from inflexible bandwidth management) when the network extends its services to VBR services. When the ring capacity is exhausted, system upgrade could become difficult and expensive.

ATM ring architectures may reduce this ring capacity concern because of their flexible bandwidth management and efficient support of various services, including CBR and VBR services. For example, Reference [28] discusses an ATM ring architecture that may be used to support SMDS-like services. A case study reported in Reference [28] indicates that, for a 10-node network, the total throughput of the ATM ring can be increased to 200 percent of the STM ring capacity without degrading delay conditions.

Major design issues for ATM rings include ATM ring transport and control architectures, dynamic bandwidth allocation (i.e., sizing and bandwidth sharing), and efficiency and economic gains over the STM/SONET ring architectures. The efficiency and economic gains are two important factors that may determine the early deployment potential of ATM technology in the SONET network infrastructure. Also, for ATM rings, ultra-high-speed ATM multiplexing (above tens of gigabits-per-second) or wavelength multiplexing of many wavelengths will be needed to provide full B-ISDN services when each customer's communication rate approaches 140 Mbps.

Current studies in ATM VP-based ring architectures involve loop networks or MANs [29-31]. The same concept is being considered for ATM interoffice networking applications. Some possible VP applications in ATM feeder rings are service separation, subscriber line identification, and nodal destination addressing for VP leased line services. Service separation can be provided via point-to-point VPs, point-to-multipoint VPs, and cycle VPs. A *cycle VP* is a VP that is shared by all connected nodes of a cycle. For example, point-to-point VPs may be used to support public switching and leased line services; point-to-multipoint VPs may be used to support distribution services, such as cable TV services [29,30]; and cycle VPs may be used to support connectionless services, such as SMDS [31].

9.6.2 An ATM VP-Based Ring Architecture

This section discusses an ATM ring architecture based on point-to-point VP add-drop multiplexing. The point-to-point VP add-drop multiplexing scheme adds-drops and passes through ATM cells at the VP level, as depicted in Figure 9.16.

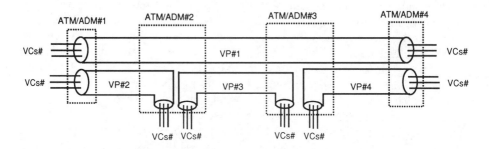

Figure 9.16. Point-to-point VP add-drop multiplexing.

Figure 9.17 depicts an ATM ring architecture using point-to-point VPs for supporting leased line services. In this ring architecture, each ring node pair is preassigned to a VP connection with different VPs assigned to different node pairs. For example, in Figure 9.17, VP#2 carries all VC connections between Nodes 1 and 3. The ATM cell add-drop or pass-through at each ring node is performed by checking the VPI value of the cells. Because only one route is available for all outgoing cells, no VP cross-connect capability is needed for this ATM VP-based ring architecture.

Note that this point-to-point VP ring architecture is very similar to SONET BSHR architectures (see Chapter 4). The SONET BSHR architectures use the concept of point-to-point "physical" paths (i.e., STS-N), whereas the ATM point-to-point VP ring uses the concept of point-to-point "virtual" paths. Thus, the evolution of the point-to-point VP ring from the SONET BSHR may occur naturally when the flexible bandwidth management capability becomes needed (e.g., supporting VBR services).

400

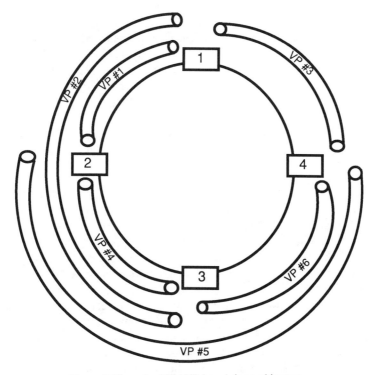

Figure 9.17. An ATM VP-based ring architecture.

To simplify the implementation (i.e., avoid the VPI table translation), the VPI values are assigned on a global basis. The global VPI value assignment is applicable to the ring architecture because the number of nodes supported by a ring is typically very limited, primarily because of traffic and reliability constraints. For example, the 12-bit VPI field in the NNI ATM cell represents 4096 VPI values available for use. Thus, the maximum number of ring nodes for this VPI global assignment is 91.[14] This maximum

---

14. Let n be the number of ring nodes. Each node pair requires one global VPI for a VP connection. Thus, the maximum number of ring nodes is the number satisfying the equation: $\frac{n \times (n-1)}{2} = 4096$, that is $n=91$.

number of ring nodes is expected to be reasonable in most practical interoffice and loop network applications.

If the point-to-point VP ring is used to support present DS1 services (via circulation emulation), each DS1 establishes a VC connection and is assigned a VPI/VCI based on the addressing information and the relative position of the DS1 group that terminates at the same source and destination on the ring. For example, VPI=2 and VCI=3 represent a DS1 that is the third DS1 of the DS1 group terminating at Nodes 1 and 3 (see Figure 9.17).

ATM/ADMs in this ring architecture can be implemented differently based on physical SONET STS-Nc terminations. The most common proposed ATM STS-Nc terminations are STS-3c, STS-12c, and STS-48c, although only the STS-3c ATM termination has been specified in current CCITT Recommendations. When ATM cells are extracted from the STS-Nc frame, dropping or passing through of the cells occurs by examining the VPI of the cells, which indicates their destination node on the ring.

Figure 9.18 depicts an ATM/ADM design with STS-3c terminations for supporting DS1 services (i.e., circuit emulation). In Figure 9.18, the ATM VP add-drop function, which is performed at the STS-3c termination level, requires three major functional modules for each STS-3c termination. The first module is the ATM/SONET interface, which converts the STS-3c payload to an ATM cell stream and vice versa. The functions performed in this module include cell delineation and self-synchronous scrambling. The scrambling guards against malicious users to simulate the SONET scrambling sequence in the information field. The second module performs header processing, which includes cell addressing (VPI in this case) and HEC. To perform cell add-drop, this module checks the VPI value of each cell to determine if it should be dropped or passed through and identifies idle cells, which can be replaced by cells added at the local office (i.e., signal adding) via a simple sequential-access protocol. This protocol passes each non-idle cell through and inserts the added cells into idle cell time slots in a sequential order.[15] The third functional module is a service-mapping module that maps ATM cells to their corresponding DS1 cards based on VPI/VCI values of ATM cells. This service-mapping module first multiplexes all ATM cells from different STS-3c payloads into a single ATM cell stream and then distributes ATM cells to corresponding DS1 groups according to their VPI values. For each DS1 group, the ATM cells are further divided and distributed to the corresponding DS1 cards by checking their VCI values. This service-mapping module essentially performs a simple VPI/VCI comparison function.

---

15. This is an example of the ATM VP add-drop capability and requires further study.

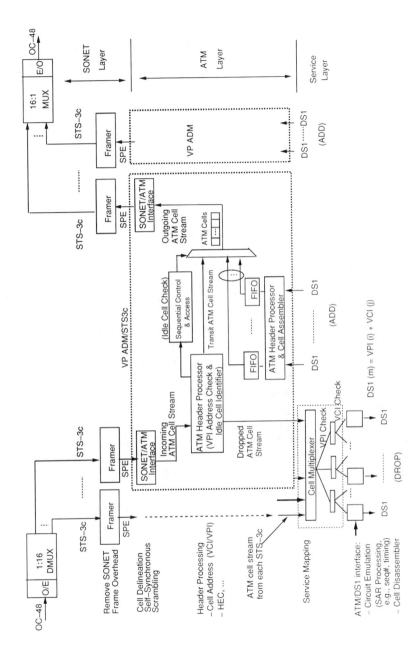

Figure 9.18. ATM/VP add-drop configuration with STS-3c terminations for supporting DS1 services.

403

For ATM termination at the STS-3c level, the ATM/ADM can be upgraded from the present SONET/STM ADM by simply replacing the SONET STS-1 line cards with ATM line cards, thus minimizing the initial capital costs for evolving to a hybrid SONET STM/ATM ring, and eventually to the ATM VP ring. Note that, for the hybrid STM/ATM ring, the STM portion is used to support non-switched DS1 and DS3 services, while the ATM portion is used to support high-speed, data-switched services (e.g., SMDS). Of course, operations for processing ATM cells will be different from those for processing SONET channels. The requirements and design for SONET rings require no change except for efficiency considerations. Thus, the ATM/ADM with STS-3c terminations could be the first candidate for transitioning from the SONET ring to the ATM ring for early deployment of ATM technology on the SONET infrastructure. If the SONET (physical) layer is chosen as the self-healing control layer in the considered ATM ring architecture, the self-healing function of the ATM ring with STS-3 terminations can be performed at the path (STS-3c) layer [32] or the line layer [33], as used in SONET ring architectures.

As described in Section 4.5.1, SONET bandwidth management can be performed in a centralized manner, using a SONET DCS, or in a distributed manner, using the TSI capability within the SONET ADM. As described in Section 4.3.1, in terms of cost and capacity, BSHRs are best suited for applications having a mesh STS-1 demand pattern, whereas unidirectional rings are best suited for applications having a centralized demand pattern. Given a mesh DS1 demand requirement, the STS-1 demand pattern usually becomes a centralized demand pattern when a centralized grooming scheme is used. The STS-1 demand pattern becomes a mesh demand pattern when a distributed grooming scheme is used. Because each ATM ring node using the ATM/ADM with STS-Nc terminations has a grooming capability similar to the SONET ring node using the ADM/TSI for distributed grooming, the BSHR architecture could be a better choice for the considered ATM ring in terms of capacity planning. Note that for a USHR architecture, two diverse routes are assigned to each duplex VP. For a BSHR architecture, one route is assigned to one VP for duplex communications.

9.7 ATM/DCS SELF-HEALING MESH NETWORK ARCHITECTURES

Like the SONET-based DCS self-healing mesh network architectures discussed in Chapter 5, the ATM self-healing mesh transport network can be implemented using ATM/DCSs. For example, current ATM self-healing network proposals [27, 34-37] use both the centralized protection control architecture with the preplanned method and the distributed protection control architecture with the divisive reconfiguration method. These are similar in design to SONET DCS self-healing networks. This section reviews an ATM/DCS functional architecture and then discusses current proposals for ATM/DCS self-healing mesh networks.

404

9.7.1 ATM/DCS Systems

ATM/DCS systems, which are used for provisioning VPs in broadband transport networks, simplify network configuration and yield a flexible routing and capacity-allocation capability. A VP used in the ATM mesh network architecture is a logical, point-to-point connection between two exchanges or users that have *fixed* routes and capacities [22,24].

To provide efficient and less expensive ATM cell transport (compared to the VC-based transport system), and also to increase transmission system utilization and minimize cell delay through the ATM DCS, the ATM/DCS must be designed in a cost-effective manner. To meet these design objectives, the ATM/DCS needs to meet the following requirements [38]:

1. Non-blocking characteristics
2. No restrictions on the usage condition of the middle link in a multi-stage switching network
3. Cell sequence integrity
4. Large system throughput (e.g., greater than tens or hundreds of gigabits-per-second)
5. Minimum cell-loss probability
6. Cross-connect capability for various VP speeds
7. VP management capabilities.

The second requirement means that switch fabrics should be determined only by the input and output connection relationship and should not be constrained by middle-stage (link) conditions. This is required because blocking may occur in non-self-routing, multi-stage switching networks if VPs are set up through the middle links and the VP bandwidths are dynamically changed. Examples of switch fabrics that meet requirements 1-3 are one-stage switching networks and self-routing multi-stage networks. The ATM/DCS also needs to handle various VP speeds uniformly with the same hardware. The VP management capabilities are also needed to efficiently operate the ATM VP-based network. These capabilities include VP maintenance signal transfer, VP transmission quality monitoring, VP rerouting against line failures, and transmission line protection switching [38].

Two experimental ATM/DCS systems have been reported [26,39]. The first [26] was designed to interconnect IEEE 802.6 MANs. It uses a self-routing, fast packet-switching fabric with output buffers. The switch port operates at approximately 50 Mbps. The second experimental ATM/DCS was designed to transport broadband signals in the VP-based mesh network [39]. This system uses a multi-stage, self-healing switching fabric with input buffers plus a contention resolution (scheduling) algorithm. The switch fabric of the ATM/DCS system proposed in Reference [39] operates at 155 Mbps and has a total capacity of 1.6 Gbps. The second generation of the ATM/DCS system based on Reference [39] was also reported in References [15,16]. The following descriptions of the ATM/DCS system are primarily taken from References [15,16].

Figure 9.19 depicts an example of the ATM/DCS functional architecture proposed in References [15,16]. The ATM/DCS consists of I/O ports, a self-routing space switch, a contention control module, and a VP management module. Each input port has its own buffer, VPI tables, and input controller. The VPI table provides the designated output port number and the new VPI for the incoming cells. The input controller and contention control module use a parallel processing architecture that dynamically assigns the sending time of input cells to prevent cell collision in the space switch. The self-routing switch receives the cells tagged with their designations from the input ports, and then the cells are transferred to the designated output ports using self-routing control. The self-routing switch module used in References [15,16] is a planar, homogeneous, self-routing switch that was designed for easy expansion of the switch size. Refer to References [15,16] for more details on this particular ATM/DCS system. It has been suggested [38] that high-speed termination (e.g., a STS-48c termination) is better than low-speed termination (e.g., an STS-3c termination) for the ATM/DCS in terms of link utilization and the transit delay within the ATM/DCS.

Table 9-4 shows a relative comparison between the ATM switch (i.e., VC switch) and the ATM/DCS [also see Figure 9.12(b)]. In general, ATM switches have better network utilization than ATM/DCSs, because the ATM VC switch can adapt various traffic conditions faster than the VP cross-connect system. However, this network efficiency for the ATM VC-switched system comes at the expense of greater system complexity. The complexity results from ATM switches setting up calls and performing path-capacity management, including capacity allocation and path-route selection. Thus, using ATM/DCSs at the intermediate nodes along the VP may reduce complexity for transporting services. The quantitative tradeoffs between network utilization and system complexity for using VP-based and VC-based ATM networks remain to be studied.

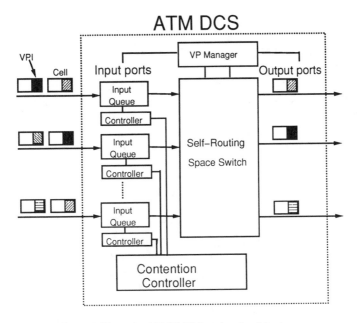

Figure 9.19. An ATM/DCS functional architecture.

Table 9-4. Design Differences between ATM Switches and ATM/DCSs

Attributes	ATM switch	ATM/DCS
Switching or cross-connect unit	VC	VP or VP/VC
Network utilization	higher	lower
Nodal complexity	higher	lower
Call setup required	yes	no
Path-capacity allocation	yes	no
Path-route selection	yes	no

407

9.7.2 ATM/DCS vs. SONET DCS

Like SONET DCSs, ATM/DCSs can also be classified as VP/VC cross-connect systems or VP cross-connect systems. The VP/VC cross-connect system terminates VPs and cross-connects VPs and VCs, as depicted in Figure 9.20(a). The VP cross-connect system terminates and cross-connects VPs only, as depicted in Figure 9.20(b). These two types of ATM cross-connections are specified in CCITT Recommendation I.311 [1]. Note that the NE used in the ATM VP-based transport network proposed in Reference [22] is primarily the VP cross-connect system, which cross-connects VPs without changing the sequence of VCs within each VP.

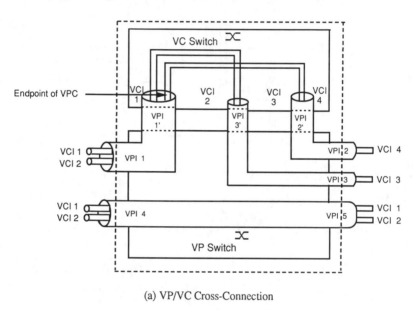

(a) VP/VC Cross-Connection

Figure 9.20.   VP cross-connect functionalities.

408

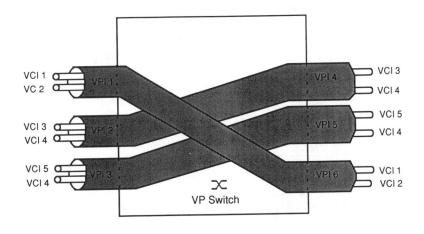

(b) VP Cross-Connection

Figure 9.20.   (Continued)

Table 9-5 shows a relative comparison between the ATM/DCS and the SONET DCS. The SONET DCS cross-connects physical paths with fixed capacities (e.g., STS-1, VT1.5, and so so), whereas the ATM/DCS cross-connects VPs with virtually any possible capacity up to the transmission line rate. To cross-connect low-speed signals from a high-speed signal stream (e.g., cross-connect VT1.5s from an STS-3 signal), the SONET DCS must demultiplex the high-speed signal to low-speed signals on a hierarchical basis. This hierarchical multiplexing/demultiplexing structure is not needed for the ATM/DCS to cross-connect VPs because each VP may accommodate any possible capacity. For the SONET DCS, any change in path capacity (i.e., from STS-1 to STS-3s) requires a change in the switching matrix configuration because the path capacity is physically associated with I/O ports of the switching matrix. In contrast, a change in the VP path capacity will not affect the switching matrix configuration in the ATM/DCS because the VP path capacity can be dynamically adjusted as long as the total capacity of all VPs being cross-connected does not exceed the ATM/DCS capacity. Due to its non-hierarchical path structure, the ATM/DCS network has better utilization than the SONET DCS network. It also has faster provisioning and restoration capabilities because its path structure is non-hierarchical and the OAM cells needed for provisioning and restoration can be conveyed within the

STS-Nc payloads in any time cycle. The SONET DCS network has limited available bandwidth of DCCs in a fixed cycle of 125 μs.

Table 9-5. Comparing SONET DCS Systems with ATM/DCS Systems

Attributes and applications	SONET DCS	ATM/DCS
Path being cross-connected	digital path (physical)	virtual path (logical)
Multiple-rate switching ports	no	yes
Paths that can be cross-connected	limited*	more*
Multiplexing hierarchy for path cross-connections	hierarchical	non-hierarchical
Switching matrix changes when path capacity changes	yes	no
Facility utilization+	lower	higher
Service provisioning (path)	slower	faster
Alternate routing (restoration)	slower	faster

* Only VT1.5, STS-1, or STS-Nc can be cross-connected by a SONET DCS, whereas any VP with a capacity ranging from zero to the line rate can be cross-connected by an ATM/DCS.

\+ Assume that service rates are DS1 and above.

9.7.3 ATM VP-Based Self-Healing System Design Concepts

The ATM VP-based self-healing system may use concepts similar to the SONET DCS self-healing system (see Chapter 5) because both systems use a similar path concept. Like its STM/SONET counterpart, the ATM VP-based self-healing function can be controlled in a centralized or distributed manner. For the centralized ATM VP-based self-healing network, one of the failed facility's end nodes detects the failure and informs the central controller via a separate data communications network or the ATM network itself, as does the SONET/DCS self-healing mesh network. For the distributed ATM self-healing network, path rerouting message broadcasting and acknowledgment message propagation may occur via OAM cells or SONET DCC channels. Several designs have been proposed to perform the self-healing function for ATM VP-based transport networks using ATM/DCSs [34-37].

Routing of VP connections is established by converting VPIs in the cell header field of input VP links into those of output VP links at each connecting node. To recover from VP faults, some backup VP routes are preassigned. When network failures are

410

detected, path recovery is performed by simply changing the VPI value at the VP originating point. Protection switching control, such as routing table reconfiguration, at intermediate nodes along the VP route is not needed. A certain VP capacity should be reserved for the backup VP route, but this spare capacity can be shared among several backup VP paths, depending on reliability factors [27].

Services carried by the ATM VP-based transport network can be restored on a physical line (called *line restoration*) or on an end-to-end VP (called *path restoration*) basis, as occurs with SONET DCS self-healing networks (see Chapter 5). The line restoration method restores all VCs in VPs that pass through the affected transmission link. The path restoration method restores VCs in each affected VP individually on an end-to-end basis.

For line restoration, the process should reestablish VPs for all affected VPs that are carried over the failed link. Standby VPs can be established on a preplanned or dynamic basis. For the preplanned method, using the ATM distributed self-healing control architecture, all preassigned VPs for each link can be stored in each node that is connected by that link. These preassigned VPs for each link can also be stored in the central controller and then downloaded to the affected node's ATM/DCS when needed. For a VP that requires protection routing, protection routes can be preestablished without reserving capacity. That is, the VPI numbers associated with each link along the protection path routes are reserved and preassigned in the path connect tables at cross-connect nodes along the paths. A certain link capacity should be reserved for sharing among many possible protection paths accommodated by the link.

When a network failure is detected, necessary path capacities are assigned along the preestablished protection path routes. This process can be done using distributed or centralized processing. Cells are then switched to the protection path route by providing new VPI values specific to the protection path. Among transmission facilities, only the originating VP terminator (for path-route switching at the originating endpoint of the path) or the ATM/DCS (for path-route switching at failed link end) is involved in routing alternation, which enables rapid restoration. (No routing table renewal or other processing at other cross-connect nodes along the protection path is required.) In this case, only two end nodes of the failed facility are involved in the line restoration process. Note that VP bandwidth reallocation is not needed in this process, as opposed to its SONET/DCS counterpart.

For the dynamic line restoration method, standby VPs for all affected VPs carried over the failed link are dynamically computed when needed [34-37]. In this case, the restoration path-finding process is similar to the one designed for SONET DCS self-healing networks. However, it is faster than its SONET counterparts because no bandwidth reallocation is needed, and OAM cells carrying protection control information are expected to have lower latency because more bandwidth is available than the SONET DCC provides. It can be expected that traffic rerouting using the

411

dynamic method is more complex than routing using the preplanned method. Path restoration requires establishing new VPs for each end-to-end working VP. These standby VPs for each VP can be preplanned or computed dynamically whenever needed.

Compared to its STM counterparts, the ATM/DCS self-healing networks may require less spare capacity and have faster restoration times due to their non-hierarchical path structure. Also, the VP bandwidth control requires no processing at cross-connect nodes along the path and no synchronization processing between the origin and the destination nodes. This reduces the time needed to control path bandwidth and, thus, increases network adaptability against unexpected traffic changes. Of course, these advantages of ATM technology all come at the expense of a more complex network control system (as compared to the SONET network).

9.7.4 Hitless Path Switching

Hitless switching has been proposed in ATM VP-based transport networks for in-service maintenance and in-service protection switching [34-37]. For facility maintenance, a VP must be moved from the original route to an alternate route without interrupting the cell stream on the original route. After maintenance is complete, the VP is switched back to the original route without service interruption. The other application is for VP protection switching. When the network component fails, the working VP is rerouted to the alternate route and switched back when the failed component is repaired. The revertive mode of protection switching allows the network to best utilize its resources as designed in the capacity-assignment phase.

The key factor that makes VP hitless switching possible is synchronization between the original route and the alternate route. The cell sequence integrity across VCs within a VP is also required for hitless path switching; however, it is not yet specified in CCITT Recommendations [1]. In ATM networks, the synchronization signals can be inserted to replace the idle cells. Several synchronization methods for hitless path switching are proposed in References [34-37]. Among synchronization alternatives, snap-shot synchronization may be the simplest one, although it requires buffers to adjust the difference of delays between the original and alternate routes if the original route is longer than the alternate route. In this case, the delay justification may be evenly performed at two ends with a total buffering delay equal to the delay difference between two routes. Let t_d and t_b be the delay difference between the original and the alternate routes and the duration of buffers at Nodes X and Y, respectively, where $t_b = \max (t_d/2)$ [34]. The following description of a hitless path switching scheme is taken from Reference [34]. Figure 9.21 shows two cases for VP hitless switching, where it is assumed that switching is to be done from the alternate route back to the original route (e.g., return to normal after restoration).

412

When the ATM/DCS at Node X receives hitless switching commands, it sets up the buffer memory based on the delay difference between the original and alternate routes. The ATM/DCS at the source node starts switching the VP from the alternate route back to the original route. At the same time, it inserts an end flag into the cell stream on the alternate route and a start flag on the original route. If the alternate route is shorter (less delay) than the original route, the ATM/DCS at the destination (i.e., Node Y) receives the end flag before the start flag. Once the ATM/DCS at Node Y receives the end flag, it removes the end flag from the ATM cell stream and switches the VP from the alternate route to the original route. No buffer requirement is needed in this case.

In the second case, the alternate route is longer than the original route, and the start flag is received after the end flag at Node Y if there is no buffering scheme imposed on the original route. If the delay difference between two routes (i.e., t_d) exceeds the duration of the buffer (t_b) at Node X, the ATM/DCS at Node X feeds the ATM cell stream on the original route via a fixed-delay buffer memory; otherwise, the buffer is not needed at Node X.

The incoming cells to the ATM/DCS at Node Y are stored in the buffer. When the ATM/DCS at Node Y detects the start flag, the incoming cells are stored in the buffer. When the ATM/DCS at Node Y detects the end flag, it removes the end flag from the cell stream and then begins transmitting cells stored in the buffer. Using the idle cell periods of the outgoing link, the cell stored in the buffer can be transmitted faster than the cell-arrival rate (delay relaxation mode). When the buffer is empty, the ATM/DCS switches the VP from the alternate route back to the original route and informs the ATM/DCS at Node Y that the switching operation at the destination is complete.

After receiving notification from Node Y, the ATM/DCS at Node X changes the operation mode of the buffer to the delay relaxation mode. After the buffer is empty, the ATM/DCS at Node X switches the VP from the alternate route to the original route, and the hitless switching process is complete.

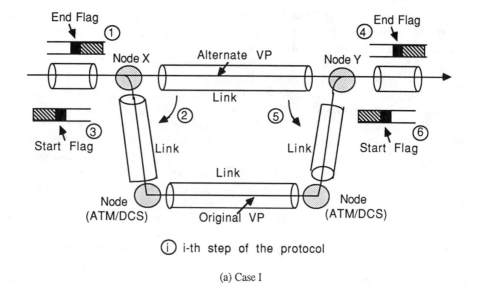

(a) Case I

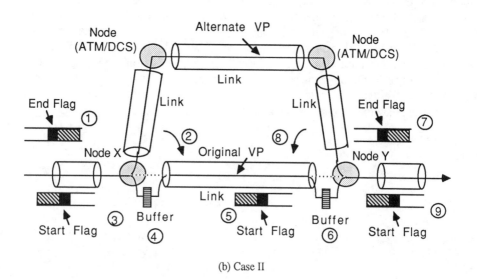

(b) Case II

Figure 9.21. Hitless protection switching for a VP.

9.7.5 Integrated ATM VP-Based Control Systems

As discussed in Chapter 5, the goals of using SONET DCS systems for fast provisioning and flexible customer control are different from the goals for network restoration. Thus, a technical challenge in the SONET DCS network is building a cost-effective DCS network that integrates fast provisioning, flexible customer control, and network restoration applications. This technical challenge also resides in the ATM/DCS network. However, because both the normal control system (for fast provisioning and flexible control) and the self-healing control system for ATM VP-based networks are in early development stages, the network restoration function may be built into the normal network control system in the early design phase. Thus, a single system could operate under both the normal and failure conditions cost-effectively and efficiently. Such integrated ATM VP-based control systems remain to be studied.

9.8 SUMMARY AND REMARKS

ATM has been recognized as a future technology that may support a wide variety of broadband services in a single and uniform network. In the past few years, many efforts have explored using ATM technology for implementing B-ISDN. These efforts include developing ATM standards [1,2,40], designing and implementing ATM switches [17,18], and studying aspects of ATM network performance [41-48]. However, in spite of these ATM research efforts, much debate remains regarding the practicality of a full-scale ATM network and the time frame in which to implement this new technology.

To stimulate early development of ATM technology, ATM transport systems based on ADMs and DCS systems using the VP concept have been proposed to provide a cost-effective and simple transport system for B-ISDN. The prototyping of ATM/ADMs and ATM/DCSs has been demonstrated in some industrial research laboratories. A preliminary study presented by NTT at GLOBECOM'90 suggested that the ATM/DCS transport network may support existing DS1 services more cost-effectively than its STM/SONET counterpart. However, further studies are still needed to prove the economic merit of the ATM/DCS network using the VP concept.

ATM VP-based network control and designs, including self-healing systems, are still in the early development stage. Technical challenges of planning ATM VP-based transport networks include economical and control complexities (both in algorithms and in hardware); tradeoffs between SONET DCS networks and ATM/DCS networks; taking advantage of the powerful management capabilities of physical SONET infrastructure and distinguishing network management functions in the ATM layer and the SONET layer; protecting the ATM transport network cost-effectively (SONET layer or ATM layer [VC or VP]); and designing a cost-effective, integrated ATM transport control system for both normal operation conditions and catastrophic failure situations. Hopefully, the review of the progress in the ATM transport control area described in this chapter will serve as useful background information for those who may be interested in research and planning in the ATM transport network control area.

REFERENCES

[1] CCITT Study Group XVIII - Report R 34, COM XVIII-R 34-E, June 1990.

[2] Sinha, R. (Editor), "Broadband Aspects of ISDN Baseline Document," T1S1.5/90-001 R2, June 1990.

[3] ISO 8208, X.25 Packet Level Protocol for Data Terminal Equipment, 1990.

[4] TA-TSY-000253, *Synchronous Optical Network (SONET) Transport Systems: Common Generic Criteria*, Bellcore, Issue 6, September 1990.

[5] Ross, F., "An Overview of FDDI: the Fiber Distributed Data Interface," *IEEE Journal on Selected Areas in Communications*, Vol. 7, No. 7, September 1989, pp. 1043-1051.

[6] "Proposed Standard Distributed Queue Dual Bus (DQDB) Metropolitan Area Network (MAN)," IEEE P802.6/D6-88/105, November 1988.

[7] TA-TSY-000772, *Generic System Requirement in Support of Switched Multi-Megabit Data Services*, Issue 3, Bellcore, October 1989.

[8] Minoli, D., *Telecommunications Technology Handbook*, Artech House, 1991.

[9] Heihanen, J., "Review of Backbone Technologies," *Computer Networks and ISDN Systems*, Vol. 21, 1991, pp. 239-245.

[10] SR-NWT-001763, *Preliminary Report on Broadband ISDN Transfer Protocols*, Issue 1, Bellcore, December 1990.

[11] CCITT SG XVIII Delayed Contribution # D.201, "Timing Recovery for CBO Services in an ATM-based Network," June 1989.

[12] Lau, R. C., and Fleischer, P. E., "Synchronous Residue-TS: A Comprise for SFET/TS," T1S1.5/91-382, November 1991.

[13] Karol, M. J., Hluchyj, M. G., and Morgan, S. P., "Input Versus Output Queueing on a Space-Division Packet Switch," *IEEE Transactions on Communications*, Vol. 35, No. 12, December 1987, pp. 1347-1356.

[14] Takeuchi, T., Suzuki, H., and Aramaki, T., "Switch Architectures and Technologies for Asynchronous Transfer Mode," *IEICE Transactions*, Vol. E 74, No. 4, April 1991, pp. 752-760.

[15] Obara, H., Sasagawa, M., and Tokizawa, I., "An ATM Cross-Connect System for Broadband Transport Networks Based on Virtual Path Concept," *Proceedings of IEEE ICC'90*, Atlanta, GA, April 1990, pp. 318.5.1-318.5.5.

416

[16] Obara, H., "Distributed ATM Cross-Connect Switch Architecture Using Transmission Scheduling Control," *Electronics and Communications in Japan*, Part 1, Vol. 74, No. 1, 1991, pp. 55-64.

[17] Prycker, M. D., *Asynchronous Transfer Mode: Solution for Broadband ISDN*, Ellis Horwood Limited, 1991.

[18] Hui, J. Y., *Switching and Traffic Theory for Integrated Broadband Networks*, Kluwer Academic Publishers, Boston, 1990.

[19] Hiramatsu, A., "ATM Communications Network Control by Neural Networks," *IEEE Transactions on Neural Networks*, Vol. 1, No. 1, March 1990, pp. 122-130.

[20] Woodruff, G. M., Rogers, R. G. H., Richards, P. S., "A Congestion Control Framework for High-Speed Integrated Packetized Transport," *Proceedings of IEEE GLOBECOM*, 1988.

[21] Sato, Y., and Sato, K-I., "Virtual Path and Link Capacity Design for ATM Networks," *IEEE Journal on Selected Areas in Communications*, Vol. 9, No. 1, January 1991, pp. 104-111.

[22] Sato, K-I., Ohta, S., and Tokizawa, I., "Broadband ATM Network Architecture based on Virtual Paths," *IEEE Transactions on Communications*, Vol. 38, No. 8, August 1990, pp. 1212-1222.

[23] Burgin, J., and Dorman, D., "Broadband ISDN Resource Management: The Role of Virtual Paths," *IEEE Communications Magazine*, Vol. 29, No. 9, September 1991, pp. 44-48.

[24] Tokizawa, I., and Sato, K-I., "Broadband Transport Techniques Based on Virtual Paths," *Proceedings of IEEE GLOBECOM'90*, San Diego, CA, December 1990, pp. 705B.4.1-705B.4.5.

[25] TR-TSY-000782, *SONET Digital Switch Trunk Interface Criteria*, Issue 2, Bellcore, September 1989.

[26] Tirtaatmadja, E., and Palmer, R. A., "The Application of Virtual Paths to the Interconnection of IEEE 802.6 Metropolitan Area Networks," *Proceedings of International Switching Symposium (ISS)*, Vol. II, Stockholm, May 1990, pp. 133-137.

[27] Sato, K-I., Hadama, H., and Tokizawa, I., "Network Reliability Enhancement with Virtual Path Strategy," *Proceedings of IEEE GLOBECOM'90*, San Diego, CA, December 1990, pp. 403.5.1-403.5.6.

[28] Imai, K., Honda, T., Kasahara, H., and Ito, T., "ATMR: Ring Architecture for Broadband Networks," *Proceedings of IEEE GLOBECOM'90*, San Diego, CA, December 1990, pp. 900.2.1-900.2.5.

[29] Maeda, Y., Kikuchi, K., and Tokura, N., "ATM Access Network Architecture," *Proceedings of IEEE ICC'91*, Denver, CO, June 1991.

[30] Ohta, N., Nomura, M., Tobuyuki, N., and Kikuchi, K., "Video Distribution on ATM-Based Optical Ring Networks," *Proceedings of IEEE ICC'90*, April 1990, pp. 323.4.1-323.4.5.

[31] Microtel Pacific Research, "Support of Connectionless Data Service Using Virtual Ring Topology," CCITT Study Group XVIII - Contribution No. D.620, December 1989.

[32] TR-TSY-000496, *SONET Add-Drop Multiplex Equipment (SONET ADM) Generic Criteria for a Unidirectional, Dual-Fed, Path Protection Switched, Self-Healing Ring Implementation,* Bellcore, Issue 2, September 1989, Supplement 1, September 1991.

[33] TA-NWT-001230, *SONET Bidirectional Line Switched Ring Criteria,* Issue 1, Bellcore, October 1991.

[34] Chujo, T., Soejima, T., and Komine, H., "Network Restoration Techniques in ATM Networks," *Joint Conference on Communications, Networks, Switching System and Satellite Communications (JC-CNSS),* Chejy Island, Korea, Dec. 1990, pp. 199-203.

[35] Kawamura, R., Sato, K., and Tokizawa, I., "High-Speed Self-Healing Techniques Utilizing Virtual Paths," *5th International Network Planning Symposium,* Kobe, Japan, May 1992.

[36] Sakauchi, H., Nishimura, Y., and Hasegawa, S., "A Network Restoration Scheme for ATM Cross-Connect Systems," *Tech. Group, CS 90-60, IEICE (Japan),* October 1990, pp. 43-48 (in Japanese).

[37] Tatsuno, H., and Tokura, N., "A Study on Hitless Path Protection Switching Technique for Asynchronous Transfer Mode Network," *Tech. Group, CS 90-47, IEICE (Japan),* October 1990, pp. 13-18 (in Japanese).

[38] Sato, K., Ueda, H., and Yoshikai, N., "The Roll of Virtual Path Crossconnection," *IEEE Magazine of Lightwave Telecommunications Systems,* Vol. 2, No. 3, August 1991, pp. 44-54.

[39] Obara, H., and Yasushi, T., "High Speed Transport Processor for Broadband Transport Networks," *Proceedings of IEEE ICC'88,* June 1990, pp. 29.5.1-29.5.6.

[40] Anderson, J., "Progress on Broadband ISDN User-Network Interface Standards," *Proceedings of IEEE GLOBECOM'90,* San Diego, CA, December 1990, pp. 900.4.1-900.4.7.

[41] SR-TSY-001453, *Broadband Industry Symposium,* Issue 1, Bellcore, September 1989.

[42] FA-NWT-001110, *Broadband ISDN Switching System Framework Generic Criteria,* Issue 1, Bellcore, December 1990.

[43] Mouftah, H. T., Ajmone Marsan, M., Kurose, J. F. (Guest Editors), "Special Issue on Computer-Aided Modeling, Analysis, and Design of Communications Networks (II)," *IEEE Journal on Selected Areas in Communications,* Vol. SAC-9, No. 1, January 1991.

[44] Bolotin, V. A., Kappel, J. G., and Kuehn, P. J. (Guest Editors), "Special Issue on Teletraffic Analysis of ATM Systems," *IEEE Journal on Selected Areas in Communications,* Vol. SAC-9, No. 3, April 1991.

[45] Coudreuse, J. P., Sincoskie, W. D., and Turner, J. S. (Guest Editors), "Special Issue on Broadband Packet Communications," *IEEE Journal on Selected Areas in Communications,* Vol. SAC-6, No. 9, December 1988.

[46] Turner, L., Aoyama, T., Pearson, D., Anastassiou, D., and Minami, T. (Guest Editors), "Special Issue on Packet Speech and Video," *IEEE Journal on Selected Areas in Communications,* Vol. SAC-7, No. 5, June 1989.

[47] Estes, G. (Guest Editor), "BISDN - Moving Toward Introductory Phases," *IEEE Magazine of Lightwave Telecommunications Systems,* Vol. 2, No. 3, August 1991.

418

[48]   Mobasser, B. (Guest Editor), "B-ISDN: High Performance Transport," *IEEE Communications Magazine,* Vol. 29, No. 9, September 1991.

APPENDIX A

Maximum Passive Protection Ring Length without Optical Signal Amplification

Assumptions:

Fiber loss for 1300-nm single-mode fiber = 0.50 dB/km
Fiber loss for 1550-nm single-mode fiber = 0.25 dB/km
Fiber splicing loss = 0.15 dB/splice/km
Connector loss per pair = 1.0 dB
Transmitting output power = 0 dBm for OC-12, OC-24, and OC-48 systems
Receiver sensitivity for BER of 10^{-9} = -35 dBm for OC-12 systems
= -32 dBm for OC-24 systems
= -29 dBm for OC-48 systems.

Ring nodes	No. of optical switches	Maximum protection ring length (miles) (for 1300 nm)			Maximum protection ring length (miles) (for 1550 nm)		
		OC-12	OC-24	OC-48	OC-12	OC-24	OC-48
3	5	27.4	24.5	21.6	44.5	39.8	35.2
4	6	26.2	23.3	20.4	42.5	37.8	33.1
5	7	24.9	22.0	19.1	40.5	35.8	31.1
6	8	23.7	20.8	17.9	38.4	33.8	29.1
7	9	22.4	19.5	16.6	36.4	31.7	27.0
8	10	21.2	18.3	15.4	34.4	29.7	25.0
9	11	19.9	17.0	14.1	32.3	27.7	23.0
10	12	18.7	15.8	12.9	30.3	25.6	20.9

APPENDIX B

A Method for Evaluating DCS Self-Healing Networks and Self-Healing Rings

B.1 Analysis Method

This appendix discusses a simple method that provides some insight into tradeoffs between the Digital Cross-connect System (DCS) self-healing network and the Self-Healing Ring (SHR). This analysis assumes that the network uses a mesh demand pattern, which is a demand pattern where any two nodes have STS-1 demands between them. The total number of demand pairs for the mesh demand pattern is n(n-1)/2 where n is the number of nodes. For simplicity, a constant demand requirement, say d STS-1s, is assumed for each demand pair. Let n be the number of nodes in the network and p (the spare capacity ratio) be the ratio of required spare capacity to corresponding working capacity. Also, let q (the transit capacity ratio) be the ratio of transit STS-1 demands to dropped STS-1 demands for each DCS. The number of STS-1s that are dropped to each node is $(n-1) \times d$. Thus, each DCS terminates a total of $[(1+q) \times (n-1) \times d] \times (1+p)$ STS-1s.

The value of q (transit ratio) depends on the degree of network connectivity. If the DCS network has a ring topology (lowest connectivity), q tends to be close to 1 assuming that a minimum hop-routing algorithm is used in the DCS network. For example, as depicted in Figure B.1(a), there are seven demand pairs terminating at Node 4, and demands from six demand pairs (i.e., 1-5, 2-5, 2-6, 3-5, 3-6, and 3-7) pass through Node 4. By contrast, q may reach 0 if a fully connected network topology is considered. The value of p (spare ratio) depends not only on network connectivity, but also on the DCS self-healing algorithm being used. As reported in References [11,14,15] of Chapter 5, p approaches 0.5 in highly connected mesh networks.

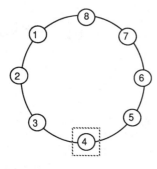 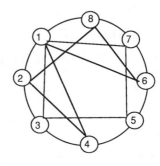

(a) Ring Topology   (b) Near Mesh Topology with Node Connectivity of Four

Figure B.1.   Examples of network topologies.

The total cost for the DCS network, denoted by C_{DCS}, is then computed as follows:

$$C_{DCS} = n \times \{C_{Dg} + C_{Dt} \times [(1+q) \times (n-1) \times d \times (1+p)]\} \tag{B-1}$$

where C_{Dg} and C_{Dt} are the startup cost and per-STS-1 terminating cost, respectively.

For simplicity, the Unidirectional Self-Healing Ring (USHR) architecture (see Section 4.2 of Chapter 4) is assumed to be the ADM ring. Let N_R be the number of rings required. It can be calculated as follows:

$$N_R = \left\lceil \frac{d \times n \times (n-1)}{2 \times N} \right\rceil$$

where N is the maximum number of STS-1s that can be carried by the OC-N ring and $\lceil x \rceil$ is the smallest integer greater than x.

For the ADM ring network, the total number of STS-1s terminated at each node is $d \times (n-1)$ because (n-1) demand pairs are terminated at each node. Here we assume that the model uses multiple overlaid rings when the capacity requirement exceeds a single ring capacity. Thus, the number of STS-1s terminated at each ring node is $\dfrac{d \times (n-1)}{N_R}$.

The total equipment cost for rings, C_R, is calculated as follows:

$$C_R = N_R \times n \times [C_{Rg}^N + C_{Rt} \times \frac{d \times (n-1)}{N_R}] \qquad (B-2)$$

where C_{Rg}^N and C_{Rt} are the startup cost and per-STS-1 cost, respectively, for an OC-N ADM.

In this study, the transport cost includes equipment, fiber material, and fiber placement costs.

B.2 Results

We first apply the above cost model to a six-node network with two extreme topologies: ring and mesh (a fully connected network topology). We then make the following assumptions:

- The considered network has eight nodes.
- Each link is 10 miles long, and the mesh demand pattern includes six STS-1s for each demand pair. The total number of demand pairs is 28.
- The maximum capacity of a SONET B-DCS is 1024 STS-1s.
- Only OC-48 ADMs are available for use in the model.
- The startup costs for a B-DCS and an OC-48 ADM are $150K and $48K, respectively.
- The per-STS-1 termination cost is $1500 for both systems.
- The fiber material and splicing cost per mile, per fiber pair is $500. The placement cost per mile is $5000.

The SONET ADM ring network requires four OC-48 USHRs because the total STS-1 demand is 168 STS-1s (6×28), and this ADM ring network has eight fiber links for each USHR. The DCS self-healing network has a near mesh topology with 16 fiber links, as depicted in Figure B.1(b). This topology has a connectivity of four for each node. According to most proposed distributed DCS self-healing algorithms, it is required that all or most of 16 links be used to gain a spare ratio of 50 percent (i.e., p=0.5), where the spare ratio is the ratio of required total spare capacity to total working capacity.

Table B-1 shows cost comparisons between the ADM ring and the DCS network for both the ring and near mesh topologies. The total ADM cost for the ADM ring is

$$C_R = 4 \times [8 \times (48,000 + 1,500 \times (8-1) \times 6/4)] = \$\,2,040,000$$

For the DCS self-healing network, the value of q is assumed to be 1.0 for the ring topology and 0.1 for the near mesh topology [see Figure B.1(b)]. Also, the spare capacity ratio, p, is assumed to be 0.5 for the DCS network. The ring and the near mesh topologies have eight and 16 fiber links, respectively. Using Equation B-1, the equipment cost for the DCS network can be obtained as shown in Table B-1.

Table B-1. Cost Comparison between the ADM Ring and the DCS Network for an Eight-Node Network

Component cost	ADM ring architecture	DCS network	
		Ring topology ($p=0.5$, $q=1.0$)	Near mesh topology* ($p=0.5$, $q=0.1$)
Fiber cost	$160K	$40K	$80K
Placement cost	$400K	$400K	$80K
Equipment cost	$2040K	$2712K	$2032K
Total transport cost	$2600K	$3152K	$2912K

* The near mesh topology has a connectivity of four for each node.

From Table B-1, we observe the following:

- Under the near mesh topology, the DCS network has lower terminating costs than the ADM ring but has higher transport costs due to higher placement costs. This implies that if the near mesh topology is embedded, the DCS network becomes more attractive than its ADM ring counterpart in terms of overall economic merit.
- Regarding equipment costs, the DCS network with a ring topology may not compete well with the ADM ring but may have better economic benefits if the DCS network topology has a near mesh structure.
- The spare capacity savings (50 percent assumed here) of the DCS network over the ADM ring does not imply transport cost savings in this particular example. It implies only that the DCS network's spare capacity savings results in lower equipment costs if a near mesh topology is used.

Figures B.2, B.3, and B.4 show some extensive results from the cost model described above. Figures B.2(a) and B.2(b) show cost ratios of the DCS network to rings versus the spare capacity ratio (p) for an eight-node network for transport costs and equipment costs, respectively. In Figure B.2, two extreme topologies (ring and mesh) and two different demand requirements (d=6 STS-1s and 10 STS-1s) are analyzed. The transit capacity ratio (i.e., q) is assumed to be 0.5 and 0.1 for the ring and mesh topologies, respectively. Results shown in Figures B.2(a) and B.2(b) suggest the following observations:

426

- Transport costs [see Figure B.2(a)]

  — For relatively low demand (d=6 STS-1s), the spare capacity savings from the DCS restoration method may not be a crucial factor in determining the economic merit of the DCS restoration method, even for the mesh topology.

  — For relatively high demand (d=10 STS-1s), the DCS network may become attractive under the ring topology with the spare capacity ratio less than 70 percent. In this case, the DCS restoration scheme may not be economical under the mesh topology, when compared to its ring counterparts, due to high fiber placement costs.

- Equipment costs only [see Figure B.2(b)]

  — For relatively low demand (d=6 STS-1s), the DCS restoration scheme may not be more economical than its ring counterpart under the ring topology, regardless of the spare capacity ratio. However, the scheme does show its economic benefits under the mesh topology with the spare capacity ratio less than 50 percent.

  — For relatively high demand (d=10 STS-1s), the DCS restoration scheme has economic merit over its ring counterpart under the mesh topology. For the ring topology, the DCS restoration scheme still shows its economic merit for a spare capacity ratio less than 50 percent.

In summary, high demand and highly embedded connectivity may be two major factors that make the DCS self-healing network more economical than its ring counterpart.

Figure B.3 depicts the cost ratio of the DCS network to the ring versus the number of nodes in the network. It shows the general trend that a large network size combined with a high demand requirement may be a major factor in making the DCS self-healing network more economical than its ring counterpart.

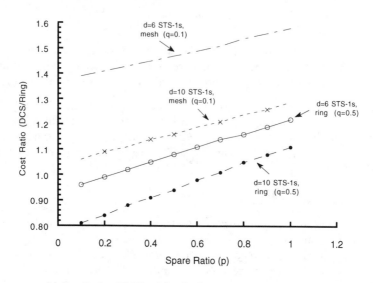

(a) Cost Ratio of DCS to Ring for 8-Node Network (Transport Costs)

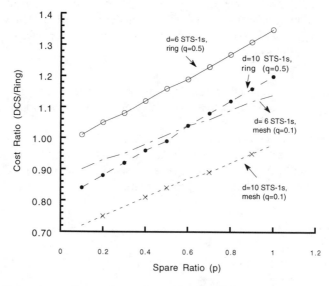

(b) Cost Ratio of DCS to Ring for 8-Node Network (Equipment Cost Only)

Figure B.2. Cost ratios of the DCS network to rings vs. spare capacity ratio for an eight-node network.

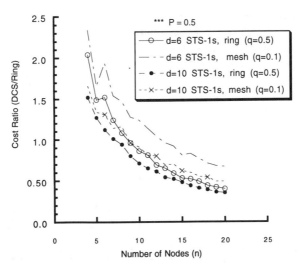

(a) Cost Ratio of DCS to Ring (Transport Costs)

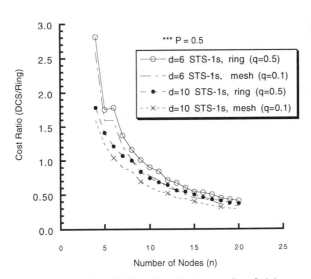

(b) Cost Ratio of DCS to Ring (Equipment Cost Only)

Figure B.3.   Cost ratios of DCS networks to rings for various networks.

429

Figure B.4 shows the impact of fiber placement cost (related to embedded network connectivity) on cost ratios for a mesh network. This case assumes that p=0.5 and q=0.1. As shown in Figure B.4, embedded (i.e., the placement cost is 0) or new connectivity may significantly impact the economic merit of the DCS network over its ring counterpart for a small network with or without a low demand requirement. However, it may not impact the economic merit of the DCS network as the network size becomes larger (e.g., n≥12 in Figure B.4).

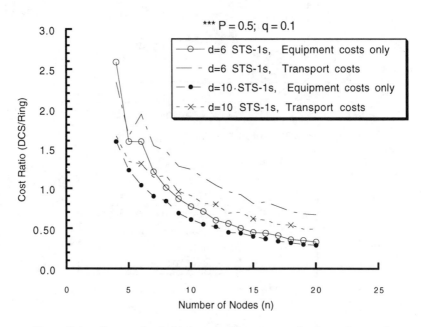

Figure B.4.   Impact of embedded connectivity on cost ratios for mesh networks.

In summary, large network size, high demand, and embedded high connectivity are three necessary and crucial conditions that ensure the economic merit of the DCS self-healing network over its ring counterpart.

APPENDIX C

Review of Disjoint Path Algorithms

This appendix reviews some algorithms needed for discussions of the survivable fiber-hubbed network design algorithm described in Section 6.3 of Chapter 6.

C.1 Shortest Path Tree Algorithm

A shortest path tree algorithm finds the shortest path from one source node to all other nodes. The most common of these algorithms is by Dijkstra (see Reference [1]). This section discusses a shortest path tree algorithm, called *Shortest Path First* (SPF) algorithm, which is modified from Dijkstra's algorithm and has been implemented in a packet-switched data network [2]. A full classification and reference list of shortest path algorithms can be found in Reference [3].

We first define variables needed for the SPF algorithm and then define the algorithm.

Definition of Variables

        SOURCE = the source node
        LIST = a variable-length list whose elements are ordered triples
        T = <PARENT, NODE, DISTANCE> where
        NODE(T) = the node currently being processed
        PARENT(T) = the parent of NODE(T) in tree T being built
        DISTANCE(T) = the smallest distance from SOURCE to NODE(T)
        LENGTH(X,Y) = the distance from Node X to Node Y

SPF Algorithm

1. If no tree exists, place <SOURCE, SOURCE, 0> on LIST.
2. Process the current LIST as follows:

    a. Search LIST for the triple T with the smallest DISTANCE and remove this T from LIST.
    b. Place NODE(T) on the shortest path tree so that its parent on the tree is PARENT(T). (If NODE(T) = SOURCE, then place it in the tree as its root).

3. For each neighbor N of NODE(T), do one of the following steps:

a.   If N is already in the shortest path tree, do nothing.
b.   If there is no triple T' on LIST such that NODE(T') = N, then place the triple <NODE(T), N, DISTANCE(T)+LENGTH(NODE(T),N)> on LIST.
c.   If there is already a triple T' on LIST such that NODE(T') = N and if DISTANCE(T') > DISTANCE(T) + LENGTH(NODE(T),N), then do the following:

   - Remove T' from LIST.
   - Place the triple <NODE(T), N, DISTANCE(T)+LENGTH (NODE(T),N)> on LIST.

4.   If LIST is non-empty, go to Step 2; otherwise, the algorithm is completed and the shortest path tree is built.

Example Using the SPF Algorithm

Figure C.1 shows how the SPF algorithm works. Figure C.1(a) shows a six-node graph and Node 1 is the source node. In Step 1, the triple <1,1,0> is placed on LIST because no tree exists from the beginning. Now NODE(T) is Node 1, where its neighbors are Nodes 2 and 3. Because LIST does not include any triple T' such that NODE(T')=2 or 3, the algorithm creates triples <1,2,0+1> (for Node 2) and <1,3,0+3> (for Node 3) and places them on LIST. It then sorts LIST so the triple with the least distance length (i.e., <1,2,1>) is placed at the head of LIST. Step 2 of the algorithm is now repeated, and Node 2 is added into the tree where its parent node is Node 1. Starting from Node 2, where its neighbors are Nodes 1, 4, and 5, Node 1 is not processed again because it is already in the tree. Step 2(b), LIST = {<2,4,2>,<1,3,3>,<2,5,6>} is repeated. Next, Node 4 is placed in the tree where its parent node is Node 2 and its neighbors are Nodes 5 and 6. Now T=<2,4,2>. Because LIST includes T'=<2,5,6>, whose NODE(T')=5, it follows Step 3(c), removing T' from LIST (since DISTANCE(T')=6 > DISTANCE(T)+LENGTH(4,5) (=2+3)), and adding the triple <4,5,5> on LIST. The same procedure is repeated until LIST is empty.

Figure C.1(b) shows a shortest path tree generated by the SPF algorithm, and Figure C.1(c) shows detailed steps of the solution tree.

C.2 Minimum-Spanning Tree Algorithm

A minimum-spanning tree is a spanning tree in which the sum of the lengths of the tree's links is as small as possible. Two well-known algorithms for computing a minimum-spanning tree are Kruskal's algorithm and Prim's algorithm. These two algorithms use the straightforward "greedy" algorithm for finding a minimum-spanning tree. Table C-1 describes these two algorithms. Reference [4] gives an historical review of the minimal-spanning tree problem.

Example Using Prim's Algorithm

A minimal-spanning tree is now needed for the network shown in Figure C.2(a). Both Prim's and Kruskal's algorithms start with the shortest link, i.e., they start with link (a,d) because the length of edge (a,d) is 1. Following Prim's algorithm, the next links to be added are (a,b) of length 2, then (d,e) of length 4, then (e,h), then (h,i), and so forth. The last link that would be added is (c,f). Figure C.2(b) shows the final tree.

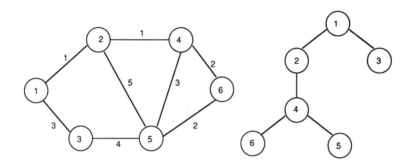

(a) Six-Node Example Network   (b) A Shortest Path Tree

Iteration	LIST	Next edge to be added	Next node to be processed	Shortest distance
1	<1,1,0>	1	1	DISTANCE(1)=0
2	<1,2,1>,<1,3,3>	1-2	2	DISTANCE(2)=1
3	<2,4,2>,<1,3,3>, <2,5,6>	2-4	4	DISTANCE(4)=2
4	<1,3,3>,<4,6,4>, <4,5,5>	1-3	3	DISTANCE(3)=3
5	<4,6,4>,<4,5,5>	4-6	6	DISTANCE(6)=4
6	<4,5,5>	4-5	5	DISTANCE(5)=5
7	Empty	-	-	-

(c) Detailed Steps of the Example

Figure C.1.   An example of an SPF algorithm.

433

Table C-1. Algorithms for Finding a Minimal Spanning Tree

Kruskal's algorithm	Prim's algorithm
Repeat the following step until tree T has n-1 edges (initially T is empty): Add to T the shortest edge that does not form a circle with edges already in T.	Repeat the following step until tree T has n-1 edges: Add to T the shortest edge between a node in T and a node not in T (initially pick up any edge of shortest length).

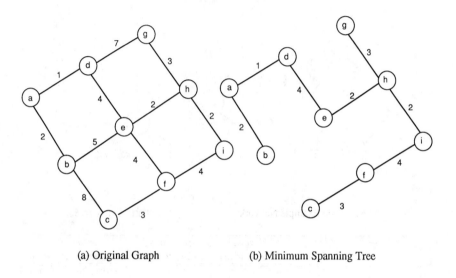

(a) Original Graph   (b) Minimum Spanning Tree

Figure C.2.   An example of Prim's algorithm.

C.3 Shortest Disjoint Path Algorithm

C.3.1 Two Shortest Link-Disjoint Paths between a Node Pair

The *Shortest Link-Disjoint Path* (SLDP) algorithm is used to find a pair of link-disjoint paths from source s to destination v of minimum total length. The following paragraphs review an algorithm discussed in Reference [5] that finds the two disjoint paths for each pair of nodes.

434

To understand the SLDP algorithm, the following definitions are needed: Let G be a directed graph containing n nodes, one of which is a distinguished *source* node s, and m links (v,w), each with a non-negative length $c(v,w)$. The distance from a node v to a node w, denoted by $d(u,w)$, is the length of the shortest path from v to w, defined only if a shortest path exists. Figure C.3 shows how the SLDP Algorithm works.

The SLDP Algorithm [5]

1. Find a shortest path tree T rooted at source s in G. Such a tree contains, for each node v, a shortest path from s to v. Compute $d(s,v)$, the shortest distance from s to v, for each node. Let P_1 be the shortest path from s to f (destination).
2. Transform the length of every link (v,w) by defining $c'(v,w)=c(v,w)-d(s,w)+d(s,v)$. [See Figure C.3(b).]
3. For destination f, create a graph G_f from G by reversing all links along the path in T from s to f. [See Figure C.3(c).] Compute the shortest path from s to f in G_f, denoted by P_2.
4. Two shortest link-disjoint paths between source s to destination f are obtained by joining P_1 and P_2 after eliminating common links. [See Figure C.3(d).]

C.3.2 Two Shortest Node-Disjoint Paths between a Node Pair

The shortest node-disjoint path pair problem can be solved by slightly modifying this algorithm. In this case, the SLDP algorithm described in Section C.3.1 can be applied after transforming the original graph G to a new graph G' by splitting each node v of G into two nodes, v_1 and v_2, joined by a link (v_1, v_2) of length 0. A link (v,w) of G becomes a link (v_2, w_1) of G'. [See Figure C.4(b).] The shortest link-disjoint path problem for graph G' is solved with source s_2. For destination v, a shortest node-disjoint path from s to v in G corresponds to a shortest pair of link-disjoint paths from s_2 to v_1 in G', and vice versa.

For the dual-homing architecture (see Chapter 5), any special CO (e.g., node s) needs to have two node-disjoint paths to its home hub (e.g., node h_1) and to its foreign hub (e.g., node h_2). In this case, a pseudo-node, p, which connects nodes h_1 and h_2 with link lengths of 0s is created. The problem of finding the two shortest node-disjoint paths from node s to h_1, and to h_2, is equivalent to the problem of finding the two shortest node-disjoint paths between nodes s and p. The SLDP algorithm can solve the latter problem.

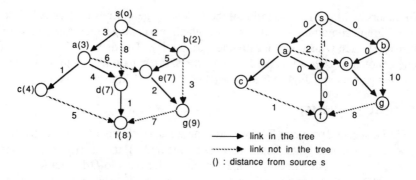

(a) Shortest-Path Tree of a Directed Graph   (b) Graph after Link Length Transformation

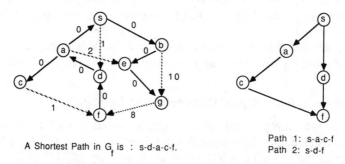

A Shortest Path in G_f is : s-d-a-c-f.

Path 1: s-a-c-f
Path 2: s-d-f

(c) Graph G_F for Graph G in (b)   (d) Shortest Disjoint Paths in Graph G

Figure C.3.   An example of an SLDP algorithm.

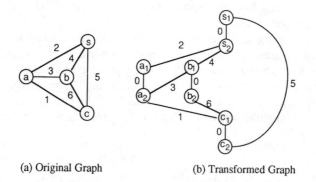

(a) Original Graph   (b) Transformed Graph

Figure C.4.   Node-splitting transformation.

436

REFERENCES

[1] Dijkstra, E. W., "A Note on Two Problems in Connection with Graphs," *Numerische Mathematik*, Vol. 1, 1959, pp. 269-172.

[2] McQuillan, J., Richer, I., and Rosen, E., "The New Routing Algorithm for the ARPANET," *IEEE Transactions on Communications*, Vol. COM-28, No. 5., May 1980, pp. 711-719.

[3] Deo, N., and Pang, C-Y., "Shortest Path Algorithms: Taxonomy and Annotation," *Networks*, Vol. 14, 1984, pp. 275-323.

[4] Graham, R. L., and Hell, P., "On the History of the Minimum Spanning Tree Problem," *Annals of the History of Computing*, Vol. 7, 1985, pp. 43-57.

[5] Suurballe, J. W., and Tarjan, R. E., "A Quick Method for Finding Shortest Pairs of Disjoint Paths," *Networks*, Vol. 14, 1984, pp. 325-336.

APPENDIX D

Demand Routing and Capacity Requirement Computation for Rings

This appendix discusses how to determine the ring capacity requirement and its required line rate, provided the end-to-end demand requirement is given. Note that the demand requirement discussed here is the STS-1 demand requirement because it is the building block for transport in SONET interoffice networks.

As discussed in Chapter 4, the SHR architectures can generally be divided into two categories: bidirectional SHRs and unidirectional SHRs [1]. The type of ring depends on the physical path (or route) traveled by a duplex communication channel between each office pair. In a *Bidirectional SHR* (BSHR), both directions of a duplex channel travel over the same path; in a *Unidirectional SHR* (USHR), the directions of a duplex channel travel over opposite paths. Figures D.1(a) and D.1(b) depict examples of the USHR and the BSHR, respectively.

For the USHR [see Figure D.1(a)], a duplex channel between Offices 1 and 3 travels over two opposite paths: path 1: 1 -> 2 -> 3, and path 2: 3 -> 4 -> 1. For BSHRs [see Figure D.1(b)], both directions of a duplex channel between Offices 1 and 3 use the same path (1 <-> 2 <-> 3), which travels through Office 2. Thus, a BSHR requires two working fibers to carry a duplex channel, and a USHR requires only one. To provide a protection capability for fiber system failures and fiber cable cuts, a BSHR may use four fibers (denoted by BSHR/4) (i.e., one working fiber pair and one protection fiber pair) or two fibers (denoted by BSHR/2) (i.e., all working fibers with spare capacity on each for protection). A USHR requires only two fibers (i.e., one working fiber and one protection fiber). The available capacity for the BSHR/2 is half the capacity of the BSHR/4.

The *ring capacity requirement* is defined to be the largest STS-1 cross-section in the ring. The STS-1 cross-section of each ring link is the same for USHRs, based on its routing rule. The OC-N line rate of the ring is selected based on its capacity requirement. Note that a limited set of line rates is used in SONET rings for most practical situations: OC-3, OC-12, and OC-48. Thus, if the capacity requirement is 15 STS-1s, either five OC-3 rings, two OC-12 rings, one OC-48 ring, or one OC-12 ring and one OC-3 ring are used.

The capacity requirement for a USHR is the sum of STS-1 demands for all demand pairs on the ring. The ring capacity requirement for BSHRs is determined based on the particular ring demand assignment algorithm used. The following discussion describes a heuristic for computing the capacity requirement for the BSHR. This algorithm assumes that demand is not split for BSHRs. In other words, all demand from one CO to another CO uses the same routing path for BSHRs. For example, in Figure D.1(b), if one STS-1 between CO-2 and CO-4 is routed through a single path, Path 2-3-4, then the remaining STS-1s for the same demand pair have to be routed via the same path.

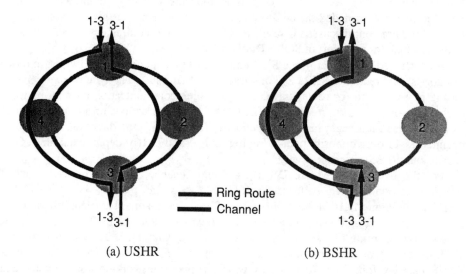

(a) USHR   (b) BSHR

Figure D.1.   Ring definition.

Demand Routing Algorithm for the BSHR

1. Rank demand requirements in descending order [see Figure D.2(a)].
2. Select demand pairs that are ring links (i.e., a link-disjoint set) and assign demands of these demand pairs to these ring links [see Figure D.2(b)]; delete these demand pairs from the demand matrix.
3. Distribute the demand requirement for each demand pair according to the order determined in Step 1, in both the clockwise and counterclockwise directions, and repeat the following process until all demand pairs have been processed:

440

a. Compare the maximum link capacity requirement when the demand requirement is routed in the clockwise and counterclockwise directions. If the maximum link capacity requirement is smaller in the clockwise (or counterclockwise) direction, route demands in the clockwise (or counterclockwise) direction.

b. If the maximum link capacity requirement is equal in both the clockwise and counterclockwise directions, compare the number of links between the origin and the destination of the considered demand pair in both the clockwise and counterclockwise directions. If there are fewer links in the clockwise (or counterclockwise) direction, route demands in the clockwise (or counterclockwise) direction. If the number of links is equal in both the clockwise and counterclockwise directions, route demands in the clockwise or counterclockwise direction alternately.

Figure D.2 illustrates an example that shows how this algorithm works. First, the demand requirement is sorted in decreasing order, with the highest demands placed in the first position. It then assigns demands to a link-disjoint set [e.g., links (1,2), (2,3), and (4,5) as shown in Figure D.2(b)]. Then, it begins distributing demands in the present routing assignment in a balanced manner and repeats the same process until all demand pairs have been processed. Figure D.2(g) shows a final demand-routing assignment that requires a ring capacity of 11 STS-1s (i.e., the link between Nodes 4 and 5). Thus, the BSHR/4 requires an OC-12 line rate, while the BSHR/2 needs an OC-48 rate because it only uses half the capacity of the ring (see Section 4.2.2 of Chapter 4).

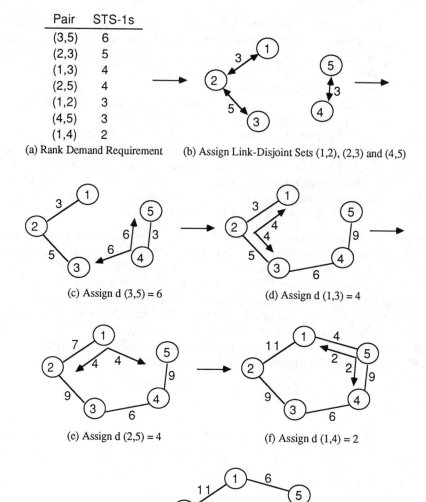

Pair	STS-1s
(3,5)	6
(2,3)	5
(1,3)	4
(2,5)	4
(1,2)	3
(4,5)	3
(1,4)	2

(a) Rank Demand Requirement   (b) Assign Link-Disjoint Sets (1,2), (2,3) and (4,5)

(c) Assign d (3,5) = 6   (d) Assign d (1,3) = 4

(e) Assign d (2,5) = 4   (f) Assign d (1,4) = 2

(g) BSHR Capacity Requirement = 11 STS-1s

Figure D.2.   Ring capacity requirement calculations for BSHRs.

442

REFERENCES

[1]   Wu, T-H., and Lau, R. C., "A Class of Self-Healing Ring Architectures for SONET Network Applications," *IEEE GLOBECOM'90,* December 1990, San Diego, CA, pp. 403.2.1-403.2.8.

443

Acronyms

1:1/ODP	1:1 Optical Diverse Protection
1:N/DP	1-for-N APS Diverse Protection
AAL	ATM Adaptation Layer
ADM	Add-Drop Multiplexer
AIS	Alarm Indication Signal
ANSI	American National Standards Institute
APS	Automatic Protection Switching
ASC	Accredited Standards Committee
ASE	Amplified Spontaneous Emission
ATM	Asynchronous Transfer Mode
ATM/DCS	ATM Cross-connect System
BCC	Bellcore Client Company
B-DCS	Broadband DCS
BER	Bit Error Ratio
BIP	Bit-Interleaved Parity
B-ISDN	Broadband Integrated Services Digital Network
BSHR	Bidirectional SHR
BSHR/2	2-fiber BSHR
BSHR/4	4-fiber BSHR
BSHR4/PPR	Passive Protected BSHR/4 architecture
BTS	Bit Transport System
CAD/CAM	Computer-Aided Design/Computer-Aided Manufacture
CBR	Constant Bit Rate
CCITT	International Telegraph and Telephone Consultative Committee
CCM	Customer Control and Management
CCN	Cross-Connect Network
CCS	Common Channel Signaling
CCW	Counterclockwise
CEPT	Conference of European Postal and Telecommunications
CEV	Controlled-Environment Vault
CLP	Cell Loss Priority
CO	Central Office
CP	Customer Premises
CPE	Customer Premises Equipment
CS	Convergence Sublayer
DCC	Data Communications Channel
DCS	Digital Cross-connect System
DQDB	Distributed Queue Dual Bus
DH	Dual Homing
DLC	Digital Loop Carrier
DMUX	Demultiplexer (or Demultiplexing)

DNHR	Dynamic Non-Hierarchical Routing
DP	Diverse Protection
DR	Dynamical Routing
DS0	Digital Signal level 0
DS1	Digital Signal level 1
DS3	Digital Signal level 3
DSX	Electronic Cross-connect System
ECF	Embedded Superframe Structure
EDFA	Erbium-Doped Fiber Amplifiers
EDSX	Electronic Digital Cross-connect System
E/O	Electrical-to-Optical
EOC	Embedded Operations Channel
FCC	Federal Communications Commission
FDDI	Fiber Distributed Data Interface
FERF	Far End Receive Failure
FFTS	Fiber Feeder Transport System
FITL	Fiber in the Loop
GFC	Generic Flow Control
GOS	Grade of Service
HDLC	High-level Data Link Control
HDT	Host Digital Terminal
HDTV	High Definition Television
HEC	Header Error Control
ICCF	Interchange Carrier Compatibility Forum
ICSR	Incremental Cost-to-Survivability Ratio
ID	Identification
IDLC	Integrated Digital Loop Carrier
I/O	Input/Output
ISO	International Standards Organization
IXC	Interexchange Carrier
LAN	Local Area Network
LATA	Local Access and Transport Area
LED	Light Emitting Diode
LOF	Loss of Frame
LOP	Loss of Pointer
LOS	Loss of Signal
LTE	Line Terminating Equipment
MAC	Media Access Control
MAN	Metropolitan Area Network
MOS	Mechanical Optical Switches
MSB	Most Significant Bit
MUX	Multiplexer (or Multiplexing)
NE	Network Element

NNI	Network-Node Interface
NTT	Nippon Telephone and Telegraph
OAM	Operations and Maintenance
OAM&P	Operations, Administration, Maintenance, and Provisioning
OBPF	Optical Bandpass Filters
OC-1	Optical Carrier Level-1
OLA	Optical Limiting Amplifier
OLTM	Optical Line Terminating Multiplexer
ONU	Optical Network Unit
OOF	Out-of-Frame
OPLDS	Optical Power Loss Detection System
OPM	Optical Protection Module
OPSC	Optical Protection Switching Controller
OS	Operations System
OSI	Open Systems Interconnection
PDN	Passive Distribution Network
PM	Physical Medium
PON	Passive Optical Network
PON/OC	PON with Optical Coupler
PON/WDM	PON with Wavelength Division Multiplexing
POP	Point of Presence
POTS	Plain Old Telephone Service
PPL	Passive Photonic Loop
PS	Power Splitter
PSC	Protection Switching Controller
PT	Payload Type
PTE	Path Terminating Equipment
RDT	Remote Digital Terminal
RN	Remote Node
RT	Remote Terminal
SAR	Segmentation and Reassembly
SD	Signal Degrade
SDH	Synchronous Digital Hierarchy
SF	Signal Failure
SH	Single Homing
SHR	Self-Healing Ring
SMDS	Switched Multi-megabit Data Service
SNR	Signal-to-Noise Ratio
SONET	Synchronous Optical Network
SPE	Synchronous Payload Envelope
SPF	Shortest Path First
SSBN	SONET Switched Bandwidth Network
STE	Section Terminating Equipment

447

STM	Synchronous Transfer Mode
STS-1	Synchronous Transport Signal-Level 1
SWF-DS1	Switched DS1/Switched Fractional DS1
SYNTRAN	Synchronous Transmission
TC	Transmission Convergence
TDM	Time Division Multiplexer (or Multiplexing)
TDMA	Time Division Multiple Access
TM	Terminal Multiplexers
TSA	Time Slot Assignment
TSI	Time Slot Interchange
UNI	User-Network Interface
UPC	Usage Parametric Control
USHR	Unidirectional SHR
USHR/L	Unidirectional line-Switched USHR
USHR/P	Unidirectional Path-Switched USHR
VBR	Variable Bit Rate
VC	Virtual Channel
VCI	Virtual Channel Identifier
VLSI	Very Large Scale Integration
VP	Virtual Path
VPI	Virtual Path Identifier
VPT	Virtual Path Terminators
VT	Virtual Tributaries
WAN	Wide Area Network
W-DCS	Wideband Digital Cross-connect System
WDM	Wavelength Division Multiplexer (or Multiplexing)
dB	Decibel

About the Author

Tsong-Ho Wu earned his BS in mathematics at National Taiwan University in 1976, and his PhD in operations research at the State University of New York at Stony Brook in 1983. From 1983-86 he was a senior network scientist/architect at United Telecommunications, Inc., where he engaged in research and project management for an enhanced nationwide packet-switched data network and its secure network architecture. In 1986, he joined Bellcore's Applied Research area, where he conducts research on fiber network survivability and broadband SONET/ATM network architecture and design, as well as on new cost-effective optical network architectures that may utilize the unique characteristics of photonic switching technology. Tsong-Ho is the author of more than forty conference and journal papers in the areas of network design, data communications, modeling and performance, and survivable broadband network architectures. He is also a member of the IEEE.

INDEX

The Artech House Telecommunications Library

Vinton G. Cerf, Series Editor